Complex Variables
for Scientists and Engineers

Complex Variables

F O R

Scientists A N D Engineers

John D. Paliouras
Professor of Mathematics
Rochester Institute of Technology

Macmillan Publishing Co., Inc.
New York

Collier Macmillan Publishers
London

MACMILLAN PUBLISHING CO., INC.
866 Third Avenue, New York, New York 10022

COLLIER-MACMILLAN CANADA, LTD.

LIBRARY OF CONGRESS CATALOGING IN PUBLICATION DATA

Paliouras, John D
    Complex variables for scientists and engineers.

    1. Functions of complex variables.   I. Title.
QA331.P164   515'.9   73-18770
ISBN 0-02-390550-6

Printing: 1 2 3 4 5 6 7 8   Year: 5 6 7 8 9 0

στή μνήμη τῶν γονέων μου

PREFACE

A FIRST course on complex variables taught to students in the sciences and engineering is invariably faced with the difficult task of meeting two basic objectives: (1) It must create a sound foundation based on the understanding of fundamental concepts and the development of manipulative skills, and (2) it must reach far enough so that the student who completes such a course will be prepared to tackle relatively advanced applications of the subject in subsequent courses that utilize complex variables. This book has been written with these two objectives in mind. Its main goal is to provide a development leading, *over a minimal and yet sound path*, to the fringes of the promised land of applications of complex variables or to a second course in the theory of analytic functions. The arrangement of the topics allows a variety of choices depending on the objectives of the course. The standard topics that are necessary, regardless of the goals of a particular course, are in the mainstream of the development, while peripheral topics are available but not stressed.

The level of the development is quite elementary, and its main theme is the *calculus of complex functions*. The only prerequisite for the study of this book is a standard course in elementary calculus. The topological aspects of the subject are developed only to the extent necessary to give the reader an intuitive understanding of these matters. However, the material contained in the exercises, in the appendices, and in Part III of the book provides ample opportunity for in-depth treatment of most of the concepts if desired. Theorems are discussed informally and, whenever possible, are illustrated via examples, but their proofs are given in the appendices at the end of each chapter. Numerous examples illustrate new concepts soon after they are introduced as well as theorems that lend themselves readily to problem solving. Most of the examples are discussed in detail although, occasionally, some less elementary steps are included which are intended to prompt the inquisitive and conscientious student to seek and provide justifications, thus affording himself the opportunity to review the underlying fundamental notions. Similar practices are followed in the proofs of theorems. Exercises are usually divided into three categories in order to accommodate problems that range from the routine type to the more formidable ones. Constant reference is made to concepts from elementary calculus that are analogous to the concepts under discussion. Thus the student is constantly reminded of the similarities between real and complex analysis. At the same time, cases in which such similarities cease to exist are pointed out.

Earlier versions of this book were used, in the form of notes, in a course given to science and engineering students at the Rochester Institute of Technology. During that period I was fortunate to have received the constructive criticism of many of my colleagues and students, as a result of which many improvements were effected; I am grateful to all of them. I am especially indebted to Professors A. Erskine, L. Fuller, C. Haines, and T. Upson, who read and corrected various parts of the final form of the manuscript. My deep appreciation goes to my wife for her infinite patience and silent encouragement.

J. D. P.

Rochester, N.Y.

CONTENTS

PART II
THE FOUNDATIONS OF COMPLEX
FUNCTION THEORY

Chapter 4   Complex Integration   151

Chapter 5   Cauchy Theory of Integration   183

Chapter 9   Some Theoretical Results   337

Answers to Exercises   351

Index   367

PART I
Preliminaries

CHAPTER 1
Complex Numbers

Section 1 Definition of a complex number. Special complex numbers. Equality, sum, difference, product, and quotient of complex numbers. Conjugation. Basic algebraic laws.

Section 2 The complex plane; real and imaginary axes. Modulus and argument of z. Distance between two complex numbers. Principal value of the argument. Properties of the modulus. Complex form of two-dimensional curves. Polar form of a complex number. Equality in polar form. Roots of complex numbers; roots of unity. Geometry of rational operations on complex numbers.

Section 1
Complex Numbers and Their Algebra

It is assumed that the reader is familiar with the system of real numbers and their elementary algebraic properties. Our work in this book will take us to a larger system of numbers, which have been given the unfortunate name of "imaginary" or "complex numbers." A historical account of the discovery of such numbers and of their development into prominence in the world of mathematics is outside the scope of this book. We only remark that the need for imaginary numbers arose from the need to find square roots of negative numbers.

The system of complex numbers can be formally introduced by use of the concept of an "ordered pair" of real numbers (a, b). The set of all such pairs with appropriate operations defined on them can be defined to constitute the system of complex numbers. The reader who is interested in this approach is referred to Appendix 1(A). Here, with due apologies to the formalists, we shall proceed to define the complex numbers in the more conventional, if somewhat incomplete, manner.

The set of **complex numbers** is defined to be the totality of all quantities of the form

$$a + ib \qquad \text{or} \qquad a + bi,$$

where a and b are real numbers and $i^2 = -1$. To the reader who may wonder what is so incomplete about this approach, we point out that nothing is said as to the meaning of the implied multiplication in the terms "ib" or "bi."

3

If $z = a + ib$ is any complex number, then a is called the **real part** of z and b is called the **imaginary part** of z; we sometimes denote them

$$R(z) \qquad \text{and} \qquad I(z),$$

respectively. We point out that both $R(z)$ *and* $I(z)$ *are real numbers.* If $R(z) = 0$ and $I(z) \neq 0$, then z is called **pure imaginary**; e.g., $z = 3i$ is such a number. In particular, if $R(z) = 0$ and $I(z) = 1$, then we write $z = i$ and we call this number the **imaginary unit.** If $I(z) = 0$, then z reduces to the real number $R(z)$; in that sense, one can think of any real number x as being a complex number of the form $z = x + 0i$.

For the remainder of this section

$$z_n = x_n + iy_n, \qquad n = 1, 2, 3,$$

are three arbitrary complex numbers.

Equality of complex numbers is defined quite naturally. Thus

$$z_1 = z_2 \qquad \text{provided that} \qquad x_1 = x_2 \quad \text{and} \quad y_1 = y_2 .$$

The **sum** of two complex numbers is defined by

$$z_1 + z_2 = (x_1 + x_2) + i(y_1 + y_2)$$

and their **product** by

$$z_1 z_2 = (x_1 x_2 - y_1 y_2) + i(x_1 y_2 + x_2 y_1).*$$

The **zero** (additive identity) of the system of complex numbers is the number

$$0 + 0i,$$

which we simply write 0, and the **unity** (multiplicative identity) is the number

$$1 + 0i,$$

which we simply write 1. It is very easy to show that for any $z = x + iy$,

$$z + (0 + 0i) = z \qquad \text{and} \qquad z(1 + 0i) = z.$$

Again, if z is any complex number, there is one and only one complex number, which we will denote by $-z$, such that

$$z + (-z) = 0;$$

---

* The **nonnegative integral powers** of a complex number z are defined as in the case of real numbers. Thus,

$$z^1 = z, z^2 = zz, z^3 = z^2 z, \ldots, z^{n+1} = z^n z,$$

and if $z \neq 0$, then $z^0 = 1$.

$-z$ will be called the **negative** (additive inverse) of z and it turns out that

$$\text{if } z = x + yi, \qquad \text{then} \qquad -z = -x - yi.$$

For any nonzero complex number $z = x + iy$ there is one and only one complex number, which we will denote by z^{-1} or $1/z$, such that

$$zz^{-1} = 1;$$

z^{-1} is called the **reciprocal** (multiplicative inverse) of z and a direct calculation yields

$$z^{-1} = \frac{x}{x^2 + y^2} - \frac{y}{x^2 + y^2}i.$$

In order to facilitate further algebraic manipulations, we now define the **difference** of two numbers by

$$z_1 - z_2 = z_1 + (-z_2),$$

which yields

$$z_1 - z_2 = (x_1 - x_2) + (y_1 - y_2)i.$$

We also define the **quotient** of two numbers by

$$\frac{z_1}{z_2} = z_1 z_2^{-1}, \qquad \text{for } z_2 \neq 0,$$

which yields

$$\frac{z_1}{z_2} = \frac{x_1 x_2 + y_1 y_2}{x_2^2 + y_2^2} + \frac{x_2 y_1 - x_1 y_2}{x_2^2 + y_2^2}i.$$

In addition to the operations defined above, we have a "new" operation, called **conjugation,** defined on the complex numbers as follows: if $z = x + iy$, then the **conjugate** of z, denoted $\bar{z}$, is defined by

$$\bar{z} = x - yi.$$

Unlike the four "binary operations" defined above, conjugation is a "unary operation": i.e., it acts on one number at a time and has the effect of negating its imaginary part.

Algebraic Properties of Complex Numbers

The operations defined above obey the following laws.

1. Commutative laws:

$$z_1 + z_2 = z_2 + z_1; \qquad z_1 z_2 = z_2 z_1.$$

2. Associative laws:

$$z_1 + (z_2 + z_3) = (z_1 + z_2) + z_3; \qquad z_1(z_2 z_3) = (z_1 z_2)z_3.$$

3. Distributive law:

$$z_1(z_2 + z_3) = z_1 z_2 + z_1 z_3.$$

4. Distributivity of conjugation:

$$\overline{z_1 + z_2} = \bar{z}_1 + \bar{z}_2; \qquad \overline{z_1 - z_2} = \bar{z}_1 - \bar{z}_2; \qquad \overline{z_1 z_2} = \bar{z}_1 \bar{z}_2;$$

$$\overline{z_1/z_2} = \bar{z}_1/\bar{z}_2.$$

5. $\bar{\bar{z}} = z.$
6. $z\bar{z} = [R(z)]^2 + [I(z)]^2.$

Some of these properties are proved in the examples that follow: the remaining ones are left for the exercises.

NOTE: The reader may have noted already that the product of two complex numbers is found by effecting an ordinary multiplication of two binomials in which use is made of the reduction formula $i^2 = -1$.

On the other hand, one finds the quotient z_1/z_2 by first multiplying numerator and denominator by $\bar{z}_2$ and then simplifying the resulting expression. Thus

$$\frac{z_1}{z_2} = \frac{z_1 \bar{z}_2}{z_2 \bar{z}_2}$$

and, in particular, for the reciprocal of z we have

$$\frac{1}{z} = \frac{\bar{z}}{z\bar{z}}.$$

EXAMPLE 1

If $z = 5 - 5i$ and $w = -3 + 4i$, find $z + w$, $z - w$, zw, z/w, $\bar{z}$, and $\bar{w}$.
Using the definitions of the algebraic operations, we find

$$z + w = (5 - 5i) + (-3 + 4i) = (5 - 3) + (-5 + 4)i = 2 - i.$$

$$z - w = (5 - 5i) - (-3 + 4i) = (5 + 3) + (-5 - 4)i = 8 - 9i.$$

$$zw = (5 - 5i)(-3 + 4i) = (-15 + 20) + (15 + 20)i = 5 + 35i.$$

$$\frac{z}{w} = \frac{5 - 5i}{-3 + 4i} = \frac{(5 - 5i)(-3 - 4i)}{(-3 + 4i)(-3 - 4i)} = -\frac{7}{5} - \frac{1}{5}i.$$

$$\bar{z} = 5 + 5i.$$

$$\bar{w} = -3 - 4i.$$

E X A M P L E 2

Prove the commutative law for addition: $z_1 + z_2 = z_2 + z_1$.

We effect this proof by employing the corresponding law for real numbers, which states that, for any two real numbers a and b, $a + b = b + a$. Thus, we have

$$z_1 + z_2 = (x_1 + iy_1) + (x_2 + iy_2)$$

$$= (x_1 + x_2) + i(y_1 + y_2) \qquad \text{definition of sum}$$

$$= (x_2 + x_1) + i(y_2 + y_1) \qquad \text{commutativity of real numbers}$$

$$= (x_2 + iy_2) + (x_1 + iy_1) \qquad \text{definition of sum}$$

$$= z_2 + z_1.$$

E X A M P L E 3

Prove that conjugation distributes over multiplication: $\overline{z_1 z_2} = \bar{z}_1 \bar{z}_2$.

On the one hand, we have

$$\overline{z_1 z_2} = \overline{(x_1 x_2 - y_1 y_2) + (x_1 y_2 + x_2 y_1)i} \qquad \text{definition of product}$$

$$= (x_1 x_2 - y_1 y_2) - (x_1 y_2 + x_2 y_1)i \qquad \text{definition of conjugate}$$

$$= (x_1 x_2 - y_1 y_2) + (-x_1 y_2 - x_2 y_1)i \quad \text{definition of negative.}$$

On the other hand,

$$\bar{z}_1 \bar{z}_2 = \overline{(x_1 + y_1 i)}\,\overline{(x_2 + y_2 i)}$$

$$= (x_1 - y_1 i)(x_2 - y_2 i) \qquad\qquad\qquad \text{definition of conjugate}$$

$$= (x_1 x_2 - y_1 y_2) + (-x_1 y_2 - x_2 y_1)i \qquad \text{definition of product.}$$

Clearly, the two sides are equal and the proof is complete.

E X A M P L E 4

Prove property 6, namely, that $z\bar{z} = [R(z)]^2 + [I(z)]^2$.

Let $z = x + iy$; then $\bar{z} = x - iy$. Therefore,

$$z\bar{z} = (x + iy)(x - iy)$$

$$= x^2 + y^2$$

$$= [R(z)]^2 + [I(z)]^2.$$

This property simply says that, for any number z, the product $z\bar{z}$ is always a nonnegative real number, since it is the sum of squares of real numbers.

It should be apparent to the reader that most of the familiar algebraic properties of the real numbers are shared by the complex numbers. There is, however, a particular property of the real numbers, namely, the property of order, which does not carry over to the complex case. By this we mean that given two arbitrary complex numbers z and w, no reasonable meaning can be attached to the expression

$$z < w;$$

discussion and proof of this fact are left as an exercise for the reader. See Review Exercise 19 at the end of the chapter.

EXERCISE 1

A

In Exercises 1.1–1.10 perform the indicated operations, reducing the answer to the form $A + Bi$.

1.1. $(5 - 2i) + (2 + 3i)$.   **1.2.** $(2 - i) - (6 - 3i)$.
1.3. $(2 + 3i)(-2 - 3i)$.   **1.4.** $-i(5 + i)$.
1.5. $i\bar{i}$.   **1.6.** $(a + bi)(a - bi)$.
1.7. $6i/(6 - 5i)$.   **1.8.** $(a + bi)/(a - bi)$.
1.9. $1/(3 + 2i)$.   **1.10.** $i^2, i^3, i^4, i^5, \ldots, i^{10}$.

1.11. From the results of the preceding exercise, formulate a rule for all the positive integral powers of i and then for the negative ones.

1.12. Show that if $z = -1 - i$, then $z^2 + 2z + 2 = 0$.

1.13. Show that the imaginary unit has the property that $-i = i^{-1} = \bar{i}$.

1.14. If $z = a + bi$, express z^2 and z^3 in the form $A + Bi$.

1.15. Reduce each of the following to the form $A + Bi$:

(a) $\dfrac{1 + i}{1 - i}$.   (b) $\dfrac{i}{1 - i} + \dfrac{1 - i}{i}$.

(c) $\dfrac{1}{i} - \dfrac{3i}{1 - i}$.   (d) $i^{123} - 4i^9 - 4i$.

B

1.16. (a) For which complex numbers, if any, is $z^{-1} = z$ true?
(b) Similarly for the equation $\bar{z} = -z$.
(c) Similarly for the equation $\bar{z} = z^{-1}$.

1.17. Prove that for any number z,

$$R(z) = \frac{1}{2}(z + \bar{z}) \quad \text{and} \quad I(z) = \frac{1}{2i}(z - \bar{z}).$$

1.18. Prove that conjugation distributes over sums, differences, and quotients. (See property 4 and Example 3.)

1.19. Prove: $z = \bar{z}$ if and only if z is a real number.*

1.20. Prove the commutative, associative, and distributive laws for complex numbers.

1.21. Prove that if $z^2 = (\bar{z})^2$, then z is either real or pure imaginary.

1.22. Prove that $\bar{\bar{z}} = z$.

1.23. Prove: For any numbers z and w, $z\bar{w} + \bar{z}w = 2R(z\bar{w})$.

C

1.24. Prove: $zw = 0$ if and only if either $z = 0$ or $w = 0$.

1.25. Prove that the zero of the complex number system is unique.

1.26. Prove that the unity of the complex number system is unique.

1.27. Prove that the negative of any complex number z is unique.

1.28. If $z = x + iy$ is a nonzero complex number, derive z^{-1} in terms of x and y and show that it is unique.

Section 2
Geometry of Complex Numbers

The reader is familiar with certain correspondences between algebraic and geometric concepts that are described in analytic geometry. Thus, for example,

1. The real numbers correspond to the x-axis.
2. $|a - b|$ corresponds to the distance between a and b.
3. Equations in two variables correspond to curves in the plane.

Similar correspondences have very important uses in the theory and applications of complex variables. The basis for this entire concept is found in the very definition of a complex number, which creates, in a natural way, a one-to-one correspondence between the set of complex numbers and the set of points in the xy-plane. Thus

the complex number $z = a + ib$ corresponds to the point (a, b) in the plane and vice versa.

The importance of this correspondence cannot be overemphasized, and it will be increasingly obvious in the subsequent developments. Indeed, the

---

* The reader is reminded that a statement of the form "*P* if and only if *Q*" actually involves two statements: (1) "If *P*, then *Q*" and (2) "If *Q*, then *P*." So, in this problem one is required to prove that (1) if $z = \bar{z}$, then z is a real number, and (2) if z is a real number, then $z = \bar{z}$.

identification of complex numbers with points in the plane is so strong that the number $a + ib$ and the point (a, b) become practically indistinguishable; as a result, we often talk about the *number* (a, b) or the *point $a + ib$*. In view of this identification, the familiar xy-plane will henceforth be referred to as the **complex plane** or **z-plane**; the x-axis and y-axis will be called **real axis** and **imaginary axis,** respectively.

Going one step further, we can also identify a complex number with a vector. Thus,

> the complex number $a + ib$ can be thought of as a vector in the plane, emanating from the origin and terminating at the point (a, b).

Now, given any number $z = a + ib$, the **modulus** of z, denoted $|z|$, is defined to be the magnitude of the vector associated with z; i.e.,

$$|z| = [a^2 + b^2]^{1/2}.$$

The **argument** of z, denoted arg z, is defined to be *any one* of the angles that the vector corresponding to z makes with the positive direction of the real axis; i.e., arg $(a + ib)$ is any angle θ such that

$$\sin \theta = \frac{b}{|z|} \quad \text{and} \quad \cos \theta = \frac{a}{|z|}:$$

see Figure 1.1.

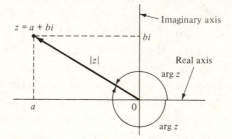

FIGURE 1.1   MODULUS OF z AND arg z

Concerning the two concepts just defined, the following remarks are of importance.

REMARK 1

It is clear from the definition that the modulus of z represents the undirected distance of z from the origin and, therefore, it is a nonnegative real number.

In particular, if $z = a + ib$ is real ($b = 0$), then

$$|z| = (a^2)^{1/2},$$

which is a definition of the absolute value of any real number a. This shows that the modulus of a complex number can be thought of as an extension of the concept of the absolute value of a real number.

REMARK 2

The notion of $|z|$ representing the linear distance between 0 and z can be extended, quite naturally, to define the **distance** between any two numbers $z = a + ib$ and $w = c + id$ to be the quantity

$$|z - w|.$$

That this is indeed the distance between the points (a, b) and (c, d) is easily shown as follows:

$$|z - w| = |(a + ib) - (c + id)|$$

$$= |(a - c) + (b - d)i|$$

$$= [(a - c)^2 + (b - d)^2]^{1/2},$$

the last expression being precisely the distance between the said points, as we know from analytic geometry.

REMARK 3

The argument of zero cannot be defined in a meaningful way. Algebraically, this is obvious, since one would have to contend with the indeterminate form $0/0$; geometrically, it is also obvious since the zero vector, to which the number $z = 0$ corresponds, has no length and hence it cannot form any angle with the positive real axis.

REMARK 4

It is clear from the definition that the argument of a number is *not* a unique quantity: in fact, every $z \neq 0$ has an infinity of distinct arguments, which differ from each other by a multiple of 2π. The situation here is identical with that encountered in analytic geometry when one expresses the co-ordinates of a point in polar form. For instance, $\arg(1 + i)$ can be taken to be $\pi/4$ or $9\pi/4$ or $-15\pi/4$ or, in general, $(\pi/4) + 2k\pi$, where k is any integer. In order to alleviate this situation, which in certain instances may be undesirable, we introduce the concept of the "principal value" of $\arg z$.

For any number $z \neq 0$, the **principal value of arg** z is defined to be the unique value of arg z that satisfies the relation

$$-\pi < \arg z \leq \pi:$$

it will be denoted Arg z. In view of Remark 4, it is easy to see that

$$\arg z = \text{Arg } z + 2k\pi, \qquad k = 0, \pm 1, \pm 2, \dots.$$

EXAMPLE 1

The number $z = -\sqrt{12} - 2i$ has been plotted in Figure 1.2. Its modulus is $|z| = |-\sqrt{12} - 2i| = [(-\sqrt{12})^2 + (-2)^2]^{1/2} = 4$. Now, denoting arg z by θ, we find that

$$\sin \theta = -\frac{1}{2} \qquad \text{and} \qquad \cos \theta = -\frac{\sqrt{3}}{2}.$$

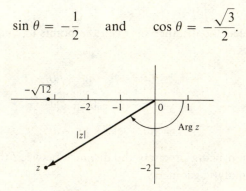

FIGURE 1.2   EXAMPLE 1

Hence

$$\arg z = \theta = \frac{7\pi}{6} + 2k\pi$$

and

$$\text{Arg } z = -\frac{5\pi}{6}.$$

EXAMPLE 2

Find z such that $|z| = 2$ and arg $z = \pi/4$.

Write $z = x + iy$. Since its argument $\theta = \pi/4$, we have

$$\sin \theta = \frac{\sqrt{2}}{2} \qquad \text{and hence} \qquad \frac{y}{|z|} = \frac{\sqrt{2}}{2}.$$

But $|z| = 2$. Therefore, $y = \sqrt{2}$. Similarly,

$$\cos \theta = \frac{\sqrt{2}}{2};$$

hence

$$\frac{x}{|z|} = \frac{\sqrt{2}}{2}$$

and therefore $x = \sqrt{2}$. It follows that $z = \sqrt{2} + \sqrt{2}i$.

Properties of |z|

For any two complex numbers z and w, the following properties are true:
1. $|z| = |-z| = |\bar{z}|$.
2. $|z - w| = |w - z|$.
3. $|z|^2 = |z^2| = z\bar{z}$, hence, if $z \neq 0$, $1/z = \bar{z}/|z|^2$.
4. $|zw| = |z| \, |w|$.
5. $|z/w| = |z|/|w|$, for $w \neq 0$.
6. $|z + w| \leq |z| + |w|$.
7. $| \, |z| - |w| \, | \leq |z - w|$.
8. $|z| - |w| \leq |z + w|$.

Property 6 is known as the **Triangle Inequality**, and its truth is established in the next example. The proofs of properties 1, 2, and 3 follow immediately from the definition of modulus; properties 4 and 5 follow by the use of 3, whereas properties 7 and 8 are corollaries of the triangle inequality. See Exercise 2.14.

EXAMPLE 3

Prove the triangle inequality.

A brief reason is given to justify each step of the proof. The reader will find it instructive to complete the justification whenever it is not entirely obvious.

We follow a procedure that is usual in proving assertions involving moduli. Specifically, instead of proving $|z + w| \leq |z| + |w|$, we prove the "squared relation" $|z + w|^2 \leq [|z| + |w|]^2$.

$|z + w|^2 = (z + w)\overline{(z + w)}$	property 3
$= (z + w)(\bar{z} + \bar{w})$	distributivity of conjugation
$= z\bar{z} + w\bar{w} + z\bar{w} + \bar{z}w$	algebra
$= |z|^2 + |w|^2 + 2R(z\bar{w})$	property 3, Exercise 1.23

$$\leq |z|^2 + |w|^2 + 2|z\overline{w}| \qquad \text{Exercise 2.14(g)}$$

$$= |z|^2 + |w|^2 + 2|z|\,|w| \qquad \text{properties 1 and 4}$$

$$= [|z| + |w|]^2 \qquad \text{algebra.}$$

But since the quantities involved in the first and last steps are nonnegative, the triangle inequality follows.

In the examples that follow, we illustrate the fact that the correspondence between concepts from analytic geometry and complex numbers can be carried one step further to give us what we may call the **complex form of equations in the plane.**

EXAMPLE 4

Show that the equation $|z + i| = 2$ represents a circle and find its center and radius.

Writing the given equation in the form

$$|z - (-i)| = 2,$$

we note that the left-hand side represents the distance from z to $-i$. So, this equation is satisfied by all points z whose distance from $-i$ is 2. Clearly, the set of all such points is the circle with center at $-i$ and of radius 2.

The same result can be obtained via algebraic manipulations as follows: Let $z = x + iy$. Then the given equation becomes

$$|x + (y + 1)i| = 2,$$

from which

$$[x^2 + (y + 1)^2]^{1/2} = 2,$$

which in turn yields

$$x^2 + (y + 1)^2 = 4.$$

The reader will recognize this as the circle with center at $(0, -1)$ and of radius $r = 2$.

EXAMPLE 5

Find the locus of all points in the plane such that $I(i + \overline{z}) = 4$.

We first simplify the left side of the given equation. Letting $z = x + iy$, we find that

$$i + \overline{z} = x + (1 - y)i.$$

Hence

$$I(i + \bar{z}) = 1 - y.$$

Therefore, substituting from the given equation, we have

$$4 = 1 - y,$$

from which

$$y = -3.$$

It follows that the locus sought is the horizontal line $y = -3$.

EXAMPLE 6

Determine first geometrically and then algebraically the locus of points z such that

$$|z - 2i| = |z + 2|.$$

Geometrically: We seek all points z whose distances from $2i$ and -2 are equal. From plane geometry we know that the locus of such points is the perpendicular bisector of the line segment joining $2i$ and -2. By inspection we find that the locus in question is the line

$$y = -x.$$

Algebraically: Letting $z = x + iy$ in the given equation, we have

$$|z - 2i| = |z + 2|$$

$$|x + i(y - 2)| = |(x + 2) + iy|$$

$$[x^2 + (y - 2)^2]^{1/2} = [(x + 2)^2 + y^2]^{1/2}.$$

Squaring both sides and simplifying we obtain, as above,

$$y = -x.$$

EXAMPLE 7

Find a complex form of the equation $x + 3y = 2$.

Letting $z = x + iy$, we recall from Exercise 1.17 that

$$x = \frac{1}{2}(z + \bar{z}) \qquad \text{and} \qquad y = \frac{1}{2i}(z - \bar{z}).$$

Then, substituting in the given equation and simplifying, we obtain

$$(3 + i)z + (-3 + i)\bar{z} = 4i,$$

which is a complex form of the given linear equation.

EXAMPLE 8

Describe by a mathematical relation the totality of all points in the plane that lie inside the circle with center z_0 and of radius r.

Paraphrasing the problem slightly, we can say that we are seeking all points z whose distance from the center z_0 is less than the radius r. But distances between points are conveniently given by the modulus of their difference. So, our locus is expressed by the relation

$$|z - z_0| < r.$$

We proceed now to introduce the "polar form" of a complex number. The reader will recall that a point (x, y) in the plane can also be expressed in terms of its polar coordinates r and θ and that the relations connecting the two coordinate systems are

$$x = r \cos \theta \quad \text{and} \quad y = r \sin \theta, \tag{1}$$

from which one obtains the inverse relations

$$r = [x^2 + y^2]^{1/2} \quad \text{and} \quad \theta = \arc \tan \frac{y}{x}. \tag{2}$$

Now, given any number

$$z = x + iy,$$

substitution from Equation (1) gives

$$z = r(\cos \theta + i \sin \theta),$$

which is called the **polar form** of z. From Equation (2) it is not difficult to see that

$$r = |z| \quad \text{and} \quad \theta = \arg z.$$

Ordinarily, we shall write

$$r \operatorname{cis} \theta$$

in place of

$$r(\cos \theta + i \sin \theta).$$

As we shall see, following the next example, among numerous other uses the polar form of a complex number provides a most convenient way of expressing products, quotients, and integral powers of such a number.

Let $z = r \operatorname{cis} \theta$ be any complex number. Since θ is actually $\arg z$, then, in view of Remark 4, θ can be assigned an infinity of distinct values, any two of which will differ by a multiple of 2π. This situation dictates a careful definition as to what we mean when we say that two numbers in polar form are equal. We define **equality of complex numbers in polar form** as follows:

$$r \text{ cis } t = \rho \text{ cis } \theta$$

if and only if

$$r = \rho \quad \text{and} \quad t = \theta + 2k\pi, \quad k = \text{integer}.$$

EXAMPLE 9

Express the number $z = -5 + \sqrt{75}i$ in polar form.
  We have

$$r = [(-5)^2 + (\sqrt{75})^2]^{1/2} = 10$$

and

$$\theta = \arctan \frac{\sqrt{75}}{-5} = \frac{2\pi}{3} + 2k\pi.$$

Hence,

$$-5 + \sqrt{75}i = 10 \text{ cis } \frac{2\pi}{3},$$

where we have chosen to use the principal value of θ, i.e., Arg z.

Now suppose that we have two arbitrary complex numbers

$$z_1 = r_1 \text{ cis } t_1 \quad \text{and} \quad z_2 = r_2 \text{ cis } t_2.$$

Then, direct calculations yield the following formulas:

$$z_1 z_2 = r_1 r_2 \text{ cis } (t_1 + t_2) \tag{3}$$

and

$$\frac{z_1}{z_2} = \frac{r_1}{r_2} \text{ cis } (t_1 - t_2), \quad \text{for } z_2 \neq 0. \tag{4}$$

These two formulas give the polar form of the product and quotient of two complex numbers. From Equation (3) one extracts the following simple rule:

The product of two complex numbers is a complex number whose modulus is the product of the two moduli and whose argument is the sum of the two arguments.

A similar rule is obtained from Equation (4).

A convenient formula for the integral powers of any complex number can now be deduced; if we set

$$z_1 = z_2 = z = r \operatorname{cis} t,$$

then repeated use of Equation (3) yields

$$z^2 = r^2 \operatorname{cis} 2t, \qquad z^3 = r^3 \operatorname{cis} 3t, \qquad z^4 = r^4 \operatorname{cis} 4t,$$

and by induction

$$z^n = r^n \operatorname{cis} nt \tag{5}$$

for any integer $n \geq 0$. The extension to the negative values of n is then immediate by use of Equation (4), with $z_1 = 1$ and $z_2 = z^n$.

With formula (5) at our disposal we are now in a position to evaluate the nth roots of any nonzero complex number c. Of course, this amounts to solving the equation

$$z^n - c = 0$$

for all its roots. So, given $c = \rho \operatorname{cis} \theta$, we set out to find all numbers $z = r \operatorname{cis} t$ such that $z^n = c$. In view of (5), the last equation becomes

$$r^n \operatorname{cis} nt = \rho \operatorname{cis} \theta.$$

Therefore,

$$r^n = \rho, \qquad \text{and} \qquad nt = \theta + 2k\pi, \qquad k = \text{integer},$$

or

$$r = \rho^{1/n} \qquad \text{and} \qquad t_k = \frac{1}{n}(\theta + 2k\pi), \qquad k = \text{integer}.$$

Since r is a nonnegative real number, it follows that $r = \rho^{1/n}$ represents the real, nonnegative nth root of ρ. On the other hand, as k takes on n consecutive integral values (preferably $k = 0, 1, \ldots, n - 1$), we obtain n distinct values of t, which along with $\rho^{1/n}$ yield the n nth roots of c:

$$z_k = \rho^{1/n} \operatorname{cis} t_k, \qquad k = 0, 1, \ldots, n - 1. \tag{6}$$

It can be shown that formula (6) indeed yields the n distinct roots of c and that any further assignment of values of k would result in repetition of roots already obtained. See Exercise 2.27.

EXAMPLE 10

Find the three cube roots of i.

In effect, we are solving the equation $z^3 = i$. Thus, expressing z and i in polar form and substituting in the above equation, we have

$$r^3 \text{ cis } 3t = 1 \text{ cis } \frac{\pi}{2}.$$

Then,

$$r^3 = 1 \quad \text{and} \quad 3t = \frac{\pi}{2} + 2k\pi,$$

and hence

$$r = 1 \quad \text{and} \quad t_k = \frac{\pi}{6} + \frac{2k\pi}{3}.$$

Letting $k = 0, 1, 2$, we obtain, respectively,

$$t_0 = \frac{\pi}{6}, \quad t_1 = \frac{5\pi}{6}, \quad t_2 = \frac{3\pi}{2}.$$

It follows that the three cube roots of i are

$$z_0 = 1 \text{ cis } \frac{\pi}{6} = \frac{\sqrt{3}}{2} + \frac{i}{2},$$

$$z_1 = 1 \text{ cis } \frac{5\pi}{6} = -\frac{\sqrt{3}}{2} + \frac{i}{2},$$

and

$$z_2 = 1 \text{ cis } \frac{3\pi}{2} = -i.$$

As a special case of the development preceding and illustrated by Example 10, one solves the equation

$$z^n = 1$$

to find the n **nth roots of unity.** See Exercise 2.28.

We close this section with a brief discussion of the geometrical equivalents of the algebraic operations on complex numbers.

Conjugation is actually a reflection in the real axis, as Figure 1.3(a) illustrates. This is easy to see, since conjugation of a complex number simply negates the imaginary part of that number.

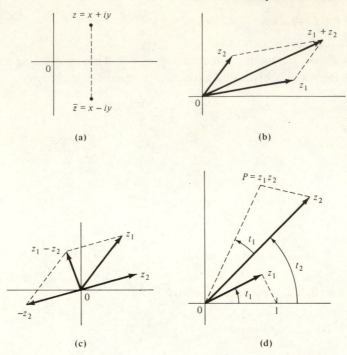

FIGURE 1.3 GEOMETRY OF OPERATIONS. (a) CONJUGATION; (b) ADDITION; (c) SUBTRACTION; AND (d) MULTIPLICATION

   Addition of complex numbers corresponds to addition of two-dimensional vectors since, by definition, the sum of such numbers is obtained by adding respective components. Consequently, the geometry of the operation of addition is the familiar "parallelogram rule" used in the addition of vectors in the plane; see Figure 1.3(b). The reader will find it instructive to argue that the geometry of the difference of two complex numbers is as suggested in Figure 1.3(c), bearing in mind that $z_1 - z_2 = z_1 + (-z_2)$.

   The geometry of the product of two complex numbers is best seen by use of the polar form of the product, from which we derive the rule that we have just given concerning the modulus and the argument of such a product. Given z_1 and z_2 [Figure 1.3(d)], choose the number $z = 1$ on the real axis and form the triangle $\triangle(0, 1, z_1)$. Then, with vector z_2 as one of its sides, form the triangle $\triangle(0, z_2, P)$ similar to the first triangle, while keeping the orientation of the equal angles t_1 the same. By similarity, we have

$$\frac{1}{|z_2|} = \frac{|z_1|}{|P|} \qquad \text{and hence} \qquad |P| = |z_1|\,|z_2|.$$

On the other hand, since by construction

$$\arg P = t_1 + t_2,$$

it follows that

$$P = z_1 z_2,$$

and the product of two complex numbers has been constructed geometrically. An analogous construction yields the quotient of two complex numbers. See Exercise 2.24.

EXERCISE 2

A

2.1. Plot the numbers $3 + 4i, 1 - i, -1 + i, 2, -3i, e + \pi i$, and $-2 + \sqrt{3}i$.

2.2. Find the distance between $2 + i$ and $3 - i$.

2.3. Is $|z^2| = |z|^2$ for all z? Prove your answer.

2.4. Write the polar form of each of the following numbers:

(a) -1.   (b) 3.   (c) $-4i$.

(d) $-2 + 2i$.   (e) $\sqrt{3}i$.   (f) $-\sqrt{27} - 3i$.

(g) $1 - i$.   (h) $2 - i$.   (i) $-\sqrt{2} - \sqrt{2}i$.

(j) $2 - 3i$.

2.5. Check your work in Exercise 2.4 by transforming your answers back to rectangular form.

2.6. Use your answers to Exercise 2.4 to perform the following operations in polar form:

(a) $(-2 + 2i)(1 - i)$.   (b) $-4i/(-2 + 2i)$.

(c) $(1 - i)^6$.   (d) $(-2 + 2i)^{15}$.

2.7. In each of the following cases, find the locus of points in the plane satisfying the given relation:

(a) $|z - 5| = 6$.   (b) $|z + 2i| \geq 1$.

(c) $R(z + 2) = -1$.   (d) $R(i\bar{z}) = 3$.

(e) $|z + i| = |z - i|$.   (f) $|z + 3| + |z + 1| = 4$.

(g) $|z + 3| - |z + 1| = \pm 1$.   (h) $-1 \leq R(z) < 1$.

(i) $I(z) < 0$.   (j) $0 < I(z + 1) \leq 2\pi$.

2.8. If c is a positive real number and z_0 is an arbitrary fixed point in the plane, show that $|z - z_0| = c$ describes a circle with center at z_0 and radius c.

2.9. Find the six sixth roots of unity and plot them. See Exercise 2.28(b).

2.10. Find all the roots of the equation $z^3 + 8 = 0$.

2.11. Solve the equation $z^2 + i = 0$ and then use your answer to solve $z^4 + 2iz^2 - 1 = 0$.

2.12. Find the three cube roots of unity. Then prove that the second and third powers of (at least) one of them yield the other two roots. See Exercise 2.28.

2.13. Use the geometric property of the roots of unity described in Exercise 2.28 to write the polar form of the 12 roots of $z^{12} - 1 = 0$ without solving the equation. Plot the roots.

B

2.14. Prove the following identities:

(a) $|z| = |-z| = |\bar{z}|$.   (b) $|z - w| = |w - z|$.

(c) $|z|^2 = z\bar{z}$.   (d) $|zw| = |z|\,|w|$.

(e) $|z/w| = |z|/|w|,\ w \neq 0$.   (f) $|\,|z| - |w|\,| \leq |z - w|$.

(g) $|R(z)| \leq |z|$ and $|I(z)| \leq |z|$.   (h) $|z| - |w| \leq |z + w|$.

2.15. Choose an arbitrary point z in the plane and then plot the points

$$-z, \quad \bar{z}, \quad -\bar{z}, \quad \frac{1}{z}, \quad -\frac{1}{z}, \quad \frac{1}{\bar{z}}, \quad -\frac{1}{\bar{z}}.$$

2.16. Prove that any point of the form $z = \text{cis } t$, for $t = $ real, lies on the circle $x^2 + y^2 = 1$.

2.17. Prove formula (4). From it then derive a rule analogous to that derived from formula (3).

2.18. Prove that, for any z, arg z + arg $\bar{z} = 2k\pi, k = $ integer.

2.19. Prove that, for any $z \neq 0$ and $w \neq 0$,

$$\text{arg } \frac{z}{w} = \text{arg } z - \text{arg } w.$$

C

2.20. Under what conditions would equality hold in each of the relations of Exercise 2.14(g)?

2.21. Prove that the equation $z^2 + 2z + 5 = 0$ cannot be satisfied by any z such that $|z| \leq 1$.

2.22. If $|z| = 1$, prove that $|z - w| = |1 - \bar{w}z|$, for any w.

2.23. Prove that if $z + 1/z$ is real, then either $I(z) = 0$ or $|z| = 1$.

2.24. Construct geometrically the quotient of two complex numbers using a method similar to that used at the end of this section to construct the product of two numbers.

2.25. If m and n are integers that have no common factors except ± 1 and if $z = r$ cis t, prove that for $k = 0, 1, 2, \ldots, n - 1$,

$$z^{m/n} = r^{m/n} \text{ cis } \frac{m}{n}(t + 2k\pi).$$

2.26. Denoting by w any one of the imaginary nth roots of unity, prove that

$$1 + w + w^2 + \cdots + w^{n-2} + w^{n-1} = 0.$$

2.27. Prove the statement following formula (6).

2.28. Prove the following two properties satisfied by the n nth roots of unity:

(a) Algebraic property

If the n roots are given by formula (6), then consecutive powers of z_1 yield $z_2, z_3, \ldots, z_{n-1}$ and z_0.

(b) Geometric property

The n nth roots of unity are the vertices of a regular polygon of n sides inscribed in the circle $|z| = 1$ and one of whose vertices is $z = 1$.

REVIEW EXERCISES — CHAPTER 1

1. Perform the following operations, writing your answer in the form $A + Bi$.

(a) $\dfrac{(3 + i)(2 - i)}{1 + i}$. (b) $\dfrac{(3 - 2i)^2}{|3 - 2i|}$. (c) $(-4)^{1/4}$.

(d) $(1 + i)^{180}$. (e) $1^{1/8}$. (f) $\dfrac{i}{(1 - i)(2 + i)}$.

(g) $I\left(\dfrac{3 + i}{-1 + 2i}\right)$. (h) $\left|\dfrac{\bar{i}}{i}\right|$, (i) $(-1 + i)^{1/3}$.

(j) $\dfrac{|26 - 36i|^2}{26 + 36i}$.

2. Mark the following statements *true* or *false*.

(a) If c is a real number, then $c = \bar{c}$.

(b) If z is pure imaginary, then $z \neq \bar{z}$.

(c) $i < 2i$.

(d) The argument of zero is zero.

(e) There is at least one number z such that $-z = z^{-1}$.

(f) If $z \neq 0$, then arg z has an infinity of distinct values.

(g) The locus of $I(2\bar{z} + i) = 0$ is a circle.

(h) For any real value of t, $|\cos t + i \sin t| = 1$.

(i) The relation $|z - w| \geq |z| - |w|$ is always true.

(j) The relation $r_1 \text{ cis } t_1 = r_2 \text{ cis } t_2$ implies that $r_1 = r_2$ and $t_1 = t_2$.

3. Under what conditions is $|z + w| = |z| + |w|$?

4. Identify all the points in the plane that satisfy $|z - 2| \leq |z|$.

5. Describe and sketch the following loci:

  (a) $\arg z = \dfrac{\pi}{4}$.   (b) $0 < \arg z < \pi$.   (c) $\pi < \arg z \leq 8\pi$.

6. Prove that if $|z| < 1$, then $R(z + 1) > 0$.

7. Show that the equation $z\bar{z} - z\bar{z}_0 - z_0\bar{z} = r^2 - a^2 - b^2$ represents a circle of radius r and center $z_0 = a + ib$.

8. If the points z, w, and v have magnitude (modulus) equal to 1 and if $z + w + v = 0$, prove that they are equidistant from each other.

9. If z, w, and v are three distinct points of a circle centered at the origin, show that

$$\arg \frac{v - w}{v - z} = \frac{1}{2} \arg \frac{w}{z}.$$

10. If z, w, and v lie on the same line, prove that $I\left(\dfrac{v - z}{w - z}\right) = 0$. Prove that the converse is also true.

11. Prove that, with the exception of zero, the relation $z = -\bar{z}$ holds only for pure imaginaries.

12. Prove the following relations:

  (a) $R(z + w) = R(z) + R(w)$.   (b) $I(z + w) = I(z) + I(w)$.

  (c) $R(zw) = R(z)R(w) - I(z)I(w)$.   (d) $I(zw) = R(z)I(w) + I(z)R(w)$.

13. Solve the equation $z^6 = \dfrac{1 - i}{\sqrt{3} + i}$ for all its roots.

14. If $z = \text{cis } t$, prove the following relations:

  (a) $z^n + \dfrac{1}{z^n} = 2 \cos nt$.   (b) $z^n - \dfrac{1}{z^n} = 2i \sin nt$.

15. Prove: If $I(z + w) = 0 = I(zw)$, then $z = \bar{w}$ or z and w are real.

16. Consider the totality of all numbers of the form

$$\alpha = \cos t + i \sin t, \qquad 0 \leq t < 2\pi.$$

Prove that any complex number can be uniquely written as $r \cdot \alpha$, where r is some real number.

17. Prove: $i\bar{z} = \overline{-iz}$.

18. If the coefficients $a_0, a_1, \ldots, a_n$ of the polynomial

$$P(z) = a_0 + a_1 z + a_2 z^2 + \cdots + a_n z^n$$

are real numbers, prove that $P(\bar{z}) = \overline{P(z)}$.

19. In Appendix 1(A), we prove that the set C of complex numbers forms a mathematical structure called a "field." In general, a field F is called an **ordered field**, provided that it contains a "positive" subset P with respect to which the following axioms hold:

(a) If x is in F, then one and only one of the following holds:

$$x = 0 \qquad \text{or} \qquad x \text{ is in } P \qquad \text{or} \qquad -x \text{ is in } P.$$

(b) If x and y are elements of P, then so is their product xy.

(c) If x and y are elements of P, then so is their sum $x + y$.

Prove that C is not an ordered field.

APPENDIX 1

Part A
A Formal Look at Complex Numbers

An **ordered pair** of real numbers is denoted (a, b). In saying that the pair is ordered, we mean that (a, b) and (b, a) are distinct entities unless $a = b$.*

The **complex numbers system** C is defined to be the totality of all ordered pairs (x, y) of real numbers, where **equality, addition,** and **multiplication** are defined, respectively, as follows:

$$(a, b) = (x, y) \qquad \text{if and only if} \qquad a = x \text{ and } b = y, \tag{1}$$

$$(a, b) + (x, y) = (a + x, b + y), \tag{2}$$

$$(a, b)(x, y) = (ax - by, ay + bx). \tag{3}$$

Any ordered pair (x, y) of real numbers will henceforth be called a **complex number.**

Under the operations defined by (2) and (3), C forms an algebraic system called a *field*; more specifically, C satisfies the following 11 laws:

For any complex numbers $z = (a, b)$, $w = (c, d)$, and $v = (e, f)$ we have:

Closure Laws:

A.1. $z + w$ is a complex number.   M.1. zw is a complex number.

* For a rigorous definition of an ordered pair, see P. R. Halmos, *Naïve Set Theory* (New York: Van Nostrand Reinhold, 1967), Sec. 6.

Commutative Laws:

A.2. $z + w = w + z$.   M.2. $zw = wz$.

Associative Laws:

A.3. $z + (w + v) = (z + w) + v$.   M.3. $z(wv) = (zw)v$.

Identities:

A.4. There is a number α in **C** such that $z + \alpha = z$.   M.4. There is a number β in **C** such that $z\beta = z$.

Inverses:

A.5. For each z in **C** there is z' in **C** such that $z + z' = \alpha$; see A.4.   M.5. For each $z \neq \alpha$ in **C** there is z'' in **C** such that $zz'' = \beta$; see M.4.

Distributive Law:

D. $z(w + v) = zw + zv$.

The number α of A.4 is called the **zero** of the system, and it is unique; see Exercise 1.25. The number z' in A.5 is called the **negative** of z, and it too is unique. The number β of M.4 is called the **unity** of the system, and z'' of M.5 is called the **reciprocal** of z; here, too, β is unique and so is z'' for a given $z \neq \alpha$.

We proceed to prove some of the above properties of **C**; the reader should provide proofs for the remaining ones.

A.1 and M.1 are clearly true, since the right-hand sides of (2) and (3) constitute an ordered pair of real numbers and hence a complex number.

A.2 is proved as follows:

$$z + w = (a, b) + (c, d) = (a + c, b + d) = (c + a, d + b)$$

$$= (c, d) + (a, b) = w + z.$$

Note that, in the process, we have used the commutativity of the real numbers.

M.2 is proved analogously.

A.3 is proved as follows:

$$z + (w + v) = (a, b) + ((c, d) + (e, f))$$

$$= (a, b) + (c + e, d + f)$$

$$= (a + c + e, b + d + f)$$

$$= (a + c, b + d) + (e, f)$$

$$= ((a, b) + (c, d)) + (e, f)$$

$$= (z + w) + v.$$

Here, again, we have used the associativity of the real numbers.

M.3 can be proved similarly.

A.4 is an "existence claim." In order to prove its truth, we must produce a complex number α with the prescribed property. We begin by assuming that such a number exists, say, $\alpha = (x, y)$. Then, for any $z = (a, b)$ we must have $z + \alpha = z$; i.e.,

$$(a, b) + (x, y) = (a, b).$$

But then $(a + x, b + y) = (a, b)$ and by Equation (1), $a + x = a$ and $b + y = b$. Therefore, $x = 0$, $y = 0$ and hence, if α is to exist, it must be of the form

$$\alpha = (x, y) = (0, 0).$$

Clearly, α has the prescribed property, as one can easily verify.

M.4 is established in a similar fashion. We begin by assuming that $\beta = (x, y)$ where the desired property exists so that, for any $z = (a, b)$,

$$(a, b)(x, y) = (a, b).$$

Then

$$(ax - by, ay + bx) = (a, b);$$

hence

$$ax - by = a \qquad \text{and} \qquad ay + bx = b.$$

Solving for x and y, we find that $x = 1$ and $y = 0$. Therefore,

$$\beta = (x, y) = (1, 0).$$

We leave the proofs of A.5, M.5, and D as an exercise for the reader.

The reader familiar with the concept of a vector space will easily recognize **C** as a two-dimensional vector space over the field of real numbers. In this context, a complex number (x, y) can be thought of as a vector with vector addition defined by (2). If one defines scalar multiplication by

$$r(x, y) = (rx, ry), \tag{4}$$

for any real number r, then it is an easy exercise to show that the postulates of a vector space are satisfied. A basis for this vector space is given by the vectors

$$\mathbf{v}_1 = (1, 0) \qquad \text{and} \qquad \mathbf{v}_2 = (0, 1). \tag{5}$$

This, of course, implies that any vector (x, y) can be written as a linear combination of the basis vectors; indeed, using (4), we find that

$$(x, y) = x(1, 0) + y(0, 1).$$

Clearly, v_1 is the unity of **C**. One would probably expect that, perhaps, v_2 is also a special type of a vector when viewed as a complex number, and this is indeed the case. For, we find that

$$(0, 1)(0, 1) = (-1, 0) = -(1, 0); \tag{6}$$

i.e., the square of v_2, *taken as a complex number*, is $-v_1$. Relation (6) plays a fundamental role in the "translation" that we are about to effect on the complex numbers.

The traditional form of the complex numbers is obtained from the above development by making the following identifications:

$$(1, 0) \leftrightarrow 1 \qquad \text{and} \qquad (0, 1) \leftrightarrow i,$$

where $i^2 = -1$. Then, by use of (5) we have the identification

$$(x, y) \leftrightarrow x + yi.$$

EXERCISE — SET 1

Prove that the identification $(x, y) \leftrightarrow x + yi$ preserves the operations on complex numbers as defined, on the one hand, by (2) and (3) in this appendix and, on the other, by the relations defining the same operations in Section 1.

Part B
Stereographic Projection

An alternative and, in some respects, very interesting way of looking at complex numbers is by means of the concept of stereographic projection.

Consider the z-plane and take a sphere $\sum$ of diameter 1, tangent to the plane at some point S: see Figure 1.4. In terms of three-dimensional coordinates, we let S be the point $(0, 0, 0)$. Then the center C of $\sum$ will be at $(0, 0, \frac{1}{2})$. We call the point S the **south pole** and the point $N(0, 0, 1)$ the **north pole** of $\sum$. The entire configuration is often referred to as the **Riemann sphere**.

It is evident that a line joining any point $z = x + iy$ in the plane to the north pole will pierce the sphere at a unique point $P(\alpha, \beta, \gamma)$. Thus, for any z in the plane, there is one and only one point on the sphere corresponding to z: conversely, given any point P on the sphere, the same projection will produce a unique point z on the plane—with one notable exception: the north pole itself. The coordinates α, β, and γ of P are related to the coordinates

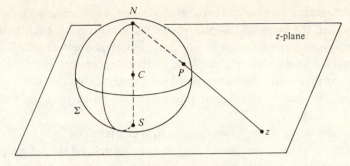

FIGURE 1.4   RIEMANN SPHERE

x and y of z by the formulas

$$\alpha = \frac{x}{1 + |z|^2}, \qquad \beta = \frac{y}{1 + |z|^2}, \qquad \gamma = \frac{|z|^2}{1 + |z|^2}. \qquad (1)$$

The inverse relations are

$$x = \frac{\alpha}{1 - \gamma}, \qquad y = \frac{\beta}{1 - \gamma}, \qquad |z|^2 = \frac{\gamma}{1 - \gamma}. \qquad (2)$$

This process of creating a one-to-one correspondence between the points of the plane and those of the sphere (except N) is called **stereographic projection.** The fact that the north pole corresponds to no point in the plane should be intuitively obvious to the reader. Algebraically, it is obvious from Equations (2), since N is the only point on $\sum$ for which $\gamma = 1$.

A closer inspection of Equations (2) will show that points in the immediate vicinity of the north pole correspond to points in remote areas of the z-plane; for, if a point is very near N, then its third coordinate γ is very near 1 and, if this is the case, then the last of Equations (2) yields a corresponding z of very large modulus. Conversely, if we take a point z in the plane with an arbitrarily large modulus $|z|$, then we see from Equations (1) that α and β will be very near 0, whereas γ will be very near 1; but this says that to such a "remote" z corresponds a point P "very close" to N on the sphere.

The preceding discussion suggests that the exception in the correspondence between the points of the sphere and those of the plane can be eliminated if we adjoin to the z-plane an ideal point with modulus larger than the modulus of any point in the plane and make it correspond to the north pole; this we proceed to do. We thus adjoin to the z-plane the **point at infinity,** denoted ∞, having the property that

$$|z| < \infty$$

for every complex number z. The complex plane augmented with this ideal point is called the **extended complex plane.** Now we can say that the stereographic projection creates a one-to-one correspondence between the Riemann sphere and the extended complex plane, without exception.

Going further with the correspondence created by the stereographic projection, the reader will find it interesting to prove that the following are true:

1. The unit circle $|z| = 1$ corresponds to the equator of $\sum$.
2. A line $y = kx$ in the plane corresponds to a circle on the sphere passing through both the north and the south poles, assuming that the ideal point at infinity is used.
3. A line $y = mx + b$, with $b \neq 0$, corresponds to a circle on $\sum$ passing through the north pole.
4. The interior of the unit circle in the plane "maps" onto the entire southern hemisphere, whereas the exterior of the same circle maps onto the northern hemisphere.
5. A spherical cap about the north pole maps onto a set of points z in the plane such that $|z| > M$, for some real M.
6. Any circle or straight line in the plane corresponds to a circle on the sphere.*

We close this appendix with some remarks concerning the concept of distance between points of the Riemann sphere as it relates to the distance of the corresponding points in the plane.

First, we note that the distance between any two points on the sphere does not exceed 1:

$$0 \leq \text{dist}\ (P_1, P_2) \leq 1.$$

Now, for any two points z_1, z_2 in the plane, let P_1, P_2 be their corresponding points on the sphere. Then, it can be shown that

$$\text{dist}\ (P_1, P_2) = \frac{|z_1 - z_2|}{(1 + |z_1|^2)^{1/2}(1 + |z_2|^2)^{1/2}}.$$

There are a number of interesting observations that one can make relative to the corresponding distances on the plane and on the sphere. For instance, it is easy to see that the distances

$$|(n + 1)^2 - n^2|, \qquad n = 0, 1, 2, \ldots$$

---

* For a proof of assertion 6 see E. Hille, *Analytic Function Theory*, Vol. 1 (Lexington, Mass.: Ginn, 1959), p. 40.

increase without bound in the z-plane, as n becomes large, whereas the corresponding distances on the sphere become infinitesimal as $n \to \infty$. As another illustration, we may consider the sequence

$$\{n \cdot i^n\}, \qquad n = 0, 1, 2, \dots .$$

It is easy to see that the sequence diverges on the plane. However, the sequence of the corresponding points on the sphere converges to the north pole.

EXERCISES — SET 2

1. Derive Equations (1) and (2).
2. Prove assertions 1–6 on p. 30.
3. Derive the formula given above for **chordal distance.**

CHAPTER 2
Analytic Functions

Section 3
Preliminaries

In this chapter we begin the study of functions of one complex variable and their calculus. At first, we consider the most general types of functions and we discuss the concept of the limit of a complex function. We then proceed to develop the hierarchy of functions according to the properties of *continuity*, *differentiability*, and *analyticity*. At each step, the functions become more restricted in the sense that they are required to satisfy more stringent conditions. On the other hand, it is precisely their restricted nature which endows these functions with desirable properties and, consequently, renders such functions more interesting and more useful.

Before we embark on a discussion of these and other items, we digress briefly to introduce certain basic concepts that are peripheral to the main development but indispensable nonetheless. We must also establish some terminology, which we will use for the remainder of this book. So, here is a minimal dose of topological preliminaries. Absorb them; we shall need them.

In what follows, we shall use the term "set" only in the specific sense of a collection of points in the z-plane and no knowledge of set theory on the part of the reader will be assumed.

Let z_0 be a point in the plane and let r be a positive real number. The **r-neighborhood** of z_0 is defined to be the totality of all points z in the plane such that

$$|z - z_0| < r;$$

we shall denote it

$$N(z_0, r).$$

The **deleted r-neighborhood** of z_0 is defined to be the totality of all z such that

$$0 < |z - z_0| < r;$$

we shall denote it

$$N^*(z_0, r).$$

It is easy to see that $N(z_0, r)$ is a circular disk centered at z_0, of radius r and not containing its circumference; $N^*(z_0, r)$ is the same disk with its center removed. When no need arises to specify the radius r, we shall simply talk about a *neighborhood* of a point.

EXAMPLE 1

(a) $N(i, 1)$, the "1-neighborhood of i," is the interior of the circle $|z - i| = 1$; i.e., it consists of all points z such that $|z - i| < 1$; see Figure 2.1(a).

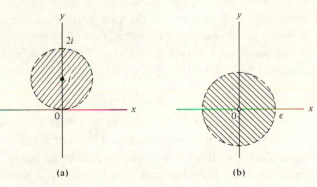

(a)   (b)

FIGURE 2.1   NEIGHBORHOODS. (a) $N(i, 1)$; (b) $N^*(0, \varepsilon)$

(b) $N^*(0, \varepsilon)$ consists of all z such that $0 < |z| < \varepsilon$; i.e., it is the interior of the circle $|z| = \varepsilon$ from which the center $z = 0$ has been removed; see Figure 2.1(b).

Let a set S be given. The **complement** of S is the set of all points in the plane that do not belong to S.

EXAMPLE 2

(a) Let S be the set of all z such that $R(z) > 1$. Clearly, S consists of all points in the plane strictly to the right of the line $x = 1$. Then, the complement of S is the set of all z on and to the left of $x = 1$, i.e., all z such that $R(z) \leq 1$.

(b) Let T be the set of all z such that $1 \leq |z| < 3$. The complement of T consists of all z such that either $|z| < 1$ or $|z| \geq 3$. Draw the configuration.

Again, let S be any set of points in the plane. A point w will be called a **boundary point** of S provided that *every neighborhood* of w contains at least one point of S and at least one point of the complement of S. The set of all boundary points of S is called the **boundary** of S.

EXAMPLE 3

(a) Let S be the disk $|z| < 2$. It is not difficult to see that the boundary of S is the circle $|z| = 2$. For, take *any* point w on the circle and draw $N(w, r)$ for *any* $r > 0$. We see that, no matter what the size of r, $N(w, r)$ contains points of S and points of the complement of S. Hence w is a boundary point of S. Furthermore, no other points have this property, except those on the circle. Note that S contains no part of its boundary.

(b) Let T be the "infinite strip" consisting of all z with $1 < I(z) \leq 3$. The boundary of T consists of the two horizontal lines $y = 1$ and $y = 3$. Note that T contains part but not all of its boundary.

(c) If V is the set of all z with $1 \leq |z - i| \leq 2$, then the boundary of V consists of the two circles $|z - i| = 1$ and $|z - i| = 2$. Here, V contains all of its boundary.

The three cases of Example 3 illustrate the fact that a set may contain no part of its boundary, or it may contain part but not all of its boundary, or it may contain all of it. If a set contains no part of its boundary, then it is called an **open set,** and if it contains all of its boundary, it is called a **closed set.** If it contains part but not all of its boundary, then the set is neither open nor closed. Again, the three cases of the preceding example illustrate, in order, an open set, a set that is neither open nor closed, and a closed set.

The concept of an open set is inseparably connected with the most important concept in complex function theory, namely, analyticity of a complex function. The latter is discussed in the last section of this chapter. The types of open sets that will be involved in our work will be rather simple. In view of this fact, for us the idea of an open set will remain a very simple idea, at least from an intuitive point of view, and so will the idea of the boundary of a set. Both of these concepts will be needed for an in-depth understanding of the development in the chapters that follow.

We shall use the term **region** to refer to a nonempty open set in the plane and the term **closed region** to refer to a region along with its boundary. See Exercise 3.21.

A set B will be called **bounded** if a circle $|z| = M$ can be found that contains all of B; i.e., B is bounded, provided that one can find a positive number M such that $|z| < M$ for every z in the set B. If no such M exists, then the set will be called **unbounded**.

EXAMPLE 4

(a) The set $R(z) > 1$ of Example 2(a) is unbounded.
(b) The set $1 \leq |z| < 3$ of Example 2(b) is bounded.
(c) The set $1 < I(z) \leq 3$ of Example 3(b) is unbounded.

EXAMPLE 5

(a) Any neighborhood or deleted neighborhood of any point z is a region.
(b) The "circular annulus" consisting of the points z with $2 \leq |z + 2| \leq 3$ is a closed region; it consists of the region between the two concentric circles $|z + 2| = 2$ and $|z + 2| = 3$ and the boundary of the region, namely, the two circles.
(c) The segment of the real axis with $-2 \leq x \leq 1$ is a closed set, but it is not a closed region, since it does not consist of a region along with its boundary. Note that the set consists entirely of boundary points and contains no interior points; see Exercise 3.16.

EXERCISE 3

A

In Exercises 3.1–3.13 find the boundary of each of the given sets; determine whether the set is open, closed, or neither and, also, whether it is bounded or unbounded.

3.1. $|z| < 1$.   **3.2.** $0 < I(z) \leq 1$.
3.3. $-2 < R(z) < 0$.   **3.4.** $|z + i| < 2$.
3.5. $1 \leq |z| \leq 3$.   **3.6.** $|z - i| \geq 3$.
3.7. The complement of the set in Exercise 3.2.
3.8. The set of points common to the sets of Exercises 3.2 and 3.3.
3.9. The set of points common to the sets of Exercises 3.1 and 3.2.

3.10. The set of points belonging to at least one of the sets of Exercises 3.1 and 3.5.

3.11. $2 \leq R(\bar{z}) < 5.$ **3.12.** $|2z - 3| \geq 1.$

3.13. $I(z^{-1}) > 3.$

B

3.14. Taken as a set with only one member, is a single point open, closed, or neither? Give reasons.

3.15. Consider the set consisting of the points

$$1, \frac{1}{2}, \frac{1}{3}, \ldots, \frac{1}{n}, \ldots$$

of the real axis. Is this set open, closed, or neither? Give reasons.

3.16. A point z is called an **interior point** of a set S if and only if z belongs to S but is not a boundary point of S. Prove that a set is open if and only if it consists exclusively of interior points.

3.17. Consider the set of all points

$$z = r \operatorname{cis} t, \quad \text{with} \quad 0 < r < 1 \quad \text{and} \quad 0 < t < 2\pi.$$

Is this set open? Closed? What is its boundary? Its complement?

C

3.18. Give an example of a set that has no boundary.

3.19. Prove that the entire plane is a set that is both open and closed.

3.20. Prove that the boundary of any set is itself a closed set.

3.21. The **empty set** is defined to be the set that contains no points.
  (a) Does the empty set have a boundary? Give reasons for your answer.
  (b) Prove that the empty set is both open and closed.

Section 4
Definition and Elementary Geometry of a Complex Function

In form, the definition of a complex function is identical with that of a function of a real variable. Thus, except for replacing the independent variable x by z and the dependent variable y by w, any definition of $y = f(x)$ can be employed to define a complex function $w = f(z)$.

By a **complex variable** z we mean a general point of some set in the plane. A formal definition of a function of a complex variable can be given in terms of ordered pairs (z, w) of complex numbers, required to satisfy certain conditions. This is actually a pairing process that assigns to each value of the variable z a unique value w. We accomplish the same objective with the following, less formal definition.

> Let D be a set of points in the plane; suppose that a rule f is given by which to each point z of D corresponds one and only one point w in the plane.

Then, f is called a **function of a complex variable** or, simply, a **complex function.** Other letters customarily used to denote functions are g, h, k, ..., and the most common way to define a function is by way of an equation of the form

$$w = f(z).$$

The set D is called the **domain** of the function f and the quantity $f(z)$ is called the **value** of f at z or the **image** of z under f. Some examples of functions of a complex variable $z = x + iy$ follow:

$$w = z, \qquad w = 5i, \qquad w = x - iy^2,$$

$$w = 3z^2 - 16z^8, \qquad w = z^{-1}, \qquad w = |z| - i\bar{z} + z^{-2}.$$

Suppose now that we have a function f having domain D and another function g with domain E. Suppose further that, for every z in D, $f(z)$ is in E; in other words, every value of the function f is in the domain of g. Then, for every z in D, $g(f(z))$ is a well-defined function with domain D and is called the **composite function** of f and g. To illustrate, consider the functions

$$f(z) = 3z + i \qquad \text{and} \qquad g(z) = z^2 + z + 1 - i.$$

Then

$$g(f(z)) = g(3z + i)$$

$$= (3z + i)^2 + (3z + i) + 1 - i$$

$$= 9z^2 + (3 + 6i)z$$

is the composite of f and g. The reader can very easily verify that *composition of functions is not commutative;* i.e., in general,

$$g(f(z)) \neq f(g(z)).$$

A very large part of the development of the theory and applications of complex functions depends on the fact that a complex function $w = f(z)$ can be thought of as, and can be decomposed into, the sum of two functions each of which is a real function of two real variables:

$$f(z) = u(x, y) + iv(x, y).$$

Of course, if instead of the rectangular form $x + iy$ of the variable z we use the polar form $z = r$ cis θ, then $f(z)$ can be decomposed into the form

$$f(z) = u(r, \theta) + iv(r, \theta).$$

EXAMPLE 1

Consider the function $w = z^2 + z + 1$.

Letting $z = x + iy$, we obtain the decomposition

$$w = (x^2 - y^2 + x + 1) + i(2xy + y).$$

On the other hand, letting $z = r$ cis t, we get the decomposition

$$w = (r^2 \cos 2t + r \cos t + 1) + i(r^2 \sin 2t + r \sin t).$$

In each case, the function has been decomposed into its real and imaginary components.

We now take a first look at the geometry of a complex function; i.e., we examine the graphical representation of a function $w = f(z)$. By definition, for each value of the independent variable $z = x + iy$ (in the domain of f), the function yields a unique value $w = u + iv$ of the dependent variable. Each of these variables has two dimensions and their combined configuration is a four-dimensional entity that is rather difficult to graph. Because of this difficulty, the graphical representation of a complex function is effected by employing two copies of the complex plane, customarily labeled

$$z\text{-plane} \quad \text{and} \quad w\text{-plane}.$$

Given a function

$$w = f(z),$$

for every $z = x + iy$ of its domain in the z-plane, we calculate the corresponding $w = u + iv$ and we locate it in the w-plane. Repetition of the same process for every point of some set S in the domain of f will yield the "image of S under f" in the w-plane. We illustrate this process in the examples that follow.

EXAMPLE 2

Consider the function $w = \bar{z}$. For each value of the independent variable $z = x + iy$ we obtain a value $w = x - iy$. For example, $z_1 = 2 + 3i$ yields $w_1 = 2 - 3i$, $z_2 = 1 - 2i$ yields $w_2 = 1 + 2i$, and so on; see Figure 2.2.

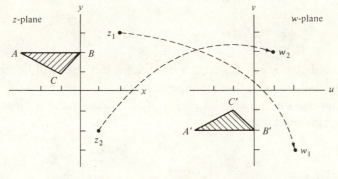

FIGURE 2.2   EXAMPLE 2

It is not difficult to see that the effect of this function on the points of the plane is a reflection in the real axis; in other words, the point w corresponding to a given z is the mirror image of z in the real axis. In more general terms, any geometric figure in the z-plane is reflected in the real axis without deformation, to yield an "inverted" congruent figure. As a result, undirected distances are preserved and so are magnitudes of angles.

EXAMPLE 3

Consider the function $w = z^2$. In Figure 2.3 we plot three points in the z-plane and the corresponding points in the w-plane. Thus we find that

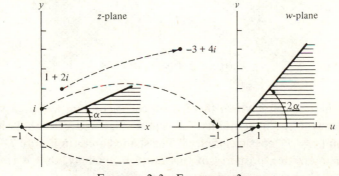

FIGURE 2.3   EXAMPLE 3

under the function $w = z^2$,

$$z = i \qquad \text{goes to} \qquad w = -1,$$

$$z = 1 + 2i \qquad \text{goes to} \qquad w = -3 + 4i,$$

and

$$z = -1 \qquad \text{goes to} \qquad w = 1.$$

The shaded areas in Figure 2.3 illustrate a fact that will be discussed later in detail, namely, that an angular region of α radians in the z-plane is transformed into an angular region of 2α radians in the w-plane, under the given function $w = z^2$.

EXAMPLE 4

The function $w = \dfrac{1 + z}{1 - z}$ is a special case of the bilinear function, which will be studied in some detail in a later section. It is easy to find that, under this function, the points $z = 0, -1, i, -i$ are mapped onto the points $w = 1, 0, i, -i$, respectively; see Figure 2.4. Of special interest is

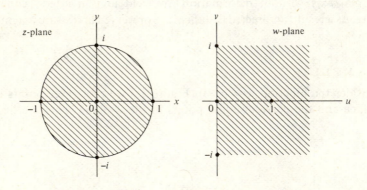

FIGURE 2.4 EXAMPLE 4

the point $z = 1$, for which there is no corresponding w. Later, when we introduce the "point at infinity," we will see that, under the above function, the point $z = 1$ "goes to infinity." The shaded areas in Figure 2.4 suggest a very interesting property of this function, namely, that it transforms the interior of the unit circle into the right half-plane.

E X E R C I S E   4

A

In Exercises 4.1–4.5 find the value of the function at each of the points indicated.

4.1. $f(z) = z^2 - 2z - 1$ at $-1, 1 + 2i$.
4.2. $f(z) = 3x^2 - i\bar{z}$ at $2i, 2 - i$.
4.3. $f(z) = (z + 1)/(z - 1)$ at $i, -i, 3i$.
4.4. $f(z) = |z|^2 - [R(z)]^2$ at $3 + i, -4 - 4i$.
4.5. $f(z) = e^x \cos y + ie^x \sin y$ at $0, 1, 2 + \pi i$.

In Exercises 4.6–4.8 decompose each of the given functions first in the form $u(x, y) + iv(x, y)$ and then in the form $u(r, \theta) + iv(r, \theta)$.

4.6. $f(z) = z^2 + 3z^3$.
4.7. $f(z) = i\bar{z} + I(i/z)$.
4.8. $f(z) = 2 + \pi i$.

In each of Exercises 4.9–4.15 plot the given z's and their corresponding w's under the respective function. Then, in each case try to generalize as to how the z-plane or parts of it are transformed by the given function.

4.9. $w = z$ $z = 1, -1, i, 0, 2 + i$.
4.10. $w = z + 1$ $z = 0, 1 + i, -1, -3 + 2i, -i$.
4.11. $w = z - 2i$ $z = 0, 2i, 1, i, -2i, 1 + i$.
4.12. $w = iz$ $z = 0, 1, 2, 3, 1 + i, 2 + 2i, i, 2i, 3i$.
4.13. $w = R(z)$ $z = 3i, -i, 2i, 1 + i, -1 - 3i, 0, 2, 2 + i$.
4.14. $w = i\bar{z}$ $z = i, 1, -i, -1, 1 + i, -1 - i, 2, -4i$.
4.15. $w = e^x \cos y + ie^x \sin y$ $z = 0, 2\pi i, -2\pi i, 1, 1 - \pi i, 2, \pi i,$
                                     $-2 + (\pi i/2), -2 + (3\pi i/2)$.

B

4.16. Consider the function $w = z^3$.

  (a) Express the function in the form $w = u(r, \theta) + iv(r, \theta)$.

  (b) Show that the angular region $0 < \arg z < \pi/3$ is mapped, under the given function, onto the upper half-plane $0 < \arg w < \pi$.

  (c) Find the area of the z-plane that will be needed to cover the entire w-plane exactly once.

  (d) Argue that if we used every point of the z-plane exactly once, then, except for $w = 0$, every point of the w-plane would be the image of three distinct points in the z-plane. How are the moduli and the arguments of these three points related?

4.17. Consider the function $w = e^x (\cos y + i \sin y)$.

   (a) Find the image, under this function, of the points $1 + \pi i$, $2 + \pi i$, $-1 + \pi i$, $-2 + \pi i$, and $3 + \pi i$. Plot these w's. Then generalize your findings by locating the image, in the w-plane, of the line $z = x + \pi i$, for all real x.

   (b) Find the image, under the above function, of the points $1 + \pi i/4$, $1 + \pi i/2$, $1 + \pi i$, $1 + 3\pi i/2$, and $1 + 2\pi i$. Plot these w's. Then, generalize your findings by locating the image of the line $z = 1 + yi$, for all real y.

   (c) Find the image of each of the lines $R(z) = -2$ and $I(z) = -\pi/2$ under the given function.

   (d) Attempt a generalization by formulating a rule that describes what happens to horizontal and vertical lines in general, under the given function.

Section 5
Limits. Continuity

In our study of the calculus of complex functions we will need some knowledge of the concepts of limit and continuity. This section is devoted to their introduction and the study of their elementary properties.

Let a function

$$w = f(z)$$

be given with domain D, and let z_0 be a fixed point that is either in D or on the boundary of D. Suppose now that the variable z approaches z_0

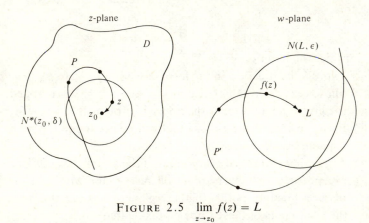

FIGURE 2.5 $\lim\limits_{z \to z_0} f(z) = L$

along some arbitrary path P that lies entirely within the domain D. Clearly, for each z along P, the function yields a point $f(z)$ in the w-plane; see Figure 2.5. If these values $f(z)$ approach a fixed number L in the w-plane, then we say that, as z approaches z_0, the **limit** of $f(z)$ is L and we write

$$\lim_{z \to z_0} f(z) = L.$$

In more formal terms, we say that

the *limit* of $f(z)$, as $z \to z_0$, is L if and only if, given any $N(L, \varepsilon)$, one can find a $N^*(z_0, \delta)$ so that whenever a point z of D is in $N^*(z_0, \delta)$, then $f(z)$ is in $N(L, \varepsilon)$.

Again, put informally, this says that *if L is to be the limit of f, as z approaches z_0, then we should be able to place $f(z)$ as close to L as we please by taking its preimage z sufficiently close to z_0.*

The following remarks constitute an important supplement to the above introduction of limits and are intended to bring out certain subtle points of the concept of a limit.

REMARK 1

The point z_0 in the above definition *need not* belong to the domain of f. Indeed, $f(z)$ may not even be defined at z_0. For instance, a function such as

$$f(z) = \frac{z^2 - 9}{z - 3}$$

can be shown to have a limit equal to 6, as $z \to 3$, even though $f(3)$ does not make sense. The definition of limit allows this situation by requiring that $z \to z_0$ but $z \neq z_0$; the latter follows from the fact that we only consider points z in a deleted neighborhood $N^*(z_0, \delta)$ of z_0. Note, however, that z_0 must "at worst" be on the boundary of the domain of f, so z can approach z_0 through allowable values, i.e., along values for which $f(z)$ is defined.

REMARK 2

The definition of limit does not specify the direction from which z must approach z_0. In fact, the definition tacitly requires that if a limit is to exist, its value L must be independent of the direction of approach. This fact is very useful in proving that a limit does not exist by showing that if $z \to z_0$ along two different paths, then the functional value approach two distinct values, i.e., that the limit depends on the path. See Example 4 and Exercises 5.10 and 5.11.

REMARK 3

Occasionally, we shall find it convenient to use the definition of limit in the following form:

$$\lim_{z \to z_0} f(z) = L \text{ if and only if}$$

for any $\varepsilon > 0$, there is a $\delta > 0$ (which usually depends on ε) so that, for any z (in the domain of f) for which $0 < |z - z_0| < \delta$ it is true that $|f(z) - L| < \varepsilon$.

The reader will find it easy to verify that this form of the definition of limit is equivalent to that given earlier.

We proceed to state a number of theorems that describe some basic properties of limits of functions. Most of these theorems should be familiar to the reader from calculus. As will be our practice throughout, the proofs of the theorems are given in an appendix at the end of the chapter.

The first theorem asserts that if a function has a limit at all, then it has exactly one limit; i.e., if a number L is a limit of a function $f(z)$, as $z \to z_0$, then no other number has that property.

Theorem 2.1 (*Uniqueness of Limit*)
If a function has a limit at a given point z_0, then its limit has a unique value.

Proof:
See Appendix 2.

In many instances, the problem of finding the limit of a function is greatly facilitated by the following theorem, which, in effect, reduces the problem to that of finding the limit of a real function of two real variables.

Theorem 2.2
Suppose that
1. *$f(z) = u(x, y) + iv(x, y)$ has domain D.*
2. *The point $z_0 = a + ib$ is in D or on the boundary of D.*
Then

$$\lim f(z) = A + iB, \quad as \ z \to z_0,$$

if and only if

$$\lim u(x, y) = A \quad and \quad \lim v(x, y) = B, \quad as \quad (x, y) \to (a, b).$$

Proof:
See Appendix 2.

In simple terms, the preceding theorem states that if a function f has a limit L, then the real and imaginary components u and v of f approach, respectively, the real and imaginary parts A and B of L, and conversely.

In view of the fact that the definitions of complex function and limit are formally the same as in the case of real functions, the next theorem should not come as a surprise. It states simply that if each of two given functions has a limit, then the sum, difference, product, and quotient of the functions have limits given, respectively, by the sum, difference, product, and quotient of the respective limits.

Theorem 2.3
Suppose that, as $z \to z_0$, $\lim f(z) = L$ and $\lim g(z) = M$. Then, as $z \to z_0$,
1. $\lim (f(z) + g(z)) = L + M$.
2. $\lim (f(z) - g(z)) = L - M$.
3. $\lim (f(z)g(z)) = LM$.
4. $\lim (f(z)/g(z)) = L/M$, *provided that $M \neq 0$.*

Proof:
The proof of this theorem is identical with that of the corresponding theorem for real functions and is, therefore, omitted. The reader can find the proof in most calculus books and adaptation from the real to the complex case requires only notational changes.

EXAMPLE 1

Consider the **identity function** $f(z) = z$.
For any point z_0 it is clear that, as $z \to z_0$, $f(z) \to z_0$, since $f(z) = z$. Hence, as $z \to z_0$, $\lim f(z) = z_0$.

EXAMPLE 2

In evaluating limits of complex functions, one may employ some of the more direct methods used in calculus. We illustrate such a method by evaluating

$$\lim_{z \to 3 - 4i} \frac{iR(z^2) - iR(z) + [I(z^2)]^2 - 1}{|z|}.$$

First, we note that

$$R(z^2) = x^2 - y^2, \qquad R(z) = x,$$

$$[I(z^2)]^2 = 4x^2y^2, \qquad |z| = (x^2 + y^2)^{1/2}.$$

Then, since $x \to 3$ and $y \to -4$ when $z \to 3 - 4i$, direct substitution yields $115 - 2i$ for the given limit.

EXAMPLE 3

$$\lim_{z \to i} \frac{z - i}{z^2 + 1} = \lim_{z \to i} \frac{z - i}{(z - i)(z + i)} = \frac{1}{2i}.$$

EXAMPLE 4

Show that if $f(z) = \dfrac{2xy}{x^2 + y^2} + \dfrac{x^2}{y + 1}i$, then, as $z \to 0$, $\lim f(z)$ does not exist.

We use the method suggested in Remark 2. So, we let z approach 0 along two different paths and we obtain two different values for the limit. First, we let $z \to 0$ along the real axis ($y = 0$) and we find that

$$\lim_{z \to 0} f(z) = \lim_{(x,0) \to (0,0)} f(z) = \lim_{x \to 0} [x^2 i] = 0.$$

On the other hand, letting $z \to 0$ along the line $y = x$, we have

$$\lim_{z \to 0} f(z) = \lim_{x \to 0} \left[1 + \frac{x^2}{x + 1} i \right] = 1.$$

Since different paths of approach yield different values, the limit does not exist.

EXAMPLE 5

If $f(z) = x^2/z$, find $\lim f(z)$ as $z \to 0$.

Since $|x| \le |z|$ [see Exercise 2.14(g)], it follows that $|x|^2/|z| \le |x|$. Hence

$$|f(z)| = \frac{|x|^2}{|z|} \le |x| \le |z|.$$

Then, as $z \to 0$, $|z| \to 0$, and therefore $|f(z)| \to 0$, since $|f(z)| \le |z|$. But if the modulus of a quantity tends to zero, then so does the quantity itself. Therefore, as $z \to 0$, $\lim f(z) = 0$.

We turn now, briefly, to the concept of continuity.

Let the function $f(z)$ be defined on a set D of the plane and let z_0 be a point in the interior of D.* Then $f(z)$ is said to be **continuous at** z_0 provided that

$$\lim_{z \to z_0} f(z) = f(z_0).$$

If a function is continuous at every point of a region R, then it is said to be **continuous on** R. It is essential to realize that the above definition requires that three conditions be satisfied if a function is to be continuous at z_0:

1. $f(z_0)$ be defined.
2. $\lim f(z)$ exist, as $z \to z_0$.
3. $\lim f(z) = f(z_0)$.

Moreover, the definition implicitly states that if $f(z)$ is to be continuous at z_0, then it must be defined throughout a neighborhood N of z_0, for it requires that z_0 be an interior point of the domain of f; see Figure 2.6.

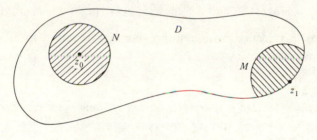

FIGURE 2.6   REMARK 4

REMARK 4

If necessary, the definition of continuity of a function f at a point z_1 can be extended to accommodate the case in which z_1, a point in the domain D of f, is a boundary point of D. This is accomplished by restricting the consideration to paths lying entirely in D. In that case, the definition would imply that f is defined in a *partial neighborhood* M of z_1 contained in D; see Figure 2.6.

The next theorem demonstrates, once again, the importance of the decomposition of a complex function into the form $u(x, y) + iv(x, y)$. It shows that the continuity of a complex function is a necessary and sufficient condition for the continuity of its component functions.

---

* The *interior* of a set S is the set of points that belong to S but not to the boundary of S. See Exercise 3.16.

Theorem 2.4

Suppose that
1. $f(z) = u(x, y) + iv(x, y)$.
2. $f(z)$ is defined at every point of a region R.
3. $z_0 = a + ib$ is a point in R.
Then

> *$f(z)$ is continuous at z_0 if and only if $u(x, y)$ and $v(x, y)$ are continuous at (a, b).*

Proof:

The proof of this theorem is an immediate consequence of Theorem 2.2, since, in essence, what must be proved here is that

$$\lim f(z) = f(z_0), \qquad \text{as } z \to z_0$$

if and only if

$$\lim u(x, y) = u(a, b) \quad \text{and} \quad \lim v(x, y) = v(a, b) \qquad \text{as } (x, y) \to (a, b).$$

We close this section with another familiar theorem as it applies to complex functions.

Theorem 2.5

Suppose that $f(z)$ and $g(z)$ are continuous at some point z_0. Then, each of the following functions is also continuous at z_0:
1. Their sum $f(z) + g(z)$.
2. Their difference $f(z) - g(z)$.
3. Their product $f(z)g(z)$.
4. Their quotient $f(z)/g(z)$, provided that $g(z_0) \neq 0$.
5. Their composite $f(g(z))$, provided that f is continuous at $g(z_0)$.

Proof:

The proof of this theorem is identical with the proof of the corresponding theorem for real functions and is, therefore, omitted. The reader can find the proof in most any calculus book; its adaptation to the complex case involves only notational changes.

EXERCISE 5

A

In Exercises 5.1–5.9 make use of direct methods (see Examples 2 and 3, this section) to find, in each case, the limit of the given function at the point indicated.

5.1. $z^2 + 3$ at $1 + i$.   **5.2.** $z^4 + 1$ at i.

5.3. $\dfrac{z^3 - 1}{z + 1}$ at $3 - 2i$.   **5.4.** $\dfrac{z^3 - a^3}{z - a}$ at a.

5.5. $\dfrac{z^2 + (3 - i)z + 2 - 2i}{z + 1 - i}$ at $-1 + i$.

5.6. $\dfrac{z^n - 1}{z - 1}$ at 1.   **5.7.** $\dfrac{i + R(z)}{|z|}$ at i.

5.8. $\dfrac{I(z^2) - 1}{z\bar{z}}$ at $3 - 4i$.   **5.9.** $\sin \pi x - e^{2xyi}$ at $1 + i$.

B

5.10. Consider the function

$$f(z) = \frac{2xy}{x^2 + y^2} - \frac{y^2}{x^2}i.$$

Find

(a) $\lim f(z)$ as $z \to 0$ along the line $y = x$.
(b) $\lim f(z)$ as $z \to 0$ along the line $y = 2x$.
(c) $\lim f(z)$ as $z \to 0$ along the parabola $y = x^2$.
What can you conclude about the limit of $f(z)$ as $z \to 0$? Justify your answer.

5.11. Use the idea suggested by the preceding exercise and Remark 2 to prove that the following limits do not exist:

(a) $\lim \dfrac{x + y - 1}{z - i}$ as $z \to i$.

(b) $\lim \dfrac{4x}{(x^2 + y^2)^{1/2}}$ as $z \to 0$.

C

5.12. Prove that any **constant function** $f(z) = c$ has a limit at any point z_0 and that the limit is c. As a consequence, prove that any constant function is continuous at any point.

5.13. Prove that, for any z_0 and any nonnegative integer n,

$$\lim z^n = z_0^n, \qquad \text{as } z \to z_0.$$

As a consequence, prove that the function $f(z) = z^n$ is continuous everywhere.

5.14. If n is a nonnegative integer and $a_0, a_1, \ldots, a_n$ are constants, then the function

$$f(z) = a_0 + a_1 z + \cdots + a_n z^n$$

is called a **polynomial.** Prove:
(a) a polynomial has a limit at any point z_0 in the plane.
(b) a polynomial is continuous at any point z_0 in the plane.

5.15. Prove that the function

$$f(z) = \ln |z| + i \, \text{Arg} \, z$$

is not continuous along the nonpositive real axis.

5.16. Use the definition of limit given in Remark 3 in conjunction with property (7) in Section 2 to prove that if $\lim f(z)$ exists (call it L), then $\lim |f(z)|$ exists and $\lim |f(z)| = |\lim f(z)|$.

5.17. Follow the suggested steps to prove that if $\lim f(z) = L$ exists and $|f(z)| \leq E$, then also $|L| \leq E$.
(a) Assume, to the contrary, that $|L| > E$.
(b) Conclude that $|L| = E + P$, for P positive.
(c) Justify each of the following assertions:

$$E \geq |f(z)| \geq |L| - |f(z) - L| = E + P - |f(z) - L|.$$

(d) Finish the proof by arguing that the above relation leads to contradiction.

Section 6
Differentiation

We have already talked briefly about the hierarchy of functions that results from the consideration of the three basic properties of continuity, differentiability, and analyticity. Continuity was briefly discussed in the preceding section, largely as part of the background needed for a discussion of the derivative of a complex function. In this section, we propose to study the stronger property of differentiability, which in turn will take us closer to our ultimate goal in this chapter, namely, the concept of analyticity.

The definition of the derivative of a complex function is formally identical with that of a real function, familiar to the reader from calculus. So, let

$$w = f(z)$$

be a complex function and take any point z_0 in the interior of the domain

D of f (see the footnote on p. 47). Let

$$z = z_0 + \Delta z \qquad (\Delta z = \Delta x + i\Delta y)$$

be any other point in D and form the **difference quotient**

$$\frac{f(z) - f(z_0)}{z - z_0}.$$

See Figure 2.7. If the limit of this quotient exists as $z \to z_0$, then we say

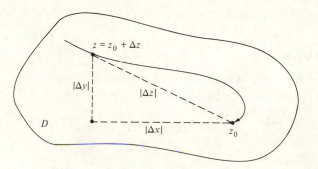

FIGURE 2.7 DERIVATIVE

that $f(z)$ is **differentiable** at z_0; the limit is called the **derivative of f at z_0** and is denoted

$$f'(z_0) \qquad \text{or} \qquad w'(z_0).$$

So, again,

$$f'(z_0) = \lim_{z \to z_0} \frac{f(z) - f(z_0)}{z - z_0},$$

provided that this limit exists.

Note that $f'(z_0)$ is a complex number; if no reference to a particular point z_0 is necessary, then we often use the notation

$$\frac{df}{dz} \qquad \text{or} \qquad \frac{dw}{dz},$$

and in that case we are referring to a function called the *derived function of f* or, simply, the **derivative of f**. The reader should convince himself that the following two expressions, which can be used alternatively to define the derivative of $w = f(z)$, differ from the one given above only in the notation used:

$$f'(z) = \lim_{\Delta z \to 0} \frac{f(z + \Delta z) - f(z)}{\Delta z} \quad \text{or} \quad w'(z_0) = \lim_{z \to z_0} \frac{w - w_0}{z - z_0},$$

where $w_0 = f(z_0)$.

Derivatives of functions may be found by direct use of the definition; the process is identical with that used in calculus and is illustrated in the examples that follow. Later, we shall develop more sophisticated and more direct methods for finding the derivatives of large families of functions.

EXAMPLE 1

Find the derivative of the constant function $f(z) = c$.

Since $f(z) = c$ for any value of z, we have

$$f'(z) = \lim_{\Delta z \to 0} \frac{f(z + \Delta z) - f(z)}{\Delta z} = \lim_{\Delta z \to 0} \frac{c - c}{\Delta z} = 0.$$

Thus the derivative of any constant function is always zero.

EXAMPLE 2

We prove that, for any integer $n \geq 0$ and any point z_0,

$$\text{if } f(z) = z^n, \quad \text{then} \quad f'(z_0) = nz_0^{n-1}.$$

The reader will certainly recognize the above formula as being the familiar "power rule" of differentiation used in calculus. We have

$$f'(z_0) = \lim_{z \to z_0} \frac{f(z) - f(z_0)}{z - z_0}$$

$$= \lim_{z \to z_0} \frac{z^n - z_0^n}{z - z_0}$$

$$= \lim_{z \to z_0} \frac{(z - z_0)(z^{n-1} + z^{n-2}z_0 + \cdots + zz_0^{n-2} + z_0^{n-1})}{z - z_0}$$

$$= \lim_{z \to z_0} (z^{n-1} + z^{n-2}z_0 + \cdots + zz_0^{n-2} + z_0^{n-1}) \quad (n \text{ terms})$$

$$= nz_0^{n-1}.$$

As special cases of the above formula we have

$$\frac{d}{dz}[z] = 1, \qquad \frac{d}{dz}[z^2] = 2z,$$

and so on.

In Example 5 we prove that the above formula holds also for negative integers.

EXAMPLE 3

We prove that the function $f(z) = \bar{z}$ has no derivative at any point, by showing that the value of the limit defining $f'(z)$ depends on the path along which $\Delta z \to 0$. See Remark 2, Section 5.

In the context of Figure 2.8, we have

$$\Delta z = |\Delta z| \operatorname{cis} \alpha, \qquad \text{hence} \qquad \overline{\Delta z} = |\Delta z| (\cos \alpha - i \sin \alpha).$$

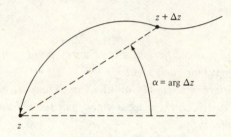

FIGURE 2.8   EXAMPLE 3

Then

$$
\begin{aligned}
f'(z) &= \lim_{\Delta z \to 0} \frac{f(z + \Delta z) - f(z)}{\Delta z} \\[2mm]
&= \lim_{\Delta z \to 0} \frac{\overline{(z + \Delta z)} - \bar{z}}{\Delta z} \\[2mm]
&= \lim_{\Delta z \to 0} \frac{\overline{\Delta z}}{\Delta z} \\[2mm]
&= \lim_{\Delta z \to 0} \frac{\cos \alpha - i \sin \alpha}{\cos \alpha + i \sin \alpha}.
\end{aligned}
$$

Clearly, this limit has no unique value, since it depends on the angle α; e.g., if $z + \Delta z \to z$ along a vertical line ($\alpha = \pi/2$), then the limit is -1, whereas if $z + \Delta z \to z$ along a horizontal line, the limit is 1.

We conclude that the given function has no derivative anywhere.

One could continue as in the preceding examples to calculate (or prove the nonexistence of) the derivative of every function that one may be interested in. However, necessary as it may prove to be in some instances, this process is tedious at best and impossible at worst. So, we direct our efforts toward a different goal. Specifically, we shall develop more general and more direct methods by which we shall be able to find the derivative of practically any given function, provided such derivative exists. Going even further, in the next section we shall develop criteria which will allow us to determine, in the first place, whether a derivative of a given function exists and, if it does, at which points it exists; then, we shall obtain a general formula for finding the derivative.

Examples 1 and 2 suggest a similarity of forms between derivatives of real and complex functions. The following theorem is a further indication in that direction. Once again, the reader will recognize the formulas in this theorem as being precisely those used in calculus.

Theorem 2.6

Suppose that f and g are differentiable at every point z of a set S and that f is differentiable at g(z) for each z in S.

Then their sum, difference, product, quotient, and composite are differentiable at every point of S at which they are defined and their derivatives are given by the following formulas:

1. $(f(z) + g(z))' = f'(z) + g'(z).$
2. $(f(z) - g(z))' = f'(z) - g'(z).$
3. $(f(z)g(z))' = f(z)g'(z) + f'(z)g(z).$

4. $\left|\dfrac{f(z)}{g(z)}\right|' = \dfrac{g(z)f'(z) - f(z)g'(z)}{(g(z))^2}.$

5. $(f(g(z)))' = f'(g(z)) \cdot g'(z).$ (*Chain Rule*)

Proof:

The proof of this theorem is identical with the proof of the corresponding theorem for real functions and it can be found in most any calculus book; it is, therefore, omitted.

We use some of the above formulas in the examples that follow to derive some more familiar rules of differentiation of complex functions that carry over from the real case.

EXAMPLE 4

We show that if c is any constant and $g(z)$ is any differentiable function, then

$$[c \cdot g(z)]' = cg'(z).$$

Taking $f(z) = c$ in formula (3) of the preceding theorem and using Example 1 of this section, we have

$$[c \cdot g(z)]' = cg'(z) + c'g(z)$$

$$= cg'(z) + 0$$

$$= cg'(z),$$

as asserted.

EXAMPLE 5

We extend the result of Example 2 by proving that

$$[z^n]' = nz^{n-1}, \qquad \text{for any integer } n.$$

Of course, in view of the aforementioned example, it suffices to prove the formula for negative integers. So, let k be a negative integer. Then, $-k$ is a positive integer, and hence the function $g(z) = z^{-k}$ has a derivative given by

$$g'(z) = -kz^{-k-1}.$$

Therefore, using formula (4) of Theorem 2.6, we have

$$[z^k]' = \left[\frac{1}{z^{-k}}\right]'$$

$$= \frac{z^{-k} \cdot 0 - (-kz^{-k-1})}{z^{-2k}}$$

$$= kz^{k-1}.$$

Thus the "power formula" holds for every integer. However, if n is negative we must exclude $z = 0$.

It is essentially at this point that the analogy between derivatives of real and complex functions ceases to exist. A classical example attesting to this fact is the pair of functions

$$f(x) = |x|^2 \qquad \text{and} \qquad g(z) = |z|^2.$$

The first function is actually $f(x) = x^2$ and, as is well known, has a derivative at every point x; the second function has a derivative at only one point, $z = 0$ (see Exercise 6.5). The situation becomes even more complicated when one attempts to find derivatives of functions such as

$$f(z) = x^2 + iy \qquad \text{or} \qquad g(z) = e^x + i \sin y \qquad \text{or} \qquad h(z) = R(z)I(z).$$

Not only do we not have any formulas giving us the derivatives of such functions, but also direct use of the definition will not always solve our problem. In fact, the situation often becomes so prohibitive of any progress that one begins to concern himself not with what the derivative is but rather with the question whether the function at hand has a derivative at all. Naturally, this raises the question to which we alluded earlier in this section:

> Is there a criterion that can be employed to determine whether a given function has a derivative at a given point?

A very satisfactory answer to this fundamental question is given in the next section.

We close this section with an example in which we illustrate the use of some of the formulas put forth in Theorem 2.6.

EXAMPLE 6

Using the rules of differentiation from Theorem 2.6 and the examples of this section, we find the following:

1. $\dfrac{d}{dz}(z^3 - 3z^2 + z^{-4} + 2) = 3z^2 - 6z - 4z^{-5}.$

2. $\dfrac{d}{dz}[(z^3 - z^{-2})(z^2 + 5)] = (z^3 - z^{-2})2z + (z^2 + 5)(3z^2 + 2z^{-3}).$

3. $\dfrac{d}{dz}(z^{-1} + 2z + 3)^4 = 4(z^{-1} + 2z + 3)^3(-z^{-2} + 2).$

4. $\dfrac{d}{dz}\left(\dfrac{z^4 - 3}{z^2 + 1}\right) = \dfrac{(z^2 + 1)4z^3 - (z^4 - 3)2z}{(z^2 + 1)^2}.$

EXERCISE 6

A

6.1. Use Theorem 2.6 and the results of the examples of this section to find f' for each of the following functions:

(a) $f(z) = z^6 + 2z^3 - 3.$

(b) $f(z) = (2z + 5)^8(1 - 2z + z^2)^{10}$.

(c) $f(z) = \dfrac{(2z + 5)^8}{(1 - 2z + z^2)^{10}}$.

6.2. Find f' by use of the definition.
(a) $f(z) = z^2 + 3z$.
(b) $f(z) = z^{-1}$.

6.3. In each of the following cases find $f'(z_0)$.
(a) $f(z) = 3z^2 - z^{-1}$ at $z_0 = i$.
(b) $f(z) = z^3 + 2z - 3$ at $z_0 = -1 + i$.
(c) $f(z) = iz^2 + (1 - i)z$ at $z_0 = \pi i$.

B

6.4. By a method similar to that of Example 3, prove that the functions

$$f(z) = R(z) \quad \text{and} \quad g(z) = I(z)$$

have no derivative at any point.

6.5. Use the identity $|z + \Delta z|^2 = (z + \Delta z)(\bar{z} + \overline{\Delta z})$ to show that

$$f(z) = |z|^2$$

has no derivative except at $z = 0$. Find $f'(0)$.

C

6.6. In the next section it will be shown that if a function

$$f(z) = u(x, y) + iv(x, y)$$

has a derivative, then f' is given by

$$f'(z) = u_x + iv_x \quad \text{or} \quad f'(z) = v_y - iu_y.^*$$

Demonstrate the truth of these relations by finding f' for
(a) $f(z) = z^2$.   (b) $f(z) = z^3$.   (c) $f(z) = z^{-1}$.

6.7. According to the result cited in the preceding exercise, if a function $f(z) = u + iv$ has a derivative, then

$$u_x = v_y \quad \text{and} \quad v_x = -u_y.$$

---

* u_x stands for $\partial u/\partial x$, v_x stands for $\partial v/\partial x$, v_y stands for $\partial v/\partial y$, and so on.

Therefore, if either one of these two equations fails to hold, then f' fails to exist. Use this fact to show that none of the following functions possesses a derivative at any point.

    (a) $f(z) = x$.   (b) $f(z) = y$.   (c) $f(z) = \bar{z}$.   (d) $f(z) = x + ix^2$.

6.8. Repeat Exercise 6.5 using facts from Exercises 6.6 and 6.7.

6.9. Prove that if $f'(z_0)$ exists, then f is continuous at z_0.

6.10. Combine Exercises 5.15 and 6.9 to show that the function

$$f(z) = \ln|z| + i \operatorname{Arg} z$$

has no derivative at any point of the nonpositive real axis.

6.11. Prove that any polynomial

$$P(z) = a_0 + a_1 z + \cdots + a_n z^n$$

has a derivative everywhere in the plane.

Section 7
The Cauchy–Riemann Equations

In this section we give a complete answer to the general types of questions that were raised in the preceding section. Specifically, we develop necessary and sufficient conditions for which a given function possesses a derivative. This is accomplished via two theorems. The first will provide us with sufficient conditions which, if satisfied by a given function, will guarantee existence of its derivative. More significantly, the theorem will tell us *where* the derivative exists; i.e., it will specify the points of the plane at which the derivative is defined and, by implication, the points at which it is not. The second theorem will then supply a formula for the derivative, provided that the latter exists.

Theorem 2.7
Given $f(z) = u(x, y) + iv(x, y)$, suppose that

1. $u(x, y), v(x, y)$ and their partial derivatives u_x, v_x, u_y, and v_y are continuous throughout some neighborhood N of some point $z_0 = (a, b)$.

2. At the point z_0, $u_x = v_y$ and $v_x = -u_y$.*

Then $f'(z_0)$ exists and

$$f' = u_x + iv_x = v_y - iu_y.$$

Proof:
See Appendix 2.

---

* See footnote on p. 57.

Theorem 2.8

Suppose that the function $f(z) = u(x, y) + iv(x, y)$ has a derivative at a point $z_0 = (a, b)$.

Then, at that point,

$$f' = u_x + iv_x = v_y - iu_y$$

and hence

$$u_x = v_y \quad and \quad v_x = -u_y.$$

Proof:
See Appendix 2.

The partial differential equations

$$u_x = v_y \quad and \quad v_x = -u_y$$

are called the **Cauchy–Riemann equations.**

It is of interest to note that the continuity of the functions $u(x, y)$, $v(x, y)$, and of their partials, which is one of the sufficient conditions for the existence of f', is not, in general, a necessary condition. Indeed, there exist functions that possess a derivative but whose component functions u and v and their partial derivatives are not all continuous.* It follows that Theorem 2.8 is only a partial converse of Theorem 2.7.

We illustrate the effectiveness of the above two theorems with a number of examples.

EXAMPLE 1

We prove that the derivative of $f(z) = z^2$ exists for all z and that $f'(z) = 2z$.

Writing f in the form $u + iv$, we have

$$f(z) = x^2 - y^2 + 2xyi;$$

hence

$$u(x, y) = x^2 - y^2, \qquad v(x, y) = 2xy,$$

$$u_x = 2x, \qquad v_x = 2y,$$

$$u_y = -2y, \qquad v_y = 2x.$$

The preceding six functions are continuous at every point $z = (x, y)$ of the plane and, clearly,

$$u_x = v_y \quad and \quad v_x = -u_y$$

---

* See, for example, E. Hille, *Analytic Function Theory*, Vol. 1 (Lexington, Mass.: Ginn, 1959), p. 79.

for all (x, y). It follows from Theorem 2.7 that $f'(z)$ exists for all z. This, in turn, implies that the hypothesis of Theorem 2.8 is satisfied for all z. Hence, according to the conclusion of the same theorem,

$$f'(z) = u_x + iv_x = 2x + i2y = 2z.$$

EXAMPLE 2

Determine the points, if any, at which $f(z) = x^2 - iy^2$ has a derivative and wherever f' exists, find it.

We have

$$u = x^2, \qquad v = -y^2,$$
$$u_x = 2x, \qquad v_x = 0,$$
$$u_y = 0, \qquad v_y = -2y.$$

The above six functions are everywhere continuous, but the Cauchy–Riemann equations are satisfied only when $y = -x$. Hence, by Theorem 2.7, f' exists only at the points of that line. Finally, by use of Theorem 2.8, we find that at the points where f' exists it is given by

$$f' = u_x + iv_x = 2x \qquad \text{or} \qquad f' = v_y - iu_y = -2y.$$

Clearly, these two expressions are equal on the line $y = -x$ and at no other points.

EXAMPLE 3

We show that the function $f(z) = \cos y - i \sin y$ has no derivative anywhere.

The six functions

$$u = \cos y, \qquad v = -\sin y,$$
$$u_x = 0, \qquad v_x = 0,$$
$$u_y = -\sin y, \qquad v_y = -\cos y$$

are continuous everywhere. However, if the Cauchy–Riemann equations are to be satisfied, we should have

$$\cos y = 0 \qquad \text{and} \qquad \sin y = 0$$

simultaneously, which is clearly an impossibility. We conclude that f' does not exist at any point.

EXAMPLE 4

We prove that the function $f(z) = e^x (\cos y + i \sin y)$ has a derivative everywhere and that $f'(z) = f(z)$.

The reader can easily verify that the functions $u = e^x \cos y, v = e^x \sin y$, and their first partials are continuous everywhere. It is also easy to verify that the Cauchy–Riemann equations are satisfied identically. Thus, by Theorem 2.7, f' exists everywhere and, indeed,

$$f'(z) = u_x + iv_x = e^x \cos y + ie^x \sin y = e^x (\cos y + i \sin y) = f(z).$$

EXERCISE 7

A

In Exercises 7.1–7.10 determine the points at which the given function has a derivative and wherever f' exists, find it.

7.1. $f(z) = x + iy^2$.

7.2. $f(z) = z^3$.

7.3. $f(z) = 3i$.

7.4. $f(z) = x^2 - iy$.

7.5. $f(z) = \sin x \cosh y + i \cos x \sinh y$.

7.6. $f(z) = R(z)$.

7.7. $f(z) = I(z)$.

7.8. $f(z) = \bar{z}$.

7.9. $f(z) = |z|^2$.

7.10. $f(z) = 2x^2 + 3y^3 i$.

7.11. In Section 5 we saw that if the limit of a function $f = u + iv$ exists, then the limit of its components u and v exists and vice versa. Similarly, we saw that if a function is continuous, then its components are continuous and conversely. Give an example to illustrate the fact that this intimate relation between a function and its components fails in the case of differentiation.

B

7.12. Using a careful definition, one can identify the function $f(z) = \text{Arg } z$ with the function $f(z) = \arctan (y/x)$, for all nonzero $z = x + iy$. Assuming this identification, show that the function $f(z) = \text{Arg } z$ has no derivative anywhere.

7.13. Suppose that $f(z) = u(x, y) + iv(x, y)$ is differentiable at a nonzero point. Then, prove that at that point the **polar form of the Cauchy–Riemann equations** is

$$r \cdot u_r = v_\theta \quad \text{and} \quad r \cdot v_r = -u_\theta.$$

7.14. It can be shown that under certain conditions (see the next exercise) if

$$f(z) = u(r, \theta) + iv(r, \theta), \qquad z \neq 0,$$

then

$$f'(z) = \frac{\bar{z}}{r}(u_r + iv_r).$$

Use this formula to find f' for each of the following functions, assuming that the necessary conditions are satisfied in each case:
(a) $f(z) = \ln r + i\theta, \qquad -\pi < \theta < \pi$.
(b) $f(z) = z^n, \qquad n = $ integer.
(c) $f(z) = r^{1/n} \operatorname{cis} \theta/n; \qquad -\pi < \theta < \pi, n = 1, 2, 3, \ldots$.

C

7.15. Prove the following: If (1) $f(z) = u(r, \theta) + iv(r, \theta)$; (2) u, v, u_r, u_θ, v_r, and v_θ are continuous in a neighborhood of a point z_0; and (3) the Cauchy–Riemann equations are satisfied at z_0, then

$$f'(z) = \frac{\bar{z}}{r}(u_r + iv_r).$$

Section 8
Analytic Functions

The concept of an analytic function is by far the single most important concept in the theory of complex variables. Functions that possess the property of being analytic are endowed with an extremely strong inner structure which manifests itself in the properties shared by such functions. Directly or indirectly, the remainder of this book is devoted to the exploration and exploitation of the properties of analytic functions, which lend themselves to far-reaching developments equally well on both the theoretical and applied fields.

A function $f(z)$ is said to be **analytic* at a point** z_0, provided that its derivative exists *throughout some neighborhood of* z_0. It is clearly evident from the definition that there is a very intimate connection between differentiability and analyticity of a function at a point. However, the two

---

* Also called **holomorphic** or **regular** or **monogenic**.

concepts are not identical, since

> analyticity at z_0 implies differentiability
> at z_0 but not vice versa.

The reason why existence of f' at a point does not imply analyticity at that point is that, in general, whereas f' may exist on any type of a set or even at an isolated point or a line segment, *analyticity is inseparably associated with open sets*; this fact follows from the definition of analyticity at a point z_0, which requires that f' exist not only at z_0 but also throughout a neighborhood of that point.

EXAMPLE 1

In Example 2, Section 7, we found that the function $f(z) = x^2 - iy^2$ possesses a derivative at the points of the line $y = -x$ *and only at those points*. Now, every neighborhood of every point on that line will contain points off the line at which f' does not exist. It follows that f is nowhere analytic, since analyticity at a point demands existence of f' throughout some neighborhood of that point.

EXAMPLE 2

From Exercises 6.5 and 7.9 we know that the function $f(z) = |z|^2$ possesses a derivative only at $z = 0$. It follows, again, that f is nowhere analytic, since f' does not exist throughout any neighborhood of any point.

If a function is analytic at every point of a set S, then it is said to be **analytic on** S. A function analytic on the entire plane is called an **entire** function. Again, it follows from the definition that if a function f is analytic at one point, then it is analytic on an open set containing the point. This fact prompts us to use the term **region of analyticity** of f to describe the totality of all points in the plane at which f is analytic.

EXAMPLE 3

In view of Exercise 6.11, a polynomial

$$P(z) = a_0 + a_1 z + a_2 z^2 + \cdots + a_n z^n$$

is an entire function, since, as we saw, $P'(z)$ exists at all z.
    Similarly, from Example 4, Section 7, we conclude that

$$f(z) = e^x (\cos y + i \sin y)$$

is also an entire function.

EXAMPLE 4

The function

$$f(z) = \frac{z^3 - z + 1}{z^2 + 1}$$

is the quotient of two entire functions, since both its numerator and denominator are polynomials. According to Theorem 2.6, $f'(z)$ exists at every point except $z = \pm i$, where f fails to be defined. Hence f is analytic at all z except at i and $-i$.

Any function that is the quotient of two entire functions is called a **meromorphic** function.

CONVENTION: Ordinarily, when we say that a function is analytic we specify the points, if any, at which it fails to be analytic. However, if such specification is of no immediate significance, we will just say that the function is analytic even though it may fail to be so at some points.

A point z_0 will be called a **singularity** or **singular point** of a function f if and only if f fails to be analytic at z_0 and *every* neighborhood of z_0 contains at least one point at which f is analytic. For instance, the function of Example 4 has two singularities, $z = \pm i$; however, the functions of Examples 1 and 2 have no singularities, although they fail to be analytic at every point in the plane.

The following three theorems are extensions of Theorems 2.6, 2.7, and 2.8, respectively.

Theorem 2.9
Suppose that
1. $f(z)$ and $g(z)$ are analytic on a set S.
2. f is analytic at every $g(z)$ for all z in S.
Then the sum, difference, product, quotient, and composite of f and g are also analytic functions at every point of S at which they are defined.

Proof:
By Theorem 2.6 and the definition of analyticity.

Theorem 2.10
Given $f(z) = u(x, y) + iv(x, y)$, suppose that
1. The functions u, v, and their first partial derivatives u_x, v_x, u_y and v_y are continuous throughout some neighborhood N of a point z_0.
2. The Cauchy–Riemann equations $u_x = v_y$ and $v_x = -u_y$ hold at every point of N.
Then $f(z)$ is analytic at z_0.

Proof:

By Theorem 2.7 and the definition of analyticity.

Theorem 2.11

Suppose that the function $f(z) = u(x, y) + iv(x, y)$ is analytic at a point z_0. Then

$$u_x = v_y \qquad and \qquad v_x = -u_y$$

at every point of some neighborhood of z_0.

Proof:

By Theorem 2.8 and the definition of analyticity.

Analytic functions possess the following most remarkable property which will be established in a later section:

If f is analytic at a point z_0, then so is f'.

We shall borrow this fact at this point to support the following development, which constitutes one of the crucial links connecting the theory and applications of complex function theory.

Let $f(z) = u + iv$ be analytic at z_0; then f' also is analytic at z_0. But since f'' is the derivative of f', then, by the same argument, f'' is analytic at z_0 and, in fact, so are all derivatives of f. Since differentiability implies continuity (see Exercise 6.9) it follows that $f, f', f'', \ldots$ are all continuous at $z_0 = (a, b)$. Now, from Theorem 2.8 we know that derivatives of complex functions are given in terms of partial derivatives of their component functions. Then, in view of Theorem 2.4, since $f', f'', \ldots$ are continuous at z_0, it follows that the partial derivatives of all orders of the functions u and v are continuous at z_0. In particular, this fact implies that the second-order cross-partials are equal:

$$u_{xy} = u_{yx} \qquad and \qquad v_{xy} = v_{yx}. \tag{1}$$

But f is analytic at z_0; hence, at that point,

$$u_x = v_y \qquad and \qquad v_x = -u_y,$$

which upon differentiation yield

$$u_{xx} = v_{yx}, \qquad v_{xx} = -u_{yx}, \qquad v_{yy} = u_{xy}, \qquad -u_{yy} = v_{xy}.$$

Appropriate substitution in (1) then yields

$$u_{xx} + u_{yy} = 0 \qquad and \qquad v_{xx} + v_{yy} = 0.$$

Either one of the last two equations is called **Laplace's equation**; any function $g(x, y)$ that satisfies Laplace's equation throughout some neighborhood of a point $z_0 = (a, b)$ is said to be **harmonic** at z_0, provided that it has continuous second-order partials there. We have thus shown that

> the real and imaginary components of an analytic function
> $f = u + iv$ are harmonic functions;

such pairs of harmonic functions are called **conjugate harmonic** functions.

As was remarked at the beginning of the preceding discussion, the fact that the components u and v of an analytic function satisfy Laplace's equation constitutes the basis for most applications of complex function theory to problems arising in a large number of fields, such as fluid dynamics, electrostatics, heat flow, and others. Crucial in this respect is also the fact that, given a harmonic function u, one can very easily find its conjugate harmonic v and thus form an analytic function $f(z) = u + iv$. The process by which the conjugate harmonic is determined is illustrated in the example that follows.

EXAMPLE 5

The function $v(x, y) = xy$ is easily seen to be harmonic, since $v_{xx} = v_{yy} = 0$. We find its conjugate harmonic; i.e., we find a function $u(x, y)$ such that $f(z) = u + iv$ is analytic.

If f is to be analytic, the Cauchy–Riemann equations must hold, and since $v_y = x$, we must also have $u_x = x$, from which, by integration,

$$u(x, y) = \tfrac{1}{2}x^2 + h(y). \tag{1}$$

Now, let us determine $h(y)$. From (1) we find that $u_y = h'(y)$. But $u_y = -v_x$ (why?) and since, from the given equation, $v_x = y$, we have

$$h'(y) = -y.$$

Then

$$h(y) = -\tfrac{1}{2}y^2 + c;$$

hence, from (1),

$$u(x, y) = \tfrac{1}{2}x^2 - \tfrac{1}{2}y^2 + c.$$

Therefore,

$$f(z) = u + iv$$
$$= \tfrac{1}{2}z^2 + c.$$

EXERCISE 8

A

In Exercises 8.1–8.10 determine the region of analyticity of the given function. Justify your answers.

8.1. z.

8.2. y^2.

8.3. z^3.

8.4. $z^2 - 1$.

8.5. $\dfrac{z - i}{z + 1}$.

8.6. $\dfrac{z^2 + z}{z(z^2 + 1)}$.

8.7. $\ln |z| + i \operatorname{Arg} z$.

8.8. $\sin x \cosh y + i \cos x \sinh y$.

8.9. $e^{x^2 - y^2}(\cos 2xy + i \sin 2xy)$.

8.10. $I(\bar{z})$.

In Exercises 8.11–8.16 demonstrate that the real and imaginary components of each function are harmonic functions.

8.11. $z^2 + z$.

8.12. z^3.

8.13. $1/z$.

8.14. $e^x(\cos y + i \sin y)$.

8.15. $\sin x \cosh y + i \cos x \sinh y$.

8.16. $e^{x^2 - y^2}(\cos 2xy + i \sin 2xy)$.

In Exercises 8.17–8.20 show that the given function is harmonic. Then find its conjugate harmonic v to form an analytic function $f = u + iv$.

8.17. $u = x$.

8.18. $u = xy$.

8.19. $u = \ln (x^2 + y^2)$.

8.20. $u = e^x \cos y$.

B

8.21. Show that if f' exists at a point, then it is given by

$$f' = u_x - iu_y = v_y + iv_x.$$

8.22. Suppose that $f(z)$ is analytic in a region R and that $f'(z) = 0$ for all z in R. Prove that f is a constant function throughout R.

8.23. Suppose that $f(z) = u + iv$ and its conjugate function $\overline{f(z)} = u - iv$ are analytic in a region R. Prove that f is constant throughout R.

C

8.24. Refer to the two paragraphs immediately following Theorem 2.11 to prove that if $f = u + iv$ is analytic, then the partial derivatives of all orders of the functions u and v are continuous functions on the region of analyticity of f.

8.25. Suppose that $f(z) = u + iv$ is analytic in a region R. Prove that if u is a real constant, then f is constant throughout R. Is the conclusion true if v is a real constant?

8.26. Suppose that $f(z)$ and $g(z)$ are analytic in a region R. Moreover, if z_0 is a point in R, suppose that $f(z_0) = g(z_0) = 0$ and $g'(z_0) \neq 0$. Prove that **L'Hospital's rule** for analytic functions holds:

$$\lim_{z \to z_0} \frac{f(z)}{g(z)} = \frac{f'(z_0)}{g'(z_0)}.$$

8.27. Study the analyticity of the function $f(z) = |x^2 - y^2| + i|2xy|$.

REVIEW EXERCISES — CHAPTER 2

1. Mark the following statements *true* or *false*.
(a) If f is continuous at z_0, then $f'(z_0)$ exists.
(b) If a set is not open, then it is closed.
(c) If $\lim f(z) = A + iB$, as $z \to z_0$, then $\lim R[f(z)] = A$.
(d) If $f'(z_0)$ exists, then f is analytic at z_0.
(e) If f is analytic at a point, then so is $-f$.
(f) If z_0 is a singularity of f and g, then it is also a singularity of the function $f + g$.
(g) If $f(z) = x^2 yi$, then $f'(z)$ does not exist anywhere.
(h) The quotient of two entire functions is an entire function.
(i) A function satisfying Laplace's equation is called *meromorphic*.
(j) The functions $u(x, y) = x + y$ and $v(x, y) = x + y$ are conjugate harmonics, since they satisfy Laplace's equation.

2. Suppose that $f(z) = u + iv$ is analytic in a region R and that $u^2 + v^2$ is constant in R. Prove that f is constant in R.

3. Find the derivative of $f(z) = 1/z^2$ by use of the definition of f'.

4. Prove that the **polar form of Laplace's equation** is

$$r^2 u_{rr} + r u_r + u_{\theta\theta} = 0.$$

5. Prove that $f(z) = |z|^4$ is differentiable at the origin. Is it analytic there?

6. Prove that each of the following functions is continuous for all z. Then give a specific reason to justify the fact that all of them fail to be analytic anywhere.
(a) $f(z) = R(z)$.   (b) $f(z) = \bar{z}$.
(c) $f(z) = |z|$.   (d) $f(z) = |z|^2$.

7. Show that, as $z \to 0$, the limit of each of the following functions does not exist; see Example 4 in Section 5.

   (a) $f(z) = \dfrac{xy}{x^4 + y^4} + 2xi.$ (b) $f(z) = \dfrac{x^3 y^2}{x^6 + y^4} - \dfrac{x}{y}i.$

8. Given $f(z) = zR(z)$, find the points, if any, at which f' exists. Is f analytic anywhere?

9. Prove that the r-neighborhood of a point z_0 is an open set.

10. Draw any geometrical figure on the z-plane and discuss the effect of the function $f(z) = \bar{z} + 1$ on that figure as a mapping from the z-plane to the w-plane.

11. Explain why the following statement is false regardless of what f may be: "f is analytic at every point z of the unit disk $|z| \leq 1$ and at no other point in the plane."

12. Determine the conjugate harmonic of the function $u = \cosh x \cos y$.

13. Form an analytic function $f = u + iv$ by finding the conjugate harmonic of the function $u(x, y) = -y$.

APPENDIX 2

Proofs of Theorems

Theorem 2.1   (*Uniqueness of Limit*)

If a function f has a limit at a given point z_0, then its limit has a unique value.

Proof:
The proof is by contradiction. Thus, we assume that the limit of f has two values:

$$\lim_{z \to z_0} f(z) = M \qquad \text{and} \qquad \lim_{z \to z_0} f(z) = L, \tag{1}$$

where $M \neq L$. Consider now the positive number

$$\varepsilon = \tfrac{1}{2}|M - L|.$$

By the definition of limit, Equations (1) imply that, given this ε, there is a $\delta > 0$ such that

$$|f(z) - M| < \varepsilon \qquad \text{and} \qquad |f(z) - L| < \varepsilon \tag{2}$$

whenever

$$0 < |z - z_0| < \delta.$$

Then, using the triangle inequality and relations (2), we have

$$|M - L| = |M - f(z) + f(z) - L|$$
$$\leq |M - f(z)| + |f(z) - L|$$
$$< \varepsilon + \varepsilon$$
$$= \tfrac{1}{2}|M - L| + \tfrac{1}{2}|M - L|$$
$$= |M - L|.$$

But, from the last sequence of relations, it follows that $|M - L| < |M - L|$, which is clearly an impossibility. Thus our assumption has led us to an absurdity. Therefore, existence of two distinct limits is impossible and the theorem is proved.

Theorem 2.2
Suppose that
1. $f(z) = u(x, y) + iv(x, y)$ has domain D.
2. The point $z_0 = a + ib$ is in D or on the boundary of D.
Then

$$\lim f(z) = A + iB, \qquad as \ z \to z_0,$$

if and only if

$$\lim u(x, y) = A \quad and \quad \lim v(x, y) = B, \qquad as \ (x, y) \to (a, b).$$

Proof:
We first prove that if $\lim f = A + iB$, then $\lim u = A$ and $\lim v = B$. We accomplish this by showing that for any $\varepsilon > 0$, there is a $\delta > 0$ such that

$$0 < |x - a| < \delta \qquad and \qquad 0 < |y - b| < \delta$$

imply that

$$|u(x, y) - A| < \varepsilon \qquad and \qquad |v(x, y) - B| < \varepsilon.$$

So, let $\varepsilon > 0$ be given. By hypothesis, there is $\eta > 0$ such that

$$0 < |z - (a + ib)| < \eta \qquad \text{implies that} \qquad |f(z) - (A + iB)| < \varepsilon.$$

Take $\delta = \eta/2$. Then (see Figure 2.9), for any $z = (x, y)$ such that

$$0 < |x - a| < \delta \qquad and \qquad 0 < |y - b| < \delta,$$

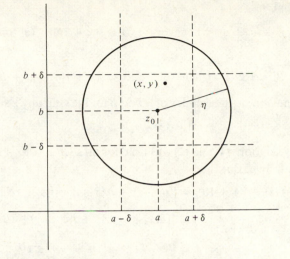

$$\text{Figure 2.9 \ Proof of Theorem 2.2}$$

we shall have, by use of the triangle inequality,

$$0 < |(x + iy) - (a + ib)|$$

$$= |(x - a) + i(y - b)|$$

$$\leq |x - a| + |y - b|$$

$$< \delta + \delta$$

$$= \eta.$$

But then, by hypothesis,

$$|f(z) - (A + iB)| < \varepsilon$$

or, which is the same,

$$|[u(x, y) - A] + i[v(x, y) - B]| < \varepsilon.$$

Finally, using Exercise 2.14(g), it follows that

$$|u(x, y) - A| < \varepsilon \qquad \text{and} \qquad |v(x, y) - B| < \varepsilon,$$

which is what we set out to·prove, namely, that

$$\text{as } (x, y) \to (a, b), \qquad \lim u(x, y) = A \quad \text{and} \quad \lim v(x, y) = B.$$

We now prove the converse. By hypothesis, $\lim u = A$ and $\lim v = B$ as $(x, y) \to (a, b)$. This, in turn implies that for any $\varepsilon > 0$ there exist $\alpha > 0$ and

$\beta > 0$ such that

$$|u(x, y) - A| < \varepsilon/2 \quad \text{whenever} \quad 0 < |(x + iy) - (a + ib)| < \alpha \quad (1)$$

and

$$|v(x, y) - B| < \varepsilon/2 \quad \text{whenever} \quad 0 < |(x + iy) - (a + ib)| < \beta. \quad (2)$$

Now, choosing δ to be the smaller of α and β and taking any z such that

$$0 < |z - z_0| < \delta,$$

then certainly relations (1) and (2) will hold for all such $z = x + iy$ and with α and β replaced by δ. But then

$$|f(z) - (A + iB)| = |[u(x, y) + iv(x, y)] - [A + iB]|$$

$$= |[u(x, y) - A] + i[v(x, y) - B]|$$

$$\leq |u(x, y) - A| + |v(x, y) - B|$$

$$< \varepsilon/2 + \varepsilon/2$$

$$= \varepsilon.$$

Therefore, $\lim_{z \to z_0} f(z) = A + iB$ and the proof is complete.

Theorem 2.7

Given $f(z) = u(x, y) + iv(x, y)$, suppose that
1. *$u(x, y)$, $v(x, y)$ and their partial derivatives u_x, v_x, u_y, and v_y are continuous throughout some neighborhood N of some point $z_0 = (a, b)$.*
2. *At the point z_0, $u_x = v_y$ and $v_x = -u_y$.*
Then $f'(z_0)$ exists and

$$f' = u_x + iv_x = v_y - iu_y.$$

Proof:
From the calculus of two real variables we know that hypothesis (1) guarantees the following*: For any point $(a + \Delta x, b + \Delta y)$ in N,

$$\Delta u = u(a + \Delta x, b + \Delta y) - u(a, b) = u_x \Delta x + u_y \Delta y + \alpha \Delta x + \beta \Delta y,$$

where α and β tend to zero as Δx and Δy tend to zero. Similarly,

$$\Delta v = v(a + \Delta x, b + \Delta y) - v(a, b) = v_x \Delta x + v_y \Delta y + \gamma \Delta x + \delta \Delta y,$$

where, again, γ and δ tend to zero with Δx and Δy. Now, by hypothesis (2), the preceding two relations become

$$\Delta u = u_x \Delta x - v_x \Delta y + \alpha \Delta x + \beta \Delta y \quad (i)$$

* See R. Courant, *Differential and Integral Calculus*, Vol. II (New York: Wiley-Interscience, 1968), pp. 59–62.

and

$$\Delta v = v_x \, \Delta x + u_x \, \Delta y + \gamma \, \Delta x + \delta \, \Delta y. \tag{ii}$$

Then,

$$f(z_0 + \Delta z) - f(z_0) = [u(a + \Delta x, b + \Delta y) + iv(a + \Delta x, b + \Delta y)]$$

$$- [u(a, b) + iv(a, b)]$$

$$= [u(a + \Delta x, b + \Delta y) - u(a, b)]$$

$$+ [v(a + \Delta x, b + \Delta y) - v(a, b)]i$$

$$= \Delta u + \Delta v \cdot i.$$

In view of (i) and (ii), the equation above yields

$$\frac{f(z_0 + \Delta z) - f(z_0)}{\Delta z} = (u_x + iv_x)\frac{\Delta x}{\Delta z} + (u_x + iv_x)\frac{\Delta y}{\Delta z}i$$

$$+ (\alpha - i\gamma)\frac{\Delta x}{\Delta z} + (\beta + i\delta)\frac{\Delta y}{\Delta z}$$

$$= u_x + iv_x + (\alpha + i\gamma)\frac{\Delta x}{\Delta z} + (\beta + i\delta)\frac{\Delta y}{\Delta z}. \tag{iii}$$

Next, we take the limit of the above relation as $\Delta z \to 0$. Clearly, the quotient on the left will yield the derivative of f at $z_0 : f'(z_0)$. On the other hand, as $\Delta z \to 0$ then also $\Delta x \to 0$ and $\Delta y \to 0$, and hence all of α, β, γ, and δ tend to zero; consequently,

$$\alpha + i\gamma \to 0 \qquad \text{and} \qquad \beta + i\delta \to 0.$$

Now, in view of Exercise 2.14(g), we have

$$\left|\frac{\Delta x}{\Delta z}\right| \le 1 \qquad \text{and} \qquad \left|\frac{\Delta y}{\Delta z}\right| \le 1.$$

We conclude then that, as $\Delta z \to 0$, the last two terms of (iii) tend to zero, and, therefore, the derivative of f at z_0 exists and, indeed,

$$f'(z_0) = u_x + iv_x.$$

The second part of the formula in the conclusion of the theorem is obtained in a similar fashion. Specifically, by use of hypothesis (2), we can express equations (i) and (ii) in terms of u_y and v_y and then proceed analogously. This will complete the proof.

Theorem 2.8

Suppose that the function $f(z) = u(x, y) + iv(x, y)$ has a derivative at a point $z_0 = (a, b)$.

Then, at that point,

$$f' = u_x + iv_x = v_y - iu_y$$

and hence

$$u_x = v_y \quad and \quad v_x = -u_y.$$

Proof:

Since $f'(z_0)$ exists, the limit defining f' must be independent of the path along which $\Delta z \to 0$. In particular, the value of f' at z_0 will be the same if we choose a horizontal path. Then, of course, $\Delta y = 0$ and hence

$$f'(z_0) = \lim_{\Delta z \to 0} \frac{f(z_0 + \Delta z) - f(z_0)}{\Delta z}$$

$$= \lim_{\Delta z \to 0} \frac{[u(a + \Delta x, b + \Delta y) + iv(a + \Delta x, b + \Delta y)] - [u(a, b) + iv(a, b)]}{\Delta x + i\, \Delta y}$$

$$= \lim_{\Delta x \to 0} \frac{[u(a + \Delta x, b) + iv(a + \Delta x, b)] - [u(a, b) + iv(a, b)]}{\Delta x}$$

$$= \lim_{\Delta x \to 0} \frac{u(a + \Delta x, b) - u(a, b)}{\Delta x} + i \lim_{\Delta x \to 0} \frac{v(a + \Delta x, b) - v(a, b)}{\Delta x}.$$

By definition, the last two limits are u_x and v_x, respectively. Therefore,

$$f'(z_0) = u_x + iv_x.$$

A choice of a vertical path ($\Delta x = 0$) and a similar argument will yield

$$f'(z_0) = v_y - iu_y,$$

and the theorem follows.

CHAPTER 3
Elementary Transformations

Section 9
Mapping

In Section 4 we acquainted ourselves with the geometric aspect of a complex function, which we can think of as a process that "sends" points of the z-plane onto points of the w-plane. More generally, a function can be thought of as a process by which entire sections of the z-plane are "mapped" onto parts of the w-plane. This aspect of a function has generated the terms **mapping** and **transformation** as alternative names for "function." For instance, we say that

the function $w = z^2 + i$ **maps** $z = 1 - i$ onto $w = -i$,

or that

the function $w = 2iz + i$ **transforms** the square $ABCD$
onto the square $A'B'C'D'$,

See Figure 3.1.

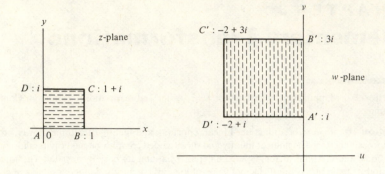

FIGURE 3.1 THE MAPPING $w = 2iz + i$

If a function f maps z_0 onto w_0, then we say that w_0 is the **image** of z_0 under f and that z_0 is a **preimage** of w_0. Note that, whereas the definition of a function forces one to speak about *the* image of a point z, a point w may have more than one preimage under a given function; for instance, under the function

$$w = z^4 + 2,$$

the point $w = 3$ has four preimages: $z = 1, -1, i, -i$.

A mapping $w = f(z)$ under which *no point* w has more than one preimage is called a **one-to-one** mapping; otherwise, it is called **many-to-one**. Put in different terms, a function f is one-to-one if *distinct points* of its domain are mapped onto *distinct points*; i.e., f is one-to-one, provided that $z_1 \neq z_2$ implies that $f(z_1) \neq f(z_2)$.

EXAMPLE 1

1. The function $w = e^x \cos y + ie^x \sin y$ is many-to-one, since, for instance, the points $z = 0, 2\pi i, 4\pi i, 6\pi i, 8\pi i, \ldots$, map onto one and the same point $w = 1$, as the reader can easily verify.

2. The function $f(z) = 3z - 5i$ is one-to-one. We prove this claim by assuming that, for some z_1 and z_2 with $z_1 \neq z_2$ in the domain of f, it is true that $f(z_1) = f(z_2)$; i.e., we assume the contrary of what we wish to prove. But this leads us to a contradiction; for, if $f(z_1) = f(z_2)$, then $3z_1 - 5i = 3z_2 - 5i$; hence $z_1 = z_2$, contradicting our assumption that $z_1 \neq z_2$ with $f(z_1) = f(z_2)$ is possible. Therefore, f is one-to-one.

Next, suppose that a function f is analytic at a point z_0 and that $f'(z_0) \neq 0$. Suppose further that two curves A and B intersect at z_0, forming an angle α, measured from A to B^*; see Figure 3.2. Now, under f, A and B have their

---

* Recall that the angle of intersection of two curves is the angle formed by their tangents at the point of intersection.

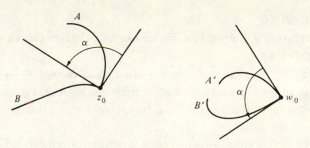

FIGURE 3.2 CONFORMALITY

images A' and B' in the w-plane, and they intersect at $w_0 = f(z_0)$. It is shown in Appendix 3(B) that the angle formed by A' and B' at w_0 and measured from A' to B' is α. This fact shows that if a function is analytic at z_0 and if it has a nonzero derivative there, then the function preserves angles both in magnitude and direction. A function having this property is called a **conformal mapping**. Certain aspects of conformal mapping are studied in some detail in Appendix 3(B). Our brief introduction of the concept at this point will enable us to appreciate some instances where conformality occurs, as we study some elementary functions in the remaining sections of this chapter.

EXERCISE 9

A

In each of Exercises 9.1–9.4, determine whether the given function is one-to-one; if it is, prove it (see Example 1). If it is not one-to-one, give at least one example illustrating your assertion.

9.1. $w = 3i$.　　　　　**9.2.** $w = z - i$.

9.3. $w = (z + 1)/(z - 1)$.　　**9.4.** $w = z^3 - 3$.

B

9.5. It is possible, in general, to change a given many-to-one function into a one-to-one function by appropriately restricting its domain. How would you restrict the domain of $w = z^2$ to obtain a one-to-one function?

9.6. Study the mapping of Example 2, Section 4, carefully. Is it a conformal mapping? Explain why.

Section 10
Elementary Complex Functions. Definitions and Basic Properties

We devote this section to the introduction of certain elementary complex functions and to the study of some of their algebraic and analytic properties. The mapping properties of these functions are studied in the remaining sections of this chapter.

At the end of the discussion of each function introduced in this section, the reader may choose to refer to the corresponding section in which we study the mapping properties of that function, or he may choose to finish this section before studying the functions as mappings; the development allows either choice.

Before we begin our study of the elementary functions, we discuss briefly the concept of the "inverse of a function." By definition, $g(z)$ is called an **inverse** of a function $f(z)$, provided that $f(g(z)) = g(f(z)) = z$. The reader may recall that the inverse of a function is not necessarily a function. However, if f is one-to-one (see p. 76), then its inverse, customarily denoted f^{-1}, is also a function; conversely, if f is many-to-one, then its inverse is not in general a function; see Exercise 9.5. The following example illustrates the two basic aspects of the concept.

EXAMPLE 1

1. Consider $f(z) = 3z - 5i$. As we showed in Example 1.2, Section 9, f is a one-to-one function; i.e., it maps distinct z's onto distinct w's. It is easy to see that $f^{-1}(z) = (z + 5i)/3$. The reader may verify that $f(f^{-1}(z)) = f^{-1}(f(z)) = z$, as prescribed by the definition.

2. The function $w = z^2$ is a many-to-one function since, for any $z \neq 0$, it maps both z and $-z$ onto the same w. Consequently, its inverse (which we will eventually define and denote $z = w^{1/2}$) is not a function. Exercise 9.5 suggests a method by which we can obtain inverses that are functions even if we start with a many-to-one function, by appropriately restricting the domain of the latter. Closely associated with this idea is the concept of "multivalued functions," which are introduced in Appendix 3(A).

THE LINEAR FUNCTION

A function of the form

$$f(z) = az + b,$$

where a and b are complex constants, is called a **linear function**. Its derivative $f'(z) = a$ is defined at every z; hence f is an entire function.

If $a = 0$, then f reduces to a **constant function**: $f(z) = b$. If $a \neq 0$, then f is a one-to-one function, since $z_1 \neq z_2$ implies that $az_1 + b \neq az_2 + b$, hence $f(z_1) \neq f(z_2)$. In this case, the inverse relation

$$z = \frac{1}{a}w - \frac{b}{a}$$

is also a linear function, which can be thought of as mapping the w-plane "back" onto the z-plane. Finally, if $a = 1$ and $b = 0$, then the linear function reduces to the **identity function** $f(z) = z$.

The mapping properties of the linear function are studied in Section 11.

THE POWER FUNCTION

For any positive integer n, the function

$$f(z) = z^n$$

is called the **power function**. In Example 2, Section 6, we prove that

$$f'(z) = nz^{n-1}$$

and is defined for all z. Therefore, f is an entire function. It is easy to see that, for $n > 1$, f is a many-to-one function. As a consequence, its inverse is not, in general, a function.

Certain aspects of the mapping $w = z^n$ are studied in Section 12.

THE RECIPROCAL FUNCTION

The function

$$f(z) = \frac{1}{z}$$

is called the **reciprocal function**. It is a one-to-one function between the z-plane, except $z = 0$, and the w-plane, except $w = 0$. In Example 5, Section 6, we showed that the derivative of f is given by

$$f'(z) = -\frac{1}{z^2},$$

and hence it exists for all $z \neq 0$. Thus, the reciprocal function is analytic on the whole plane, save the origin. This fact follows also from Theorem 2.9.

So far, we have excluded from our discussion the point $z = 0$, which has no image under the reciprocal function, as well as the point $w = 0$, which has no preimage. In what follows, we will find it convenient and useful to eliminate these exceptions by introducing the "point at infinity."

> The *point at infinity*, denoted ∞, is an ideal point
> which has the property that, for any z, $|z| < \infty$.

The z-plane augmented with this ideal point is called the **extended complex plane**; see also Appendix 1(B). Although we will find it expedient to use expressions such as "the point $z = \infty$," the point at infinity is not to be treated as a number, especially when it comes to using algebraic operations on it. We emphasize once again that ∞ is an *ideal point* whose only property we know at this stage of our discussion is that *it is larger in magnitude than any number z*; i.e., for all z, $|z| < \infty$.

The above definition of the point at infinity is motivated by the fact that, under the function $w = 1/z$, if we let $z \to 0$, then the corresponding w's will be numbers of arbitrarily large moduli; i.e., as $z \to 0$, $|w|$ "tends to infinity."

With the point at infinity at our disposal, we can now say that, under the reciprocal function, the image of $z = 0$ is $w = \infty$ and the preimage of $w = 0$ is $z = \infty$.

The natural question that arises here is whether we can consider, in general, the behavior of any given function at the point $z = \infty$. The answer to this question is in the affirmative if we adopt, as we do, the following convention:

> The behavior of a function $f(z)$ at $z = \infty$ will be identified with
> the behavior of $f\left(\dfrac{1}{z}\right)$ at the point $z = 0$.

Again, this convention is motivated by the same limiting process:

$$\frac{1}{z} \to 0 \text{ as } z \to \infty, \text{ and vice versa.}$$

We illustrate this notion in the following

EXAMPLE 2

We examine the behavior of the function $f(z) = \dfrac{z}{1 + z}$ at $z = \infty$.

According to the above convention, we examine instead the behavior of $f\left(\dfrac{1}{z}\right)$ at $z = 0$. We find that

$$f\left(\frac{1}{z}\right) = \frac{1}{z + 1},$$

which, at $z = 0$, yields $w = 1$. Hence, at $z = \infty$, the given function is assigned the value 1.

Mapping properties of the reciprocal function are studied in Section 13.

THE BILINEAR FUNCTION

If n is a nonnegative integer and $a_0, a_1, \ldots, a_n$ are complex constants, then the function

$$P(z) = a_0 + a_1 z + \cdots + a_n z^n$$

is called a **polynomial**. In view of Example 3, Section 8, a polynomial is an entire function. Now let $P(z)$ and $Q(z)$ be two polynomials. Then the function

$$F(z) = \frac{P(z)}{Q(z)},$$

defined for all z such that $Q(z) \neq 0$, is called a **rational function**. According to Theorem 2.9, a rational function is analytic at every point where its denominator is not zero.

Of special interest to us is the rational function

$$f(z) = \frac{az + b}{cz + d}, \qquad (ad - bc \neq 0) \tag{1}$$

which is called a **bilinear function**.* Since it is a rational function, it is analytic everywhere except $z = -d/c$. Clearly, if $c = 0$, then the bilinear map reduces to a linear function.

For the remainder of our discussion of the bilinear function we assume that $c \neq 0$. Under this condition, Equation (1) represents a one-to-one function from the extended z-plane to the extended w-plane. In particular, the point $z = -d/c$ maps onto the point $w = \infty$ and the point $z = \infty$ maps onto $w = a/c$.

A simple algebraic manipulation yields the inverse of the bilinear function:

$$z = \frac{-dw + b}{cw - a};$$

it is easy to see that the above is not only a function but, in fact, a bilinear function that can be pictured as mapping the extended w-plane "back" onto the extended z-plane in a one-to-one fashion.

A discussion of some of the mapping properties of the bilinear function appears in Section 14.

---

* This function is also called **Moebius transformation** or **linear fractional map**.

THE EXPONENTIAL FUNCTION

Undoubtedly, one of the most important functions in all of mathematics is the **exponential function**, which, for the case of a complex variable $z = x + iy$, is defined by

$$e^z = e^x(\cos y + i \sin y).$$

We will see shortly that, in a certain sense, the function just defined is a "natural extension" of the function e^x to the complex case. We note, for instance, that if z is a real number, in which case $y = 0$, then $e^z = e^x$; this is an indication that the manner in which the complex exponential was defined above constitutes a generalization of the real exponential.

If z is pure imaginary ($x = 0$), we have

$$e^{iy} = \cos y + i \sin y,$$

which is known as **Euler's formula**. Originally studied by Euler about two centuries ago, this form can be employed to write the polar form

$$z = r(\cos t + i \sin t)$$

of a complex number in the compact form

$$z = re^{it}.$$

We have already proved in Example 4, Section 7, that the exponential is an entire function and, indeed, that

$$\frac{d}{dz}(e^z) = e^z.$$

This fact is a further indication that our choice of definition for e^z preserves all the "usual" properties of the real exponential, familiar to the reader from calculus.

Following is a list of the most basic algebraic properties of e^z. Again, most of these properties should be familiar to the reader. The truth of some of these properties is established in the examples that follow; the reader will benefit by establishing the remaining ones.

Properties of e^z

For any complex quantities z and w the following hold:
1. $e^z \neq 0$.
2. $e^0 = 1$.
3. $e^{z+w} = e^z e^w$.
4. $e^{z-w} = e^z/e^w$.
5. $\overline{e^z} = e^{\bar{z}}$.
6. $e^z = e^{z+2\pi i}$ (periodicity of the exponential).
7. If $z = x + iy$, then $|e^z| = e^x$ and arg $(e^z) = y$.

We hasten to remark that property 7 is an immediate consequence of the definition of e^z, since any complex quantity written in the form

$$R(\cos T + i \sin T),$$

where R and T are real quantities, has modulus R and argument T.

EXAMPLE 3

We prove that, for any z, $e^z \neq 0$.

The proof is by contradiction. So, suppose that a number $z = a + ib$ exists such that

$$e^z = 0.$$

Then

$$e^a \cos b + ie^a \sin b = 0;$$

hence

$$e^a \cos b = 0 \quad \text{and} \quad e^a \sin b = 0$$

simultaneously. But since the real exponential e^a is never zero, it must be that

$$\cos b = 0 \quad \text{and} \quad \sin b = 0.$$

But this is an impossibility for every value of b. It follows that no such z exists; hence, $e^z \neq 0$ for all z.

EXAMPLE 4

We prove the periodicity of the exponential: $e^z = e^{z + 2\pi i}$, for all $z = x + iy$.

We have

$$e^{z + 2\pi i} = e^{x + (y + 2\pi)i}$$

$$= e^x \operatorname{cis} (y + 2\pi)$$

$$= e^x \operatorname{cis} y$$

$$= e^{x + iy}$$

$$= e^z.$$

Note that, in proving the periodicity of the exponential, we made use of the periodicity of $\sin y$ and $\cos y$ in order to effect the third step of the above proof.

The periodicity of the exponential has a very interesting geometrical interpretation, which will be discussed in Section 15. In the following example, the reader will find an algebraic illustration of this property of the complex exponential. It should be pointed out that after we introduce the logarithm of complex numbers, problems of the type discussed in the next example will be much easier to handle. However, the reader will profit greatly by studying this next example carefully.

EXAMPLE 5

Find all roots of the equation $e^z = -i$.

We provide the main steps of the process; the reader should supply the missing details and justifications.

The given equation can be written

$$e^x \cos y + ie^x \sin y = 0 - i.$$

Hence,

$$e^x \cos y = 0 \qquad \text{and} \qquad e^x \sin y = -1.$$

From the first of these two equations we obtain

$$y = \frac{\pi}{2} + k\pi, \qquad k = \text{integer}.$$

But then, the second of the two equations above becomes

$$\pm e^x = -1,$$

from which we have, as the only possibility, $e^x = 1$; hence, $x = 0$.

Now, it is easy to verify that, if $y = \pi/2 + k\pi$ for $k =$ even integer, then $e^x \sin y = -1$ is impossible to satisfy; consequently, the allowable values of y are restricted to those obtained when k is odd:

$$y = -\frac{\pi}{2} + 2k\pi, \qquad k = \text{integer}.$$

It follows that the roots of the given equation are

$$z = 0 + \left(-\frac{\pi}{2} + 2k\pi \right) i,$$

In turn, this means that

$$e^{(-\pi/2 + 2k\pi)i} = -i,$$

which is an illustration of the periodicity of the exponential function.

Some of the basic mapping properties of the complex exponential are studied in Section 15.

THE LOGARITHMIC FUNCTION

In our discussion of the complex exponential we saw that our definition of that function was chosen so as to preserve some of the familiar properties of the real exponential and to extend them to the complex case. As we are about to introduce the logarithm of a complex quantity and, subsequently, the logarithmic function, we are motivated by the same desire, namely, to preserve as many of the familiar properties of logarithms as possible.

Suppose now that, given any complex number z, we have, somehow, attached a specific and unambiguous meaning to the symbol

$$\log z.$$

Suppose further, that the concept represented by this symbol constitutes an extension of the real (natural) logarithm $\ln x$; i.e., suppose that

$$\textit{if z is a positive real number,} \qquad \textit{then} \log z = \ln z. \tag{1}$$

Finally, suppose that the familiar properties of $\ln x$ are shared by $\log z$ and, in particular, that

$$\log (zw) = \log z + \log w \tag{2}$$

and

$$\log (z^z) = \alpha \cdot \log z \tag{3}$$

for any complex numbers z, w, and α. In this *hypothetical context*, let $z = re^{it}$ be the complex number given above. Then

$$
\begin{aligned}
\log z &= \log (re^{it}) \\[4pt]
&= \log r + \log (e^{it}) & \text{by (2)} \\[4pt]
&= \log r + it \cdot \log e & \text{by (3)} \\[4pt]
&= \ln r + it \cdot \ln e & \text{by (1)} \\[4pt]
&= \ln r + it \\[4pt]
&= \ln |z| + i \arg z.
\end{aligned}
$$

Thus, on the assumption that some of the properties of real logarithms are shared by $\log z$, we find that

$$\log z = \ln |z| + i \arg z. \tag{4}$$

It turns out that (4) is an excellent candidate for a definition of the logarithm of a complex quantity other than zero. Thus we define the **logarithm** of z by

$$\log z = \ln |z| + i \arg z, \qquad \text{for all } z \neq 0. \tag{5}$$

The following remarks are in order at this point.

REMARK 1
In definition (5), we have chosen to distinguish between "log" and "ln" in order to avoid using the symbol being defined in its own definition. We thus use "log" to denote the logarithm of an arbitrary complex number z (which may well be a real number) as defined by (5), and we use "ln" to denote the unique value of the natural logarithm of the positive real number $|z|$.

REMARK 2
Note that "the logarithm" of z is actually "the logarithms" of z; indeed, the presence of arg z in (5) tells us that log z has an infinite number of distinct values. However, any two of these values differ by an integral multiple of $2\pi i$; see, also, Remark 4, Section 2.

REMARK 3
In spite of our original intention, the logarithm of a positive real number, as defined by (5), is not exactly the same as the natural logarithm "ln"; for, if R is a positive real number, then arg $R = 2k\pi$. Hence

$$\log R = \ln R + 2k\pi i, \qquad k = \text{integer}.$$

However, this is only a superficial shortcoming, which will be rectified very shortly when we define the "principal value" of log z.

EXAMPLE 6

We find the logarithms of the numbers $z = i$, 2, $-ei$, and -1.
    We have

$$\log i = \ln |i| + i \arg i = \ln (1) + i\left(\frac{\pi}{2} + 2k\pi\right) = i\left(\frac{\pi}{2} + 2k\pi\right).$$

$$\log 2 = \ln |2| + i \arg 2 = \ln 2 + i(2k\pi) = \ln 2 + 2k\pi i;$$

see Remark 3.

$$\log (-ei) = \ln (e) + i\left(\frac{3\pi}{2} + 2k\pi\right) = 1 + \left(\frac{3\pi}{2} + 2k\pi\right)i.$$

$$\log (-1) = (\pi + 2k\pi)i.$$

As we pointed out at the beginning of our discussion of complex logarithms, we would like to have log z defined in such a way that familiar properties of the real logarithms will be preserved. It turns out that log z, as defined by (5), shares most of those properties. However, before we state and prove some of these properties, it is imperative to understand the following: Since the logarithm of a complex number has an infinity of distinct values, any two of which differ by $2k\pi i$, each side of the equalities involved in these properties is not a single number but a set of numbers. In view of this fact, we will take each of these equalities to mean that every number of either side is equal to every number of the other side give or take an integral multiple of $2\pi i$.

It is in the sense of the preceding paragraph and only in that sense that we will understand the validity of the following.

Properties of log z

For any nonzero numbers z and w, the following hold:
1. $\log(zw) = \log z + \log w$.
2. $\log(z/w) = \log z - \log w$.
3. $\log e^z = z$.
4. $e^{\log z} = z$.
5. $\log(z^p) = p \cdot \log z$, for any rational number* p in lowest terms.

EXAMPLE 7
We prove property 1.

$$\log(zw) = \ln|zw| + i \arg(zw)$$
$$= \ln[|z||w|] + i[\arg z + \arg w]$$
$$= \ln|z| + \ln|w| + i \arg z + i \arg w$$
$$= [\ln|z| + i \arg z] + [\ln|w| + i \arg w]$$
$$= \log z + \log w.$$

EXAMPLE 8
We prove property 3.

Since $|e^z| = e^x$ and $\arg(e^z) = y$, we have

$$\log e^z = \ln|e^z| + i \arg(e^z) = \ln e^x + iy = x + iy = z.$$

---

* A real number p is called **rational**, provided that it is the quotient of two integers: $p = m/n$, where m and n are integers and $n \neq 0$.

EXAMPLE 9

Solve the equation $e^{z+1} = -2$ for its roots.

Taking logarithms of both sides, we have

$$\log [e^{z+1}] = \log (-2).$$

Then

$$z + 1 = \ln 2 + (\pi + 2k\pi)i$$

and, therefore,

$$z = (\ln 2 - 1) + (2k + 1)\pi i.$$

Since the logarithm of every nonzero complex number has an infinity of distinct values, it is obvious that log z as defined by (5) cannot be used to define a function, since, by definition, to each allowable value of z must correspond *one and only one* image. We overcome this obstacle by use of the **principal value of log** z, which we define by

$$\text{Log } z = \ln |z| + i \text{ Arg } z, \quad z \neq 0, \qquad -\pi < \text{Arg } z \leq \pi$$

We then define the **logarithmic function** to be the function

$$f(z) = \text{Log } z.$$

Clearly, the polar form of the logarithmic function is

$$\text{Log } z = \ln r + it, \qquad -\pi < t \leq \pi.$$

From the definitions of log z and Log z, it is evident that these two forms are related by the formula

$$\log z = \text{Log } z + 2k\pi i, \qquad k = \text{integer}.$$

The derivative of the logarithmic function is precisely what we have expected it to be:

$$\frac{d}{dz}[\text{Log } z] = \frac{1}{z}. \tag{6}$$

We may prove this fact either by use of Theorems 2.7 and 2.8, or, in polar form, by use of Exercises 7.13 and 7.14; the reader is asked to establish Equation (6) in the exercises at the end of this section. In view of Exercise 6.10, the derivative of the logarithmic function exists everywhere except at the points along the nonpositive real axis. We conclude, then, that the function $f(z) = \text{Log } z$ is analytic at every nonzero z such that $-\pi < \text{arg } z < \pi$; see Figure 3.3.

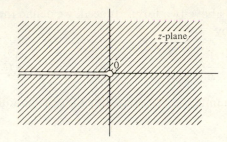

FIGURE 3.3 REGION OF ANALYTICITY OF Log z

In our study of the "multivalued functions" in Appendix 3(A), we will see that our choice of definition for Log z was one of many which we could have made. We could have chosen, for instance, Log $z = \ln r + it$ with $0 \leq t < 2\pi$, or $\pi/2 < t \leq 5\pi/2$, or $\alpha \leq t < \alpha + 2\pi$ for any angle α, and still have a logarithmic function defined for every $z \neq 0$. The only stipulation that we would have to observe would be to restrict the argument of z to precisely *one cycle*. The fact that we chose the ray arg $z = \pi$ as the "cutoff" ray of our cycle is only a matter of preference and nothing more.

The properties of log z are, of course, shared by Log z; however, the equalities in those properties will now have to be taken with the understanding that, whenever necessary, the arguments will be altered by the appropriate multiple of $2\pi i$ to comply with the restriction

$$-\pi < \arg z \leq \pi.$$

Finally, we note that the properties

$$\text{Log } e^z = z \qquad \text{and} \qquad e^{\text{Log } z} = z$$

tell us that the functions e^z and Log z are inverses of each other. The geometrical manifestation of this fact will emerge most clearly during our discussion of the logarithmic function as a mapping in Section 15.

TRIGONOMETRIC AND HYPERBOLIC FUNCTIONS

By use of Euler's formula, it is easy to verify that if x is a real number, then

$$\sin x = \frac{1}{2i}(e^{ix} - e^{-ix}) \qquad \text{and} \qquad \cos x = \frac{1}{2}(e^{ix} + e^{-ix}).$$

These two formulas can be said to represent the *complex form of the real sine and cosine functions*. The definitions of the sine and cosine functions of a complex variable z that we are about to adopt are nothing but natural extensions of the above two forms. Indeed, we define the two basic **trigono-metric functions** by

$$\sin z = \frac{1}{2i}(e^{iz} - e^{-iz}) \quad \text{and} \quad \cos z = \frac{1}{2}(e^{iz} + e^{-iz}),$$

for all complex numbers z. The remaining four trigonometric functions are defined as usual:

$$\tan z = \frac{\sin z}{\cos z}, \quad \cot z = \frac{\cos z}{\sin z}, \quad \sec z = \frac{1}{\cos z}, \quad \csc z = \frac{1}{\sin z}.$$

It is clear from their definitions that $\sin z$ and $\cos z$ are entire functions; the other four trigonometric functions are meromorphic and they fail to be analytic precisely at the points where their respective denominators vanish.

As in the case of the exponential and the logarithmic functions, a large number of properties of the real trigonometric functions carry over to the complex case. Following is a partial list of such properties.

Properties of $\sin z$ and $\cos z$

For any two complex numbers z and w, the following hold:
1. $\sin z = 0$ if and only if $z = k\pi$, k = integer.
2. $\cos z = 0$ if and only if $z = \pi/2 + k\pi$, k = integer.
3. $\sin(-z) = -\sin z$.
4. $\cos(-z) = \cos z$.
5. $\sin^2 z + \cos^2 z = 1$.
6. $\sin(z + w) = \sin z \cos w + \sin w \cos z$.
7. $\cos(z + w) = \cos z \cos w - \sin z \sin w$.
8. $|\sin z|^2 = \sin^2 x + \sinh^2 y$, where $z = x + iy$.
9. $|\cos z|^2 = \cos^2 x + \sinh^2 y$, where $z = x + iy$.
10. $\frac{d}{dz}[\sin z] = \cos z$.
11. $\frac{d}{dz}[\cos z] = -\sin z$.

EXAMPLE 10
We prove property 1 as follows.
First, if $z = k\pi$, then, clearly,

$$\sin z = \frac{1}{2i}(e^{k\pi i} - e^{-k\pi i})$$

$$= \frac{1}{2i}(\cos k\pi + i \sin k\pi - \cos k\pi + i \sin k\pi) = \sin k\pi = 0.$$

Conversely, suppose that sin $z = 0$. Then,

$$\frac{1}{2i}(e^{iz} - e^{-iz}) = 0; \qquad \text{hence} \qquad e^{iz} = e^{-iz}.$$

It follows that

$$e^{2iz} = 1,$$

which, by use of logarithms, yields

$$2iz = 2k\pi i, \qquad k = \text{integer},$$

and, therefore,

$$z = k\pi.$$

EXAMPLE 11

We prove property 10 as follows.

$$\frac{d}{dz}[\sin z] = \frac{d}{dz}\left[\frac{1}{2i}(e^{iz} - e^{-iz})\right]$$

$$= \frac{1}{2i}(ie^{iz} + ie^{-iz})$$

$$= \frac{1}{2}(e^{iz} + e^{-iz}) = \cos z,$$

where, in taking the derivative of the exponentials, we have made use of Exercise 10.19.

EXAMPLE 12

Decompose $\cos z$ in the form $u + iv$.

Letting $z = x + iy$, we have

$$\cos z = \frac{1}{2}(e^{iz} + e^{-iz})$$

$$= \frac{1}{2}(e^{-y}e^{ix} + e^{y}e^{-ix})$$

$$= \frac{1}{2}[e^{-y}(\cos x + i \sin x) + e^{y}(\cos x - i \sin x)]$$

$$= \frac{1}{2}(e^{y} + e^{-y}) \cos x - \frac{i}{2}(e^{y} - e^{-y}) \sin x.$$

Therefore,

$$\cos z = \cos x \cosh y - i \sin x \sinh y.$$

Similarly, one finds that the decomposition of $\sin z$ is given by

$$\sin z = \sin x \cosh y + i \cos x \sinh y.$$

We close this section with a brief acquaintance with the hyperbolic functions of a complex variable whose definitions should look familiar to the reader.

The **hyperbolic sine** is defined by

$$\sinh z = \tfrac{1}{2}(e^z - e^{-z})$$

and the **hyperbolic cosine** by

$$\cosh z = \tfrac{1}{2}(e^z + e^{-z}).$$

Clearly, $\sinh z$ and $\cosh z$ are entire functions and their derivatives are given by

$$\frac{d}{dz}[\sinh z] = \cosh z \qquad \text{and} \qquad \frac{d}{dz}[\cosh z] = \sinh z,$$

as the reader may easily show by simply differentiating the exponentials that define these two functions.

The remaining four hyperbolic functions are defined as usual:

$$\tanh z = \frac{\sinh z}{\cosh z}, \qquad \coth z = \frac{\cosh z}{\sinh z}, \qquad \operatorname{sech} z = \frac{1}{\cosh z},$$

$$\operatorname{csch} z = \frac{1}{\sinh z}.$$

A comparison of the expressions used to define $\sin z$, $\cos z$, $\sinh z$, and $\cosh z$ immediately suggests a close functional relation among these four functions. Indeed, there is a number of identities connecting these functions. Some of these identities appear in the exercises at the end of this section.

Certain mapping properties of the trigonometric and hyperbolic functions are examined briefly in Section 16.

EXERCISE 10

A

10.1. Write each of the following in the form $A + iB$:

(a) $e^{i\pi/2}$.   (b) $e^{1-\pi i}$.   (c) $e^{-7\pi i}$.

(d) $e^{\ln 2 + \pi i/3}$.   (e) $e^{2-2\pi i}$.

10.2. Establish the truth of the relation $e^{i\pi} + 1 = 0$, which connects the eight most important symbols and numbers of our number system:

$$0, 1, i, e, \pi, \cdot, +, \text{ and } =.$$

10.3. (a) Use the definition of e^z to prove properties 3 and 5 on p. 82.
(b) Use property 3 to establish properties 2 and 4.

10.4. Use $z = re^{it}$ and the appropriate properties of the exponential to prove formulas (3) and (4), p. 17.

10.5. Find all z satisfying each of the following equations:
(a) $e^z = -3i$. (b) $e^z = 1 - i$.

10.6. Prove that if $z = re^{it}$, then $\bar{z} = re^{-it}$.

10.7. Find the logarithm of each of the following numbers:
(a) $-i$. (b) 1. (c) $1 + i$. (d) $3 + 4i$. (e) $2 - i$.

10.8. Find $\log \sqrt{-i}$ first by using property 5, p. 87, and then by determining the two values of $\sqrt{-i}$ and finding their logarithms. Comment on the apparent discrepancies in your answers.

10.9. Find all z satisfying the equation $\log (z + 1) = i\pi$.

10.10. Find all roots of the equation $e^{2z} = -i$.

10.11. Use the definitions of the *complex* functions involved to write the following numbers in the form $A + iB$:
(a) $\cos \pi$. (b) $\sin (\pi/2)$. (c) $\sinh (\pi i)$.
(d) $\cosh (2i)$. (e) $\tan (-\pi i)$. (f) $\sin (i)$.
(g) $\cos (-i)$. (h) $\coth (i)$. (i) $\sin (1 + i)$.

10.12. Find the derivatives of the six trigonometric functions in terms of trigonometric functions.

10.13. Find the derivatives of the six hyperbolic functions in terms of hyperbolic functions.

10.14. Prove properties 3 and 4 on p. 90.

10.15. Prove property 2, p. 90, using the method of Example 10.

10.16. Prove: (a) $\cos (iz) = \cosh z$. (b) $\sin (iz) = i \sinh z$.

10.17. Prove that, just as in the real case, the complex sine and cosine functions are periodic of period 2π.

B

10.18. Prove that the following functions are not analytic anywhere.
(a) $e^{\bar{z}}$. (b) $e^{R(z)}$. (c) $e^{iI(z)}$.

10.19. Prove: For any complex constant c, $d[e^{cz}]/dz = ce^{cz}$, for all z.

10.20. Prove: $f(z) = e^{z^2}$ is an entire function and $f'(z) = 2ze^{z^2}$.

10.21. Show that, for any real number x,

(a) $\sin x = \dfrac{1}{2i}(e^{ix} - e^{-ix})$. (b) $\cos x = \frac{1}{2}(e^{ix} + e^{-ix})$.

10.22. Verify that if $I(z) > 0$, then $|e^{iz}| < 1$.

10.23. Prove properties 2, 3, 4, and 5 on p. 87; for property 5 use Exercise 2.25.

10.24. Use Exercise 7.14 and the polar form of Log z to show that the derivative of the logarithmic function is $1/z$, as asserted on p. 88.

10.25. Assuming the identification of Arg z with the inverse tangent function as described in Exercise 7.12, and using the definition of $|z|$ in terms of x and y (see p. 10), repeat the proof of the preceding exercise in terms of rectangular coordinates.

10.26. Prove properties 5, 6, and 7 on p. 90.

10.27. Prove properties 8 and 9 on p. 90.

10.28. Prove:

(a) $\sinh z = 0$ if and only if $z = n\pi i$.

(b) $\cosh z = 0$ if and only if $z = (n + \frac{1}{2})\pi i$.

10.29. Prove that the familiar property $|\sin x| \leq 1$ of the real sine function is not shared by $\sin z$; similarly for $\cos z$.

10.30. Prove that, for any z,

(a) $\overline{\sin z} = \sin \bar z$.   (b) $\overline{\cos z} = \cos \bar z$.   (c) $\overline{\tan z} = \tan \bar z$.

10.31. Prove that $f(z) = \sin \bar z$ is nowhere analytic.

10.32. Use the Cauchy–Riemann equations and the results of Example 12 to show that $\sin z$ and $\cos z$ are entire functions.

10.33. Find the fallacy in the following "proof":

$$0 = 0$$

$$\text{Log}\,(1) = \text{Log}\,(1)$$

$$\text{Log}\,(-1)^2 = \text{Log}\,(1)$$

$$2\,\text{Log}\,(-1) = \text{Log}\,(1).$$

Therefore,

$$2\pi i = 0.$$

C

10.34. If $z \neq 0$ and w are arbitrary complex numbers, we define the **general power** z^w by use of the formula

$$z^w = e^{w \log z}.$$

Use this formula to find all values of the following "numbers," thus demonstrating the fact that, in general, an expression of the form z^w has an infinite number of values.

(a) i^i.   (b) $(1 + i)^{1 - i}$.

Note that the definition of the exponential e^z must be considered as an exception to the above definition, if e^z is to be a (single-valued) function, i.e., if e^z is to have one and only one value for each value of z.

10.35. The **principal value of the general power** is defined by

$$z^w = e^{w\,\mathrm{Log}\,z}.$$

(a) Find the principal values of the two "numbers" in the preceding exercise.

(b) Use the formula given in this exercise to show that for any $z \neq 0$ and any w,

$$\mathrm{Log}\,(z^w) = w\,\mathrm{Log}\,z.$$

Compare with property 5, p. 87.

Section 11
The Linear Transformation

Some basic algebraic and analytical properties of the **linear transformation**

$$w = az + b \tag{1}$$

have been studied in Section 10.

The study of the mapping properties of (1) is best effected by examining separately the mappings

$$\zeta = az \quad\text{and}\quad w = \zeta + b$$

and then taking their composite,

$$w = \zeta + b = az + b.*$$

The first of the above two mappings,

$$\zeta = az, \tag{2}$$

will be referred to as a **rotation stretching**; the reason for this terminology becomes apparent when one examines the relations

$$|\zeta| = |a||z| \quad\text{and}\quad \arg \zeta = \arg a + \arg z,$$

which follow from (2). From these two relations, one derives the following:

---

* In terms of mappings, we shall see that "composite $f(g(z))$" means that the mapping $g(z)$ "is followed by" the mapping $f(\zeta)$.

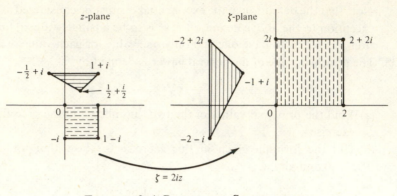

FIGURE 3.4 ROTATION STRETCHING

Under the mapping $\zeta = az$, the image of a point z is the point ζ, whose modulus is $|z|$ "stretched" by a factor $|a|$ and whose argument is arg z rotated through an angle arg a.

If $|a| = 1$, then (2) is a **pure rotation** and if arg $a = 0$, then it is a **stretching** (or "shrinking" if $|a| < 1$). Of course, if $|a| = 1$ and arg $a = 0$, then $a = 1$ and (2) reduces to the **identity transformation** $\zeta = z$. Figure 3.4 illustrates the case of a rotation stretching. We note that such a mapping preserves similarity (it is a *similarity transformation*), since it rotates every point through the same angle, namely, arg a, and multiplies the modulus of every point by the same factor, $|a|$.

The transformation

$$w = \zeta + b,$$

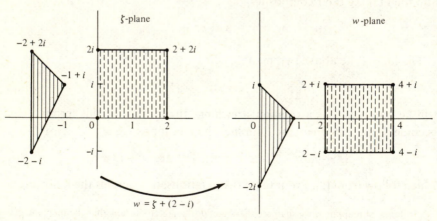

FIGURE 3.5 TRANSLATION

called **translation**, has the property that it "shifts" or "translates" every point ζ by the constant vector b. Clearly, a translation preserves congruence and it involves neither rotation nor stretching. Figure 3.5 illustrates the case of a translation.

It is now a simple matter to study the linear transformation $w = az + b$ as the combined effect of the rotation stretching $\zeta = az$ followed by the translation $w = \zeta + b$, *in that order*. Figure 3.6 shows the combined result of the mappings illustrated in Figures 3.4 and 3.5.

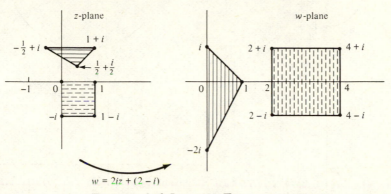

$$w = 2iz + (2 - i)$$

FIGURE 3.6 LINEAR TRANSFORMATION

EXAMPLE 1

The function

$$w = 2iz + 1 + i$$

is clearly a linear transformation. Under this mapping, every point is rotated through an angle $\arg 2i = \pi/2$, magnified by a factor $|2i| = 2$ and then translated by a vector $1 + i$. In Figure 3.7, we illustrate the effect of these three transformations for each of the following cases. The reader will find it instructive to supply the missing details.

1. The point $P : 1 + 2i$ is rotated into the point $P' : -2 + i$, magnified into $P'' : -4 + 2i$ and finally shifted into $P''' : -3 + 3i$.

2. The line segment

$$S : \arg z = \frac{\pi}{4}, \qquad 1 < |z| < 2,$$

is rotated into

$$S' : \arg \zeta = \frac{3\pi}{4}, \qquad 1 < |\zeta| < 2,$$

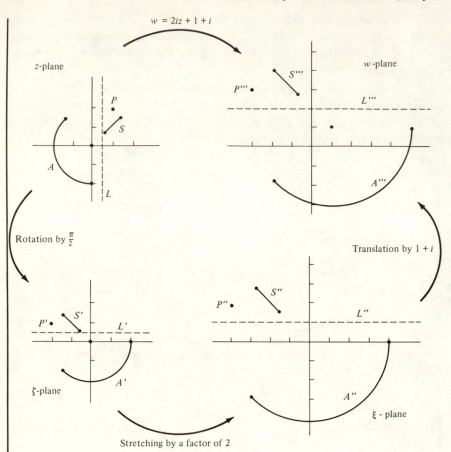

FIGURE 3.7 EXAMPLE 1

magnified into

$$S'' : \arg \zeta = \frac{3\pi}{4}, \qquad 2 < |\zeta| < 4,$$

and finally shifted into

$$S''' : \arg [w - (1 + i)] = \frac{3\pi}{4}, \qquad 2 < |w - (1 + i)| < 4.$$

3. The circular arc

$$A : |z| = 2, \frac{3\pi}{4} \leq \arg z \leq \frac{3\pi}{2},$$

is mapped, by means of the same three transformations, onto the circular arc

$$A''' : |w - (1 + i)| = 4, \frac{5\pi}{4} \leq \arg [w - (1 + i)] \leq 2\pi.$$

4. The vertical line

$$L : R(z) = \tfrac{1}{2}$$

is rotated into

$$L' : I(\zeta) = \tfrac{1}{2}, \qquad \text{in the } \zeta\text{-plane,}$$

stretched into

$$L'' : I(\xi) = 1, \qquad \text{in the } \xi\text{-plane,}$$

and, finally, translated into

$$L''' : I(w) = 2, \qquad \text{in the } w\text{-plane.}$$

The reader may verify that if we let $z = \tfrac{1}{2} + iy$ in the given function, then $w = (1 - 2y) + 2i$, which, as y varies in the z-plane, yields precisely the line $L''' : v = 2$.

The nonconstant linear transformation

$$w = az + b, \qquad a \neq 0,$$

is conformal everywhere (see p. 77), since its derivative $w' = a$ exists and is nonzero for all z. Thus, under the above mapping, angles are preserved both in magnitude and direction. This is easy to see, in this case, since the linear transformation is a similarity transformation. Looking at this fact differently, we note that, under the linear transformation, every point in the plane, and hence every curve, is rotated through the same angle and therefore angles are preserved.

EXERCISE 11

A

11.1. Find the images of the curves $\arg z = \pi/3$, $|z| = 2$, $R(z) = -1$, and $I(z) = 2$ under each of the following functions; sketch the given curves and their images in each case.

(a) $w = iz$.   (b) $w = -iz + 2i$.
(c) $w = (-1 + i)z$.   (d) $w = z + 1$.
(e) $w = (1 - i)z + (1 - i)$.   (f) $w = 2i(z + 1 + i)$.

11.2. Describe geometrically and algebraically the image of the angular region

$$0 < \arg z < \frac{\pi}{6}$$

under the function $w = -2z - 2i$.

11.3. Find the image of the parabola $y = x^2$ under the mapping $w = -2z + 2i$.

B

11.4. Find a linear transformation that maps the half-plane $I(z) > 0$ onto the region $R(w) > 1$ and the point $z = i$ onto $w = 2 - i$.

11.5. A point z_0 is called a **fixed point** of a function $w = f(z)$, provided that $f(z_0) = z_0$. Thus, in order to find the fixed points of a function, we solve the equation $z = f(z)$. Use this fact to find the fixed points, if any, of each of the following functions.

(a) $w = iz$.   (b) $w = 2z - 1 + i$.
(c) $w = z - 2i$.   (d) $w = \frac{5}{2}z + i$.

11.6. Find a linear transformation that has more than one fixed point.

11.7. Find a linear transformation that will map $z = i$ and $z = -i$ onto $w = 0$ and $w = 2$, respectively.

C

11.8. Prove that if $b \neq 0$, then the mapping $w = z + b$ has no fixed points.

11.9. Prove that the mapping $w = z + b, b \neq 0$, is the only linear mapping with no fixed points.

11.10. Prove that the only linear transformation with more than one fixed point is the identity map $w = z$.

11.11. Prove the assertions made in the last two sentences of the closing paragraph of this section.

Section 12
The Power Transformation

Certain mapping properties of the **power transformation**

$$w = z^n, \qquad n = 2, 3, 4, \ldots,$$

are more easily studied in polar form (see also p. 79). Thus, expressing the above function in polar form, we have

$$w = r^n(\cos nt + i \sin nt), \tag{1}$$

from which we see readily that if

$$|z| = r \quad \text{and} \quad \arg z = t,$$

then

$$|w| = r^n \quad \text{and} \quad \arg w = nt.$$

In words,

> the power transformation maps a point z with modulus r and argument t onto a point with modulus r^n and argument nt.

For example,

$$\text{under } w = z^3, \quad z = 2 \operatorname{cis}\left(\frac{\pi}{3}\right) \text{ maps onto } w = 8 \operatorname{cis} \pi.$$

In general, under (1), a ray emanating from the origin with angle of inclination α maps onto a ray having angle of inclination $n\alpha$; going one step further, it is easy to see that, as suggested in Figure 3.8, a sector of a circle of radius r subtending a central angle ϕ is transformed into a sector of a circle of radius r^n subtending a central angle $n\phi$. As a consequence, for instance, under $w = z^2$, the first quadrant of the z-plane maps onto the upper half of the w-plane, the upper half of the z-plane maps onto the whole w-plane, and, if we used up all the z-plane, then we would cover the w-plane twice. Again, generalizing this special case, we see that under the power transformation $w = z^n$, the z-plane maps onto the w-plane n times; i.e., every point of the w-plane, except $w = 0$, is the image of n distinct points of the z-plane. This fact, of course, is the geometric manifestation of the fact that every nonzero number has n distinct nth roots; see Exercise 2.27.

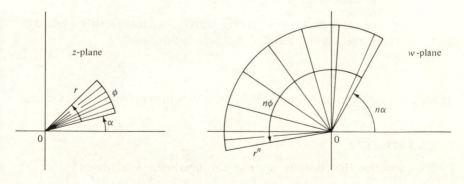

FIGURE 3.8 THE MAPPING $w = z^n$

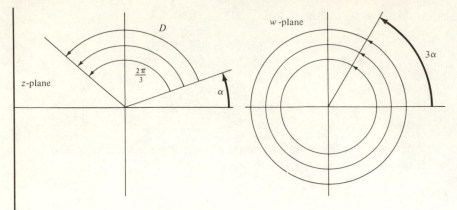

FIGURE 3.9 EXAMPLE 1: $w = z^3$

EXAMPLE 1

Consider the function $w = z^3$ and restrict its domain D as follows: D is to be the set of all z such that

$$\alpha \le \arg z < \alpha + \frac{2\pi}{3} \qquad \text{or} \qquad z = 0,$$

where α is an arbitrary angle. Clearly, if $z = 0$, then $w = 0$. For any other z in D, the given function cubes its modulus and triples its argument:

$$|z| \to |z|^3 \qquad \text{and} \qquad \arg z \to 3(\arg z).$$

In other words, the w-plane is covered "three times as fast" as is the z-plane. Figure 3.9 illustrates how the domain D, which is "one third" of the z-plane, is mapped onto the entire w-plane. It is easy to see that if z varied over the whole of the z-plane, then every w, except $w = 0$, would have three distinct preimages.

Although certain aspects of the power function are more easily studied in polar form, the rectangular form of the "square function"

$$w = z^2$$

reveals some interesting relations. We examine them in the following example.

EXAMPLE 2

We know that the decomposition of the function $w = z^2$ yields

$$u(x, y) = x^2 - y^2 \qquad \text{and} \qquad v(x, y) = 2xy.$$

Consider now, in the z-plane, the rectangular hyperbola

$$x^2 - y^2 = c, \qquad c \neq 0.$$

Then, clearly, $u = c$ and as x and y are assigned all the allowable values v ranges from $-\infty$ to $+\infty$. It follows then that, under $w = z^2$, the above hyperbola maps onto the vertical line $u = c$.

Next, consider the hyperbola

$$2xy = k, \qquad k \neq 0.$$

Then, as above, it is not difficult to see that, under the given function, its image is the horizontal line $v = k$.

The reader may show that, under the same function, horizontal and vertical lines in the z-plane map onto parabolas in the w-plane. See Exercise 12.7.

Since the power function is entire (see Example 3, Section 8) and since its derivative is nonzero, except at $z = 0$, it follows that $w = z^n$ is conformal at every $z \neq 0$. Exercises 12.8–12.10 illustrate this fact.

EXERCISE 12

A

12.1. Find the image of the sector $0 < \arg z < \pi/2$ under $w = z^2$.

12.2. Find the angle α so that, under the mapping $w = z^4$, the image of the sector $0 < \arg z < \alpha$ will be the upper half of the w-plane.

12.3. Show that, under the mapping $w = z^4$, the sectors

$$0 < \arg z < \frac{\pi}{4} \qquad \text{and} \qquad \frac{\pi}{2} < \arg z < \frac{3\pi}{4}$$

have the same image in the w-plane.

12.4. Find the image of each of the following curves or areas of the z-plane under the mapping $w = z^2$.
   (a) $x^2 - y^2 = 3$.   (b) $y = 0$.
   (c) $x = 0$.   (d) $x = 2$.
   (e) $y = 3$.   (f) $y = 1 - x$.
   (g) $|z| > 2$.   (h) $|\arg z| < \pi/2$.
   (i) $1 < R(z) < 2$.
   (j) The set of all z such that $1 < |z| < 2$ and $|\arg z| < \pi/4$.

12.5. Determine the angle α so that, under the mapping $w = z^6$, the sector $\pi/2 < \arg z \leq \alpha$, including $z = 0$, will cover the entire w-plane exactly once.

12.6. Under the mapping $w = z^5$, $z = 1$ maps onto $w = 1$. Find four more distinct points that will also map onto $w = 1$ under this function.

B

12.7. (a) Using the fact that the decomposition of the square function $w = z^2$ yields $u = x^2 - y^2$ and $v = 2xy$, show that, for this function, $v^2 = 4x^2(x^2 - u) = 4y^2(u + y^2)$.

     (b) Use the result from (a) to show that, under $w = z^2$, horizontal $(x = c \neq 0)$ and vertical $(y = k \neq 0)$ lines map onto parabolas.

12.8. Illustrate the conformality of the mapping $w = z^2$ by proving that the parabolas of the preceding exercise (which are the images of lines intersecting at right angles) intersect at right angles.

12.9. Give a further illustration of the conformality of $w = z^2$ by proving that the two sets of hyperbolas involved in Example 2 are orthogonal and then noting that their images are also orthogonal families. (Recall that two families of curves are said to be orthogonal if each member of the one family intersects every member of the other family at right angles.)

12.10. Give an example to illustrate the fact that the mapping $w = z^n$ is not conformal at $z = 0$.

Section 13
The Reciprocal Transformation

Some basic algebraic and analytical properties of the **reciprocal transformation**

$$w = \frac{1}{z} \tag{1}$$

as well as its use in introducing the point at infinity were discussed in Section 10. As a mapping, the reciprocal transformation has some very interesting properties.

It is intuitively evident that, under Equation (1), points in the vicinity of $z = 0$ are mapped onto points in the remote regions of the w-plane, and points "far away" from $z = 0$ are mapped onto points "close" to $w = 0$. But let us be a little more precise. Writing z and w in polar form, we see that if

$$z = r \text{ cis } t,$$

then

$$w = \frac{1}{r} \cos(-t).$$

From the latter we read the following:

Under the reciprocal function a point with modulus r and argument t maps onto a point with modulus $\dfrac{1}{r}$ and argument $-t$.

The transformation described by the preceding statement can be effected geometrically as follows (see Figure 3.10).

Take a point $z \neq 0$ having modulus r and argument t and lying inside the unit circle. Draw line L through z perpendicular to the ray R from 0 through z, and then from the points at which L intersects the unit circle draw tangents S and T (to the unit circle), which, as the reader may show, will intersect at some point ζ on R. One may then show that $|\zeta| = 1/r$ and that, in fact, $\zeta = 1/\bar{z}$. Next, locate the point $w = \bar{\zeta}$. It is now easy to show that

$$|w| = \frac{1}{r} \qquad \text{and} \qquad \arg w = -t$$

and hence

$$w = \frac{1}{z}.$$

We have thus constructed the reciprocal of an arbitrary complex number $z \neq 0$ by purely geometric means; a similar construction will yield the reciprocal of an arbitrary number z outside the unit circle. In Exercise 13.5 the reader is asked to discuss what happens to points on the unit circle under the reciprocal function. The process by which we located ζ in the above construction is called **inversion in the unit circle.**

The above geometric construction can be summed up by saying that the reciprocal transformation is the composite of the two functions

$$\zeta = \frac{1}{\bar{z}} \qquad \text{and} \qquad w = \bar{\zeta};$$

i.e., inversion in the unit circle followed by conjugation.

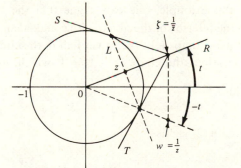

FIGURE 3.10 CONSTRUCTION OF $w = 1/z$

Exercise 13.2 brings out the basic property possessed by the reciprocal mapping with respect to the unit circle; the reader is urged to answer that exercise as early as possible.

In the following example we discuss some special cases of another remarkable property demonstrated by the reciprocal map; namely, that it maps lines and circles onto lines or circles.

EXAMPLE 1

We find the images of various curves under the mapping $w = 1/z$.
1. Consider the vertical line $x = 1$. From the decomposition

$$\frac{1}{z} = \frac{x}{x^2 + y^2} - \frac{y}{x^2 + y^2}i,$$

we have

$$u = \frac{x}{x^2 + y^2} \quad \text{and} \quad v = -\frac{y}{x^2 + y^2}. \tag{2}$$

Then, since every point on the given line is of the form $z = 1 + yi$, we find that

$$u = \frac{1}{1 + y^2} \quad \text{and} \quad v = -\frac{y}{1 + y^2}.$$

By squaring and adding the last two equations, we obtain

$$u^2 + v^2 = u,$$

which, upon completion of the square, yields the circle

$$|w - \tfrac{1}{2}| = \tfrac{1}{2},$$

which is the image of the given line $R(z) = 1$. Using the development preceding this example, one can argue that the half-plane $R(z) > 1$ maps into the interior of the above circle and, also, that the upper half of the half-plane maps into the lower half of the circle, and vice versa.
2. Next, we consider the circle $|z - 1| = 1$, which, upon simplification, is reduced to

$$x^2 + y^2 = 2x.$$

Substitution in (2) yields

$$u = \frac{1}{2} \quad \text{and} \quad v = -\frac{y}{2x}.$$

Now, as $z = (x, y)$ varies along the given circle, v takes on all real values, while u remains constant at $\frac{1}{2}$. We conclude, then, that the image of the given circle is the vertical line $u = \frac{1}{2}$. Again, using the development preceding this example, the reader may wish to show that the interior of the given circle maps on the half-plane to the right of the image line $u = \frac{1}{2}$ but with their upper and lower halves reversed.

3. Consider now the circle $|z| = 2$. From the given function $w = 1/z$ we obtain $|w| = 1/|z|$, and substituting from the equation of the given circle we get $|w| = \frac{1}{2}$. So, in this case, a circle maps onto a circle. Are their upper and lower halves interchanged in this case?

The three cases of the preceding example are instances of a more general result, which we state and then proceed to prove:

Under the reciprocal transformation lines and circles map onto lines or circles.

The proof is based on the following two elementary facts:

(A) $\dfrac{1}{z} = \dfrac{x}{x^2 + y^2} - i\dfrac{y}{x^2 + y^2}.$

(B) The equation $a(x^2 + y^2) + bx + cy + d = 0$ represents a circle (if $a \neq 0$) or a line (if $a = 0$) and, conversely, any line or circle is representable by an equation of the above form.

Now, let a circle or a line be given; call it K. Then, for some constants a, b, c, and d,

$$K : a(x^2 + y^2) + bx + cy + d = 0. \tag{3}$$

From (A) we have

$$u = \frac{x}{x^2 + y^2}, \qquad v = -\frac{y}{x^2 + y^2}, \qquad u^2 + v^2 = \frac{1}{x^2 + y^2}. \tag{4}$$

Then dividing Equation (3) by $x^2 + y^2$ and substituting from Equations (4), we obtain

$$d(u^2 + v^2) + bu - cv + a = 0, \tag{5}$$

which is a line (if $d = 0$) or a circle in the w-plane; our proof is complete.

A good number of exercises at the end of this section refer to some special cases of the above development.

The reciprocal transformation is conformal at every $z \neq 0$ (see the illustration in the next example). Its inverse,

$$z = \frac{1}{w},$$

is also a reciprocal transformation.

EXAMPLE 2

Consider the line

$$L_1 : x - y + 2 = 0.$$

In the notation of Equation (3) $a = 0$, $b = 1$, $c = -1$, and $d = 2$. Hence, under $w = 1/z$, L_1 maps onto a line or circle given by (5). So, L_1 maps onto the circle

$$C_1 : 2(u^2 + v^2) + u + v = 0.$$

Similarly, we find that the line

$$L_2 : x + y - 2 = 0$$

maps onto the circle

$$C_2 : 2(u^2 + v^2) - u + v = 0.$$

Now, the lines intersect at right angles at the point $z = 2i$, as one can easily verify. Since $w = 1/z$ is conformal at that point, it follows that the circles C_1 and C_2 must intersect at right angles in the w-plane. We verify that this is indeed the case.

First, we find that the circles intersect at $w = 0$ and $w = -i/2$. Then, finding the derivatives of the two circles and evaluating them at either point of intersection, we find that

$$\text{for } C_1, \frac{dv}{du} = 1 \quad \text{and} \quad \text{for } C_2, \frac{dv}{du} = -1,$$

or vice versa. This, of course, says that the circles intersect at right angles.

EXERCISE 13

A

13.1. Find and plot the image of each of the following points under the reciprocal transformation.

(a) 1.   (b) -1.   (c) i.

(d) $-i$.   (e) $1 + i$.   (f) $5 - 12i$.

(g) $-3 + 4i$.   (h) $-3i$.   (i) $1/(1 - i)$.

13.2. The unit circle $|z| = 1$ and the coordinate axes subdivide the plane into eight regions. Find the image of each of these regions under $w = 1/z$ and specify what happens to their boundaries under this transformation.

13.3. Find the image of each of the following lines or circles under the reciprocal map.

(a) $y = 1$.   (b) $y = x - 1$.

(c) $x = -1$.   (d) $x + y = 1$.

(e) $x^2 + (y - 1)^2 = 1$. (f) $(x + 1)^2 + (y - 2)^2 = 4$.

(g) The real axis.   (h) The imaginary axis.

13.4. Verify that the angle of intersection at each of the points at which $|z + 1 + 2i| = 2$ and $y = x + 1$ intersect is preserved under $w = 1/z$.

13.5. Find the image of the unit circle under the reciprocal map and specify what happens to its upper and lower halves.

13.6. Find the image of the circle $|z + 1| = 1$ under $w = 1/z$.

13.7. Find the fixed points of the reciprocal map; see Exercise 11.5.

B

13.8. Show that any vertical line $x = c \neq 0$ maps onto the circle

$$\left| w - \frac{1}{2c} \right| = \frac{1}{2|c|}$$

under the reciprocal transformation.

13.9. Discuss the image of a horizontal line $y = k \neq 0$ under the reciprocal map and derive an equation of that image, similar to that of Exercise 13.8.

C

13.10. Use the composite of the functions $\zeta = z + 1$ and $w = 1/\zeta$ to discuss some general mapping properties of the mapping

$$w = \frac{1}{z + 1}.$$

13.11. Refer to Example 2, this section, and give a sound explanation of the "unusual" fact that whereas L_1 and L_2 have only one point of intersection in the z-plane, their images have two points of intersection in the w-plane.

13.12. Find the image of the infinite strip $0 < y < \frac{1}{2}$ under the reciprocal map.

Section 14
The Bilinear Transformation

If a, b, c, and d are complex constants, then

$$w = \frac{az + b}{cz + d}, \qquad \text{for } ad - bc \neq 0, \tag{1}$$

is called the **bilinear transformation**. The reason for the condition $ad - bc \neq 0$ imposed on the coefficients is that (as the reader is asked to verify in Exercise 14.8) if this inequality does not hold, then (1) reduces to a constant function.

Certain fundamental properties of the bilinear function are discussed in Section 10.

To avoid reducing Equation (1) to the linear case (which we studied in Section 11), we assume, for the remainder of this section, that $c \neq 0$. Under this assumption, Equation (1) maps the extended z-plane in a one-to-one fashion onto the extended w-plane; the "exceptional" points of this mapping are $z = -d/c$, which maps onto $w = \infty$, and $z = \infty$, which maps onto $w = a/c$.

The bilinear transformation possesses a number of very interesting mapping properties, some of which are also very useful in applications. Here we choose to discuss two of the most basic of these properties.

The first one consists of the fact that, as in the case of the reciprocal map,

> under the bilinear transformation lines and circles map onto lines or circles.

We give an outline of a proof of this assertion; the proof is based on the following two facts:

(A) The bilinear map is the composite of the following three functions *taken in the given order*:

$$\zeta = cz + d, \qquad \xi = \frac{1}{\zeta}, \qquad w = \frac{a}{c} + \frac{bc - ad}{c}\xi.$$

So, the bilinear map is the composite of a linear map followed by a reciprocal map and then, again, by a linear map.

(B) The linear map is a similarity transformation (see p. 96) and the reciprocal transformation maps lines and circles onto lines or circles.

The reader may now use the above facts appropriately to argue in detail that given a line or a circle, say, K in the z-plane, the first of the three functions in (A) will rotate it, magnify it, and translate it into a line or a circle K'; the second function in (A) will invert it into a line or a circle K''; and, finally, the third function will rotate it, magnify it, and translate it into a line or a circle K''' in the w-plane.

Example 1, below, illustrates a special case of the preceding discussion; this particular case of mapping a half-plane onto a unit disk is of interest to scientists in the field of circuit analysis and stability of linear systems.

EXAMPLE 1

Show that the special bilinear map $w = \dfrac{z - 1}{z + 1}$ transforms the half-plane $R(z) > 0$ onto the unit disk $|w| < 1$.

We effect this mapping in three gradual steps, each of which is accompanied by a drawing in Figure 3.11. By use of (A), we know that the given map is obtained by successive application of the maps

$$\zeta = z + 1, \qquad \xi = \frac{1}{\zeta}, \qquad w = -2\xi + 1.$$

We thus have:

1. Under the map $\zeta = z + 1$, every point of the given half-plane is rotated through 0 radians (no change), magnified by a factor of 1 (no change), and finally shifted by a vector 1 to yield the half-plane $R(\zeta) > 1$; see Figure 3.11(a).

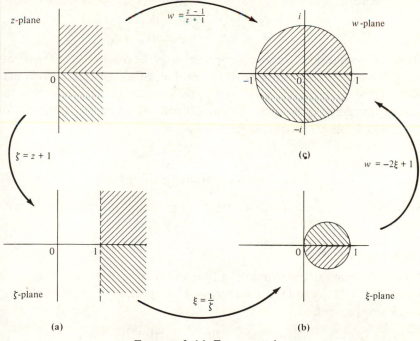

FIGURE 3.11 EXAMPLE 1

2. Under the map $\xi = 1/\zeta$ the half-plane $R(\zeta) > 1$ is mapped onto the interior of the circle $|\xi - \frac{1}{2}| = \frac{1}{2}$, but with the upper and lower halves of the half-plane and of the circle interchanged, as suggested in Figure 3.11(b). The details of this mapping may be found in Example 1(1), Section 13.

3. Finally, under the map $w = -2\xi + 1$, the interior of the circle found in (2) will be rotated through $-\pi$ radians onto the interior of the circle $|\xi + \frac{1}{2}| = \frac{1}{2}$; note that the rotation will interchange the positions of the upper and lower halves of the disk. Then the rotation will be followed by a stretching by a factor of 2 onto the interior of the circle $|\xi + 1| = 1$, and, lastly, a shift through the vector 1 will yield the disk $|w| < 1$, as asserted.

The second property of the bilinear transformation to which we wish to address ourselves may be described as follows:

> Given any triple z_1, z_2, z_3 of distinct points in the z-plane and any triple w_1, w_2, w_3 of distinct points in the w-plane, there is a unique bilinear transformation mapping z_j onto $w_j, j = 1, 2, 3$.

In order to find the mapping whose existence is asserted above, we make use of the form

$$\frac{(w - w_1)(w_2 - w_3)}{(w - w_3)(w_2 - w_1)} = \frac{(z - z_1)(z_2 - z_3)}{(z - z_3)(z_2 - z_1)}, \tag{2}$$

whose right-hand side is known, in geometry, as the **cross-ratio** of the points z_1, z_2, z_3, and z. It is not difficult to see that, by simplification, one may reduce (2) to the form of (1). Of course, the essential claim here is that the bilinear transformation resulting from (2) will indeed map z_j onto w_j, for $j = 1, 2, 3$. From (2) we obtain the expression

$$(w - w_1)(w_2 - w_3)(z - z_3)(z_2 - z_1)$$
$$= (w - w_3)(w_2 - w_1)(z - z_1)(z_2 - z_3); \tag{3}$$

then, letting $z = z_j$, we obtain $w = w_j$ for $j = 1, 2, 3$, and the claim is justified. (See Exercise 14.12.)

EXAMPLE 2

Find the bilinear transformation that will map $z_1 = 0$, $z_2 = i$, $z_3 = -1$ onto $w_1 = 12$, $w_2 = 11 + i$, and $w_3 = 11$, respectively.

Using (2), we find

$$\frac{(w - 12)i}{(w - 11)(-1 + i)} = \frac{z(1 + i)}{(z + 1)i},$$

which yields the bilinear map

$$w = \frac{10z - 12}{z - 1}.$$

The bilinear mapping (remember our assumption: $c \neq 0$) has *at most* two fixed points, which are the roots of the equation

$$cz^2 + (d - a)z - b = 0.$$

Since the derivative of the bilinear map exists and is nonzero for all $z \neq -d/c$ (verify this!), it follows that (1) is a conformal transformation for all $z \neq -d/c$.

EXAMPLE 3

We find the fixed points of the mapping $w = \dfrac{z - 1}{z + 1}$.

Setting $w = z$ (see Exercise 11.5), we find that

$$z^2 + z = z - 1,$$

which, upon solution, yields the fixed points $z = \pm i$.

EXERCISE 14

A

14.1. Find the image of each of the points $z = 0, 1, -1, i, -i, \infty$, under the map $w = \dfrac{-iz + 2}{z + i}$.

14.2. Find the fixed points of the transformations

(a) $w = \dfrac{-iz + i}{z - i}$,   (b) $w = \dfrac{z - 2}{z - 1}$.

14.3. Find the bilinear transformation that maps $0, 1$, and i onto $-1, 0$, and i, respectively.

14.4. Find the image of the line $I(z) = \frac{1}{2}$ under the map $w = \dfrac{4z}{2iz + i}$.

14.5. Give an example of a bilinear map with exactly one fixed point.

14.6. Give an example of a bilinear map with no fixed points.

14.7. Show that Equation (2) of this section reduces to (1).

B

14.8. Prove that if $ad - bc = 0$, then (1) reduces to a constant map.

14.9. Prove that the bilinear map is continuous at all $z \neq -d/c$.

14.10. Find the image of the half-plane $I(z) \geq 0$ under the map $w = \dfrac{z}{z-1}$.

14.11. As in 14.10, for $I(z) > 0$ under $w = \dfrac{z-i}{z+i}$.

14.12. Verify the claim made immediately preceding Example 2. Also, prove that the function obtained by means of (2) is unique, that is, that it is the only bilinear function mapping the given z_j's onto the given w_j's, as specified.

c

14.13. Supply the details in the proof of the assertion that the bilinear mapping transforms lines and circles onto lines or circles; p. 110.

14.14. A certain bilinear transformation is known to carry the point X of the adjacent configuration to ∞. Sketch the resulting configuration in the w-plane, specifying the images of the given points and curves.

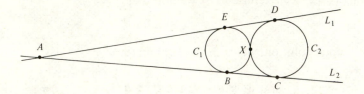

Section 15
The Exponential and Logarithmic Transformations

We begin our study of the **exponential transformation**

$$w = e^z \tag{1}$$

by examining two particular cases, which, in turn, will lead us into an obvious generalization that completely describes Equation (1) as a mapping.

EXAMPLE 1

1. We find the image, under $w = e^z$, of a horizontal line $y = b$. First, we recall that if $w = e^z$, then $|w| = e^x$ and $\arg w = y$. Now, every point on the given line has the form

$$z = x + ib, \qquad -\infty < x < \infty;$$

hence, as x varies from $-\infty$ to $+\infty$, e^x varies from 0 to $+\infty$ while y remains fixed at $y = b$. In other words, as x varies from $-\infty$ to $+\infty$,

$|w|$ varies from 0 to $+\infty$ while arg w remains fixed at arg $w = b$. It follows that, as z varies on the given line, w describes a ray emanating from (but not containing) the origin and whose angle of inclination is b radians; see Figure 3.12.

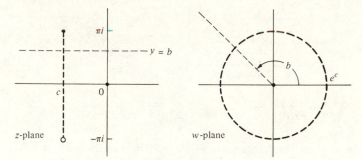

FIGURE 3.12 EXAMPLE 1

2. We find the image, under Equation (1), of the vertical line segment

$$x = c, \qquad -\pi < y \le \pi.$$

Again, every point on the given line segment (see Figure 3.12) has the form

$$z = c + iy, \qquad -\pi < y \le \pi;$$

hence, as y varies from $-\pi$ to $+\pi$, $\cos y + i \sin y$ describes a complete circle, while $|w|$ remains fixed at e^c. In other words, as z varies on the given line segment, w describes a circle centered at $w = 0$ and having radius e^c.

It is very important to note that if y were allowed to vary over a larger domain (but, always, on the same vertical line), then w would repeat its trace on the same circle, and if we took the entire vertical line $x = c$, then the circle $|w| = e^c$ would be repeated infinitely many times.

We summarize our findings in the above example as follows:

Under $w = e^z$, horizontal lines map onto rays emanating from $w = 0$ and vertical lines map onto circles centered at $w = 0$.

Using these two basic facts, we now argue as follows: If we take *all* the horizontal lines between $y = -\pi$ (not included) and $y = \pi$ (inclusive), their images will be *all* the rays with angles of inclination ranging from $-\pi$ to π. But the totality of all such rays exhausts all the points in the w-plane, except $w = 0$; see Figure 3.13. On the other hand, if we take *all* the vertical line segments, as in the above example, contained between the lines $y = -\pi$

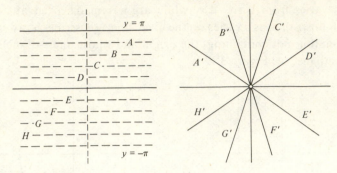

FIGURE 3.13 $w = e^z$: LINES ONTO RAYS

and $y = \pi$, then their images will be *all* the circles with positive radius centered at $w = 0$; see Figure 3.14. But, again, the totality of all such circles will cover every $w \neq 0$.

From the preceding discussion we conclude that the **fundamental strip**

$$S : -\pi < y \leq \pi, \qquad -\infty < x < +\infty,$$

maps onto the entire *w*-plane save its origin. Going one step further, one may argue in a similar fashion that the horizontal strip (see Figure 3.15)

$$S_1 : \pi < y \leq 3\pi, \qquad -\infty < x < +\infty,$$

has exactly the same "fate," under the exponential map, as the fundamental strip S. In fact, the same is true about the horizontal strips

$$S_2 : -3\pi < y \leq -\pi, \qquad -\infty < x < +\infty,$$

$$S_3 : 3\pi < y \leq 5\pi, \qquad -\infty < x < +\infty,$$

and, in general, about any horizontal strip

$$S_\alpha : \alpha < y \leq \alpha + 2\pi, \qquad -\infty < x < +\infty,$$

for any real number α.

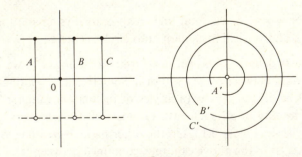

FIGURE 3.14 $w = e^z$: LINES ONTO CIRCLES

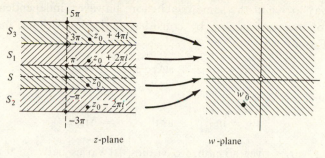

z-plane w-plane

FIGURE 3.15 THE MAPPING $w = e^z$

The effect of the exponential mapping on the z-plane should now be all but obvious. Evidently, the mapping is many-to-one in a "vertical sense" as Figure 3.15 suggests; for instance, if w_0 is the image of a point z_0 under the exponential map, then w_0 is also the image of $z_0 + 2\pi i$ and of $z_0 - 2\pi i$, and of $z_0 + 4\pi i$, and so on. This pattern is, of course, nothing but the geometrical manifestation of the periodicity of the exponential

$$e^z = e^{z + 2k\pi i},$$

which is proved in Example 4, Section 10. The exponential transformation can be considered as a one-to-one mapping if we restrict its domain to the fundamental strip S, or to any other horizontal strip of width 2π; in that case, the exponential function is capable of possessing an inverse that is a function, as we will see shortly.

Since the derivative of the exponential is never zero and exists for all z, it follows that $w = e^z$ is conformal everywhere. A simple illustration of the conformality of the exponential was given in Example 1, where we saw that the images of orthogonal curves are orthogonal curves.

In Section 10, we saw that the equation

$$\text{Log}(e^z) = z \tag{2}$$

holds for every z and that this property expresses the fact that the function $\text{Log } z$ is the inverse of the function e^z. If we examine Equation (2) carefully, we may read in it the following: If we find e^z, for any given z, and then we apply the function Log to e^z, we return to z. In short: "$\text{Log } z$ undoes what e^z does to any z." The preceding sentence completely describes the **logarithmic transformation**

$$w = \text{Log } z$$

as a mapping, since it simply says that $\text{Log } z$ takes the z-plane (minus its origin) and maps it onto the fundamental strip

$$-\pi < v \le \pi, \qquad -\infty < u < \infty,$$

of the w-plane.

One may arrive at the same conclusion, however, independently of the concept of inverse function, by considering the logarithmic mapping as follows. From

$$w = \ln |z| + i \operatorname{Arg} z, \qquad z \neq 0,$$

we obtain

$$u = \ln |z| \qquad \text{and} \qquad v = \operatorname{Arg} z.$$

Now, as z varies over all nonzero values, $|z|$ varies between 0 and $+\infty$; hence $\ln |z|$ varies from $-\infty$ to ∞, and, therefore,

$$-\infty < u < \infty.$$

On the other hand, since, by definition, $-\pi < \operatorname{Arg} z \leq \pi$, we have

$$-\pi < v \leq \pi.$$

Obviously, the last two relations involving u and v represent the fundamental strip of the w-plane; see Figure 3.16.

In view of our discussion, in Section 10, concerning the derivative and the region of analyticity of $\operatorname{Log} z$, it follows that the logarithmic mapping is conformal everywhere except at every point of the nonpositive real axis.

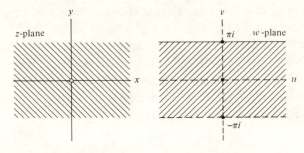

FIGURE 3.16 THE MAPPING $w = \operatorname{Log} z$

EXERCISE 15

A

15.1. Find the image of each of the following curves under $w = e^z$.
   (a) Ray: $y = 1, x > -2$.
   (b) Ray: $y = 1, x > -1$.
   (c) Ray: $y = 1, x > 0$.
   (d) Ray: $y = 1, x > 1$.

(e) Segment: $x = -2, -\pi/2 < y < \pi$.
(f) Segment: $x = 0, -3\pi/2 < y < 3\pi/2$.
(g) Segment: $x = 1, 0 \le y < \pi$.
(h) Segment: $x = 2, 0 \le y < 2\pi$.
(i) Line: $y = 3$.
(j) Line: $x = -8$.

15.2. Find the image of each of the following curves under $w = \text{Log } z$.

(a) Circle: $|z| = c, c > 0$.
(b) The ray emanating from (but not including) the origin having an angle of inclination $\alpha = -\pi/4$.
(c) As in (b) with $\alpha = 3\pi/4$.
(d) As in (b) with $\alpha = \pi/6$.

B

15.3. Find the image under the exponential function of the rectangle with vertices at $-1, 3, 3 + 2i$, and $-1 + 2i$.

15.4. Find the image under the exponential function of the polygon formed by joining by straight lines the points $0, 2, 2 + i, -2 + i, -2, -2 - 2i, -2i$, and 0, in that order.

C

15.5. Prove that, under $w = e^z$, a line $y = mx + b$, with $m \ne 0$, maps onto a logarithmic spiral, by proving and then combining the following two assertions:

(a) The line $y = mx + b$ is representable in the form

$$z = (1 + im)t + bi, \qquad -\infty < t < \infty.$$

(b) If $\rho = |w|$ and $\phi = \text{arg } w$, then

$$\rho = ce^{\theta/m},$$

where c is a constant and θ is a function of ϕ.

Section 16
The Transformations $w = \sin z$ and $w = \cos z$

In this brief section we examine some special mapping properties of the **trigonometric transformations**

$$w = \sin z \qquad \text{and} \qquad w = \cos z.$$

Certain algebraic and analytical properties of these functions are studied in Section 10.

From the decomposition

$$\sin z = \sin x \cosh y + i \sinh y \cos x$$

we obtain the equations

$$u = \sin x \cosh y \qquad \text{and} \qquad v = \cos x \sinh y, \tag{1}$$

which will form the basis for our study of some mapping attributes of the sine function. We begin with two examples.

EXAMPLE 1

Consider the interval of the real axis described by

$$-\frac{\pi}{2} \le x \le \frac{\pi}{2}, \qquad y = 0.$$

Recall that if $y = 0$, then $\cosh y = 1$ and $\sinh y = 0$. Then, for any point on the given interval, Equations (1) become

$$u = \sin x \qquad \text{and} \qquad v = 0.$$

Now, as x varies between $-\pi/2$ and $\pi/2$, $\sin x$ and therefore u varies between -1 and 1. Hence it follows that, under $w = \sin z$, the given interval maps onto the interval

$$-1 \le u \le 1, \qquad v = 0,$$

in the w-plane; see Figure 3.17.

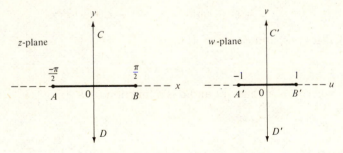

FIGURE 3.17 EXAMPLES 1 AND 2

Using precisely the same argument, one may show that the real axis of the z-plane maps onto the same interval of Example 1,

$$-1 \le u \le 1, \qquad v = 0,$$

of the w-plane; in fact, the interval in question is covered an infinite number of times. See Exercise 16.2.

EXAMPLE 2

We show that the imaginary axis $x = 0$ is mapped onto the imaginary axis $u = 0$, under the sine mapping.

Again, the basis for our arguments is provided by Equations (1). For, when $x = 0$,

$$u = 0 \qquad \text{and} \qquad v = \sinh y.$$

Then we see that, as y varies from $-\infty$ to ∞ on the imaginary axis, $\sinh y$ and hence v varies from $-\infty$ to ∞, while u remains fixed at 0.

It follows then that the axis $x = 0$ is mapped onto the axis $u = 0$ and, in fact, it is mapped so that the upper half is mapped onto the upper half and the lower half onto the lower half; see Figure 3.17.

In the exercises at the end of this section, the reader is asked to argue the following two facts:

(A) Under $w = \sin z$, any horizontal segment $y = b \neq 0$, $2k\pi - \pi/2 \leq x \leq 2k\pi + \pi/2$ maps onto either the upper or the lower half of the ellipse,

$$\frac{u^2}{\cosh^2 b} + \frac{v^2}{\sinh^2 b} = 1,$$

depending on whether $b > 0$ or $b < 0$.

(B) Under $w = \sin z$, any vertical line $x = c$, $c \neq k\pi/2$, $k =$ integer, [see Exercise 16.4(b)] is mapped onto either the right-hand half or the left-hand half of the hyperbola

$$\frac{u^2}{\sin^2 c} - \frac{v^2}{\cos^2 c} = 1,$$

depending on whether $2k\pi < c < 2k\pi + \pi$ or $2k\pi - \pi < c < 2k\pi$.

We now turn, briefly, to the transformation $w = \cos z$. Using the decomposition

$$\cos z = \cos x \cosh y - i \sin x \sinh y,$$

one may proceed as in the case of $\sin z$ to consider several special cases. However, this is hardly necessary here, in view of the fact that we have some knowledge of the mapping properties of $w = \sin z$. For, by use of the identity

$$\cos z = \sin \left(z + \frac{\pi}{2} \right), \tag{2}$$

we can translate what we know about sin z into mapping properties of cos z. This we do by noting that, in view of (2), $w = \cos z$ can be expressed as the composite of the mappings

$$\zeta = z + \frac{\pi}{2} \qquad \text{and} \qquad w = \sin \zeta,$$

the first one of which is a pure translation. We illustrate this process in the following.

EXAMPLE 3

Take the interval

$$T : -\pi \leq x \leq 0, \qquad y = 0.$$

Under the translation

$$\zeta = z + \frac{\pi}{2},$$

T is mapped onto the interval

$$T' : -\frac{\pi}{2} \leq R(\zeta) \leq \frac{\pi}{2}, \qquad I(\zeta) = 0.$$

Then, using the result of Example 1, we find that, under $w = \sin \zeta$, T' is mapped onto

$$T'' : -1 \leq u \leq 1, \qquad v = 0.$$

We conclude that, under $w = \cos z$, the interval T is mapped onto the interval T''.

EXERCISE 16

A

16.1. Imitating the process of Example 1, show that each of the following intervals is mapped, under $w = \sin z$, onto the interval

$$S : -1 \leq u \leq 1, \qquad v = 0.$$

(a) $-3\pi/2 \leq x \leq -\pi/2, y = 0.$ (b) $-\pi \leq x \leq 0, y = 0.$
(c) $0 \leq x \leq \pi, y = 0.$ (d) $\pi/2 \leq x \leq 3\pi/2, y = 0.$

16.2. Generalize Exercise 16.1 by arguing that, under $w = \sin z$, the real axis, $y = 0$, is mapped onto the interval S, in a many-to-one fashion.

16.3. Using either method suggested on p. 121, show that $w = \cos z$ has the following mapping properties:

(a) The half-line $x = \pi$, $y \le 0$, is mapped onto the ray $v = 0$, $u \le -1$, in a one-to-one fashion.

(b) The half-line $x = \pi$, $y \ge 0$, is mapped onto the ray $v = 0$, $u \le -1$, in a one-to-one fashion.

(c) The line $x = 0$ is mapped onto the ray $v = 0$, $u \ge 1$, in a "two-to-one" fashion.

16.4. (a) From Equations (1), obtain the equations appearing in (A) and (B), with b replaced by y and c by x.

(b) For what values of x and y do the equations obtained in (a) *fail to hold*, respectively?

(c) Prove assertion (A).

(d) Prove assertion (B).

B

16.5. Account for the exceptions made in assertions (A) and (B), by finding the image of each of the following lines under $w = \sin z$.

(a) $y = 0$.   (b) $x = \pi/2$.   (c) $x = -\pi/2$.

(d) $x = \pi$.   (e) $x = \pi/2 + k\pi$.   (f) $x = k\pi$.

16.6. (a) Argue that the mapping in (A) is many-to-one.

(b) Restrict the domain of the function in (A) so that the mapping there is one-to-one.

16.7. Repeat the two parts of Exercise 16.6 for the mapping in (B).

C

16.8. Find the image, under $w = \sin z$, of the rectangle in the adjacent figure, specifying precisely the images of the six points $A, B, \ldots, F$.

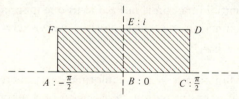

REVIEW EXERCISES — CHAPTER 3

1. Find the image of the line $y = 2x$ under the mapping $w = z^2$.

2. Find the image of the hyperbola $y = -3/x$ under $w = z^2$.

3. Find the image of the region $0 < |z| < 3, 0 < \arg z < \pi/2$, under each of the following functions:
   (a) $w = z + 1$.   (b) $w = -iz$.
   (c) $w = z^2$.   (d) $w = z^3$.
   (e) $w = z^4$.   (f) $w = \text{Log } z$.

4. Imitate the method used in Example 1, Section 11, to find and sketch the various mapping stages of the square with vertices $+1, -1, +i$, and $-i$ under the mapping $w = -3iz - 1 + i$.

5. Find the fixed points, if any, of each of the following mappings:
   (a) $w = 2 - 1/z$.   (b) $w = 3z - 4$.
   (c) $w = z + e^z$.   (d) $w = z + 1/z$.

6. What is the value of the function

$$w = \frac{z + i}{z - i}$$

   at the point $z = \infty$?

7. Find the image of $|z - i| = 1$ under the reciprocal function.

8. Find the bilinear transformation that will map the points $-1, 0$, and 1 onto $-1, i$, and 1, respectively.

9. Solve each of the following equations for all values of z:
   (a) $e^{2z-i} = i$.   (b) $\cosh z = 2$.
   (c) $\sin z = 2i$.   (d) $\log (z - 1) = 3$.

10. Establish property 11 on p. 90.

11. Prove:
    (a) $\tan^2 z + 1 = \sec^2 z$, for all z.
    (b) $\sin 2z = 2 \sin z \cos z$, for all z.

12. Find the modulus and argument of the quantity e^{z^2}.

13. Mark the following statements *true* or *false*.
    (a) The period of e^z is 2π.
    (b) For every z, it is always true that $|\cos z| \leq 1$.
    (c) The exponential function is not a one-to-one mapping.
    (d) A constant function $w = c$ is conformal everywhere.
    (e) $\cos (iz) = \cosh z$, for all z.
    (f) $\text{Log } 0 = 0$.
    (g) The power function $w = z^n$ is conformal everywhere for $n > 0$.
    (h) The bilinear transformation is a one-to-one mapping of the extended z-plane onto the extended w-plane.
    (i) The linear mapping $w = az + b$ is a rotation stretching followed by conjugation.
    (j) A constant function $w = c$ has no fixed points.

APPENDIX 3

Part A

Multivalued Functions

In our definition of a function (Section 4), we specified that for each value of the independent variable in the domain of the function, we obtain *one and only one* value of the dependent variable. As a result, one often talks about a "single-valued function" not only to emphasize the "only one" of the definition, but also to distinguish such functions from the "multivalued functions," which we now introduce.

We have the following definition.

Let D be a set of points in the plane and suppose that a rule g is given by which

> to each z in D corresponds at least one point w in the plane and there is at least one point in D to which correspond at least two distinct w's.

Then g is called a **multivalued function** with domain D. As in the case of single-valued functions, the notation

$$w = g(z)$$

will be used to denote the value(s) that g attains at z. At this point, the reader will find it extremely helpful to review the entire Section 4.

Most of the multivalued functions with which we shall be concerned result from the inverse relations of single-valued functions, which are many-to-one. Again, a review of Section 9, and Section 10 through Example 1, is strongly recommended at this point. In this section, we propose to study two such multivalued functions to introduce some of the problems that accompany such functions.

THE ROOT FUNCTION

The inverse relation

$$w = z^{1/n}, \qquad n = 2, 3, 4, \ldots, \tag{1}$$

that results from the power function will be called the **root function**. From our work in Section 2 we know that, for every nonzero z, Equation (1) yields n distinct values of w and hence defines a multivalued function with the entire plane as its domain.

We now restrict our considerations to the case $n = 3$ and we discuss some aspects of

$$w = z^{1/3}. \tag{2}$$

Each step of our discussion may then be easily adapted to the general case defined by (1).

Take a nonzero complex number

$$z = re^{it}.$$

Then Equation (2) yields three distinct image points, given by

$$w_k = r^{1/3}e^{i(t + 2k\pi)/3}, \qquad k = 0, 1, 2. \tag{3}$$

The modulus of each of these points is, clearly, $r^{1/3}$, while their arguments differ from each other by $2\pi/3$. See Figure 3.18. Going a step further, consider a ray $Z : \arg z = \alpha$. Then, under $w = z^{1/3}$, Z has three images, each of which is also a ray:

$$W_0 : \arg w = \frac{\alpha}{3},$$

$$W_1 : \arg w = \frac{\alpha + 2\pi}{3},$$

$$W_2 : \arg w = \frac{\alpha + 4\pi}{3}.$$

Now, if we let Z revolve about the origin, say, in the counterclockwise sense, the rays W_0, W_1, and W_2 will revolve similarly, and when Z has completed one revolution in the z-plane, each W_k will have covered "one-third" of the w-plane. Compare this situation (illustrated in Figure 3.18) with that described in Example 1, Section 12.

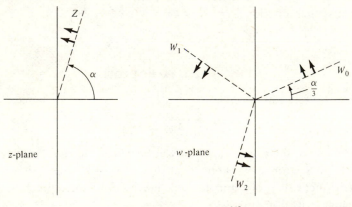

z-plane   w-plane

FIGURE 3.18 $w = z^{1/3}$

The characteristic behavior of the root function outlined in the preceding paragraph suggests that the multivalued function

$$w = z^{1/3}$$

may be split into the following three parts:

$$w_0 = r^{1/3} e^{it/3}, \qquad\qquad 0 \le t < 2\pi,$$

$$w_1 = r^{1/3} e^{i(t + 2\pi)/3}, \qquad 0 \le t < 2\pi, \qquad\qquad (4)$$

$$w_2 = r^{1/3} e^{i(t + 4\pi)/3}, \qquad 0 \le t < 2\pi.$$

It is easy to see that each of these expressions defines a *single-valued function* that maps the entire z-plane onto "one-third" of the w-plane. Specifically,

w_0 maps the z-plane onto the wedge $0 \le \arg w < \dfrac{2\pi}{3}$ plus $w = 0$,

w_1 maps the z-plane onto the wedge $\dfrac{2\pi}{3} \le \arg w < \dfrac{4\pi}{3}$ plus $w = 0$,

w_2 maps the z-plane onto the wedge $\dfrac{4\pi}{3} \le \arg w < 2\pi$ plus $w = 0$.

See Figure 3.19. Moreover, one may show that all three of the functions above are analytic everywhere except along the nonnegative real axis. Next, we take the functions defined by Equations (4), we restrict their domain by deleting from it the nonnegative real axis, and we define

$$f_0(z) = r^{1/3} e^{it/3}, \qquad\qquad 0 < t < 2\pi, r \ne 0,$$

$$f_1(z) = r^{1/3} e^{i(t + 2\pi)/3}, \qquad 0 < t < 2\pi, r \ne 0,$$

$$f_2(z) = r^{1/3} e^{i(t + 4\pi)/3}, \qquad 0 < t < 2\pi, r \ne 0.$$

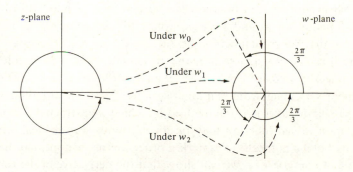

FIGURE 3.19 THE BRANCHES OF $w = z^{1/3}$

Each of these functions is called a **branch** of the multivalued function $w = z^{1/3}$. The ray consisting of the nonnegative real axis is called a **branch cut**, and the point $z = 0$ from which the branch cut emanates is a **branch point** of each branch. It follows from their definition that each of the three branches is a single-valued function that is analytic at every point of its domain. One of the branches, preferably f_0, is usually designated as the **principal branch** of the root function.

It is very important to realize that the cut which we effected along the positive real axis can be replaced by any ray emanating from the origin or, in general, by any nonself-intersecting continuous curve emanating from the origin and extending without bound. As a consequence we can say that the failure of the function $w = z^{1/3}$ to be analytic along the cut is artificial in the sense that the only point at which the root function fails to be analytic is $z = 0$. For, if z_0 is any point other than zero, we can choose a cut away from z_0 so that z_0 will belong to one of the branches of $w = z^{1/3}$ and hence analyticity at z_0 will be guaranteed. The freedom of choice that one has concerning the branch cut is not available when it comes to branch points; in fact, there is no choice at all. The branch points of the root function are $z = 0$ and $z = \infty$.

The preceding discussion of the "cube root" can be adapted, quite easily, to the general case of the root function

$$w = z^{1/n}$$

which will be found to have n branches. As a consequence of the discussion above we may say that the root function is analytic at every point $z_0 \neq 0$, and in so saying we mean that the root function can be represented by any one of its branches, provided that we choose a cut that will not contain z_0.

As we noted above, following the definition of a multivalued function, there is a very intimate connection between this concept and the concept of the inverse function. In the opening remarks of Section 10 we note that if a function is one-to-one, then its inverse is also a function. Moreover, we note there that even if a given function is many-to-one we may still obtain an inverse function for it, and we accomplish this by properly restricting its domain so that the given function (or, rather, a "part of it") becomes one-to-one. See Exercise 9.5. In essence, the process by which we extracted the functions w_0, w_1, and w_2 from $w = z^{1/3}$ is an illustration of this fact. The reader will benefit very much from making a serious effort to justify the last statement; he will discover that the method by which we extracted the branches from a multivalued function may be an ingenious way to deal with such functions, but it is not really as esoteric a process as might appear at the outset.

By use of Exercise 7.15, we can show that the derivative of the root function (more precisely: the derivative of each branch of the root function) is given by

$$\frac{dw}{dz} = \frac{1}{n}z^{(1/n)-1}$$

and exists for all $z \neq 0$.

THE LOGARITHM FUNCTION

The expression

$$w = \log z, \qquad z \neq 0, \tag{5}$$

defined by the relation

$$\log z = \ln|z| + i \arg z,$$

will be called the **logarithm function**; it is, obviously, a multivalued function, since to every nonzero z correspond infinitely many w's, any two of which differ by a multiple of $2\pi i$. The logarithm function is not to be confused with the *logarithmic function* (see p. 88).

We have seen in Section 10 that the logarithm function is the inverse of the exponential function. If we now think of the mapping aspects of the exponential function and the fact that inverse functions, taken as mappings, "undo" what the functions from which they result do, it is easy to see that the logarithm function maps the z-plane (except $z = 0$) onto the w-plane an infinite number of times, each time onto an infinite horizontal strip of width 2π; see Figure 3.20 and compare it with Figure 3.15. Also, review the discussion of the logarithmic mapping in Section 15.

In view of the relation $\arg z = \text{Arg } z + 2k\pi$ (see p. 12), the multivalued function defined by (5) can be written

$$w = \ln|z| + i\,(\text{Arg } z + 2k\pi).$$

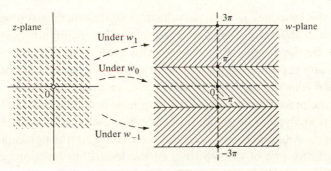

FIGURE 3.20 THE MAPPING $w = \log z$

If we set $k = 0$, we obtain a single-valued function

$$w_0 = \ln |z| + i \text{ Arg } z,$$

which is actually the *logarithmic function* introduced on p. 88. If we set $k = 1$, we obtain another single-valued function

$$w_1 = \ln |z| + i(\text{Arg } z + 2\pi)$$

defined for all $z \neq 0$. Similarly, for $k = -1$, we obtain

$$w_{-1} = \ln |z| + i (\text{Arg } z - 2\pi),$$

also single-valued and defined for all $z \neq 0$. In general, any and every integral value of k will yield such a function. Note that the effect of setting $k = 0, 1, -1, 2, \ldots$, is that, respectively, arg z is restricted to

$$-\pi < \text{arg } z \leq \pi, \qquad \text{which is the domain of } w_0,$$

$$\pi < \text{arg } z \leq 3\pi, \qquad \text{which is the domain of } w_1,$$

$$-3\pi < \text{arg } z \leq -\pi, \qquad \text{which is the domain of } w_{-1}$$

and so forth. Clearly, each of these domains has a "cut" along the negative real axis. We now proceed to remove the negative real axis from each of these domains in order to define the *branches* of the logarithm function:

$$f_0(z) = \ln |z| + i \text{ arg } z, \qquad -\pi < \text{arg } z < \pi,$$

$$f_1(z) = \ln |z| + i \text{ arg } z, \qquad \pi < \text{arg } z < 3\pi,$$

$$f_{-1}(z) = \ln |z| + i \text{ arg } z, \qquad -3\pi < \text{arg } z < -\pi,$$

$$f_2(z) = \ln |z| + i \text{ arg } z, \qquad 3\pi < \text{arg } z < 5\pi,$$

$$f_{-2}(z) = \cdots \qquad\qquad\qquad\qquad \cdots$$

$$\vdots \qquad\qquad\qquad\qquad\qquad \vdots$$

Each of the branches is a single-valued analytic function throughout the domain of its definition. The *branch cut* of the logarithm function has been *chosen* to be the nonpositive real axis. However, as in the case of the root function, any ray emanating from the origin can be used as the branch cut. The *branch points* of the logarithm function are zero and infinity.

From the preceding discussion we may now conclude that the logarithm function is analytic at every point of the plane except the origin. For, given any point $z_0 \neq 0$, we choose a cut not containing z_0, thus placing z_0 in the domain of any one of the branches of the logarithm function, and hence guaranteeing analyticity at z_0. It is very important to realize that once a cut is decided upon, then the function fails to be analytic at every point of that cut; however, no point of the z-plane other than zero can be called a singularity of $w = \log z$.

By use of Exercise 7.15 we can show that the derivative of the logarithm function (i.e., the derivative of each of its branches) is given by

$$\frac{dw}{dz} = \frac{1}{z}$$

and exists for all $z \neq 0$.

A very interesting concept intimately connected with multivalued functions is that of a *Riemann surface*. An introduction to this concept is given in Part C of this appendix.

In the preceding two special cases we pointed out the branches and branch points of the particular multivalued functions at hand. In general, given a multivalued function $g(z)$ with domain D, by a **branch** of g we mean any (single-valued) analytic function $w_0 = f(z)$ with domain D_0 contained in D and such that, for every z_0 in D_0, $f(z_0)$ is one of the values of $g(z_0)$. A precise definition of a **branch point** would necessitate consideration of numerous items that we choose not to include here. However, there is a basic characteristic of a branch point that we may describe informally as follows. If z is a branch point, then any circle centered at z and of "small" enough radius that we may draw starting at a point of one of the branches of the multivalued function will end at a point of another branch. On the other hand, if z is not a branch point, then we can draw a small enough circle around and centered at z all of which will belong to the same branch. An illustration follows.

EXAMPLE

From our discussion of the root function we may conclude that the square root function can be broken up into the following branches:

$$f_0(z) = r^{1/2}e^{it/2}, \qquad 0 < t < 2\pi, r \neq 0,$$

$$f_1(z) = r^{1/2}e^{it/2}, \qquad 2\pi < t < 4\pi, r \neq 0.$$

Here, again, we have chosen the nonnegative real axis as the branch cut.

Now take $z = 0$ and draw a circle around it starting, say, at the point $P : 2e^{i\pi/4}$; clearly, P belongs to the domain of f_0. As P describes a circle around $z = 0$, the argument of P increases from $\pi/4$ to $9\pi/4$ so that, by the time P returns to the initial position, it now belongs to the domain of f_1. Thus the process of going around a branch point interchanges branches of the multivalued function $w = z^{1/2}$.

On the other hand, as the reader may easily argue, if we take any other point, say $Q : i - 1$, we can describe a small enough circle about Q (small enough to leave $z = 0$ out) without having to leave the branch of the function from which we start the circular motion. Note that we have a choice of putting Q in the domain of f_0 (by writing Q with an argument of $3\pi/4$) or in the domain of f_1 (with argument $11\pi/4$).

EXERCISES—SET 1

1. Find the images of each of the following points under $w = z^{1/2}$:
   (a) 1.   (b) i.
   (c) $-i$.   (d) $2 + 2\sqrt{3}\,i$.
2. Find the images of the ray arg $z = 2\pi/3$ under $w = z^{1/4}$.
3. Find the images of the following curves under $w = \log z$:
   (a) the circle $|z| = 1$.   (b) the circle $|z| = \pi/2$.
   (c) the ray arg $z = \pi/4$.   (d) the ray arg $z = \pi/2$.
   (e) the sector $\pi/4 < \arg z < \pi/2$.
4. Find all possible branches of each of the following multivalued functions using the nonnegative imaginary axis as a branch cut:
   (a) $w = z^{1/6}$,   (b) $w = \log z$.
   Draw the resulting configuration in each case.
5. Verify that the derivatives of the root and logarithm functions are as asserted on p. 129 and p. 131, respectively.
6. Choose an appropriate cut and use it to define all the single-valued functions that can be extracted from the multivalued function $w = \arg z$. (Here we cannot talk about branches for this multivalued function because we have no analyticity at any point. See Exercise 7.12.)
7. Justify the statement following Equations (4).
8. Show that each function defined by Equations (4) is discontinuous at every point of the positive real axis.
9. Combine the preceding exercise with Exercise 6.9 to prove that the functions defined by Equations (4) fail to be analytic along the nonnegative real axis.
10. Discuss the branch points, branch cut, and the branches of the multivalued function $w = (z - a)^{1/2}$.
11. Study several aspects of the multivalued function
    $$w = (z^2 - 1)^{1/2}$$
    by working out the details of the following problems.
    (a) Show that by letting
    $$z - 1 = re^{it}, \qquad \text{for } -\pi < t \le \pi,$$
    $$z + 1 = \rho e^{i\tau}, \qquad \text{for } 0 \le \tau < 2\pi,$$

the given function can be broken up into two single-valued functions given by

$$w_0 = \sqrt{r\rho}\; e^{i(t+\tau)/2}$$

and

$$w_1 = -\sqrt{r\rho}\; e^{i(t+\tau)/2}.$$

(b) Show that w_0 and w_1 are discontinuous at every point of the real axis to the right of $z = 1$, by showing that as z approaches any such point the lim $(t + \tau)$ does not exist.

(c) As in (b), for the real axis to the left of $z = -1$.

(d) Show that w_0 and w_1 are continuous at every other point in the plane.

(e) From (b) and (c) conclude that w_0 and w_1 fail to be analytic at all real numbers x with $|x| \geq 1$.

(f) Impose appropriate restrictions on the domain of each of the functions w_0 and w_1 so that the resulting functions will be analytic on their domain and hence will define branches of the given function.

(g) Locate the branch points of the given multivalued function.

12. Derive the **inverse cosine function** as follows:
    (a) Write $w = \cos z$ in its exponential form.
    (b) Simplify your answer in (a) to obtain the equation

$$e^{2iz} - 2we^{iz} + 1 = 0.$$

Then use the quadratic formula to solve for z.

(c) Interchange w and z in your answer in (b) to obtain the desired multi-valued function

$$w = -i \log [z + (z^2 - 1)^{1/2}],$$

customarily denoted $w = \text{arc cos } z$ or $w = \cos^{-1} z$.

13. By a method analogous to that suggested in the preceding exercise, start with the function $w = \sin z$ and derive the multivalued **inverse sine function**

$$w = -i \log [iz + (1 - z^2)^{1/2}].$$

14. Repeat the process of Exercise 12 to obtain the **inverse tangent function**

$$w = \frac{i}{2} \log \frac{1 - iz}{1 + iz} = \frac{i}{2} \log \frac{i + z}{i - z}.$$

Part B
Conformality

The general concept of conformality encompasses a wide spectrum of topics that are directly or indirectly related to, perhaps, the most important link between the theory and applications of complex function theory, namely, conformal mapping. We propose here to acquaint ourselves with some of these topics. No in-depth discussion of any of these topics is intended, although a number of formal proofs will be included.

CONFORMAL MAPPING

The concept of conformal mapping can be thought of as a geometrical interpretation of analyticity. The primary geometric aspect of conformality, *isogonality in magnitude and direction*, was introduced in Section 9. Here we will take a closer look at this property and examine certain related ideas.

Let the function $w = f(z)$ be analytic at a point z_0 and suppose that $f'(z_0) \neq 0$. Then, by definition, f is analytic in a neighborhood N of z_0. The following development is realized within N.

Suppose that a *smooth curve** A passes through z_0 and that a variable point z approaches z_0 along A. Under f, A has an image A' in the w-plane and, as z approaches z_0 along A, $w = f(z)$ approaches $w_0 = f(z_0)$ along A'. Then, since $f'(z_0)$ exists, we know that

$$f'(z_0) = \lim_{z \to z_0} \frac{w - w_0}{z - z_0},$$

which we can express by saying that

$$\text{as } z \to z_0, \qquad \frac{w - w_0}{z - z_0} \to f'(z_0), \tag{6}$$

from which it follows that

$$\arg \frac{w - w_0}{z - z_0} \to \arg f'(z_0).$$

This, in turn, implies that

$$[\arg (w - w_0) - \arg (z - z_0)] \to \arg f'(z_0), \tag{7}$$

where the last relation is taken to hold within a multiple of 2π.

---

* A curve A is called **smooth** if A is representable parametrically by two continuously differentiable functions $x = \phi(t)$, $y = \psi(t)$ on an interval $\alpha \leq t \leq \beta$; for more details see Section 17.

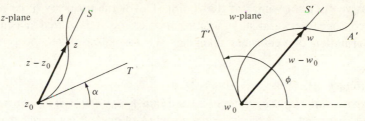

FIGURE 3.21 $\phi = \alpha + \arg f'(z_0)$

Now, as $z \to z_0$, the secant S (see Figure 3.21) tends to T, the tangent to A at z_0, whose existence is guaranteed by the definition of a smooth curve; see Remark 1, Section 17. Denoting by α the angle of inclination of T, we see that

$$\text{as } z \to z_0, \qquad \arg (z - z_0) \to \alpha. \tag{8}$$

In the meantime, in the w-plane,

$$\text{as } z \to z_0, \qquad w \to w_0,$$

and hence the secant S' tends to T', which is the tangent to A' at w_0. Denoting by ϕ the angle of inclination of T', we have that

$$\text{as } z \to z_0, \qquad \arg (w - w_0) \to \phi. \tag{9}$$

Therefore, combining (7), (8), and (9), we see that

$$\text{as } z \to z_0, \qquad (\phi - \alpha) \to \arg f'(z_0)$$

and in the limit

$$\phi = \alpha + \arg f'(z_0).$$

We summarize the preceding development in the following theorem.

Theorem
Suppose that $f(z)$ is analytic at z_0 and that $f'(z_0) \neq 0$. Let A be any smooth curve passing through z_0 and let A' denote the image of A under f.
Then, if the angle of inclination of A at z_0 is α, then the angle of inclination of A' at $f(z_0)$ is $\alpha + \arg f'(z_0)$.

Put in different terms, the preceding theorem says that *every* smooth curve passing through z_0 is rotated through the same angle $\arg f'(z_0)$.

An immediate consequence of the above theorem is the fact that the angle between any two smooth curves intersecting at z_0 will be preserved in magnitude and direction by a mapping f which is analytic at z_0 and whose derivative does not vanish at that point. This is the property of *isogonality* referred to earlier, and any function possessing it is called a **conformal transformation** or **conformal mapping**. We formalize this concept in the following.

Corollary

Suppose that $f(z)$ is analytic at z_0 and that $f'(z_0) \neq 0$. Let A and B be two smooth curves intersecting at z_0 and forming an angle θ measured from A to B.

Then their images, A' and B', under f form an angle that measured from A' to B' is θ. In short, under f, the angle at which A and B intersect at z_0 is preserved in magnitude and direction.

The proof of this corollary is immediate, since, according to the theorem above, each of the curves A and B is rotated through the same angle arg $f'(z_0)$. See Figure 3.22.

We continue with a brief look at another geometrical aspect of conformal mapping.

From (6) we have that

$$\text{as } z \to z_0, \qquad \left| \frac{w - w_0}{z - z_0} \right| \to |f'(z_0)|;$$

hence

$$\text{as } z \to z_0, \qquad |w - w_0| \to |f'(z_0)||z - z_0|.$$

The last expression says that *in the limit,* $|w - w_0|$ is a constant multiple of $|z - z_0|$, where the constant is the positive real number $|f'(z_0)|$; the proportionality constant $|f'(z_0)|$ is often called the *ratio of magnification* at z_0. Thus every length in a small neighborhood of z_0 is magnified by the same positive factor and we say that f is a **scale-preserving mapping** at z_0 *in the infinitesimal sense*. It is emphasized that the scale-preserving property of conformal mapping fails when a large region of the plane is considered.

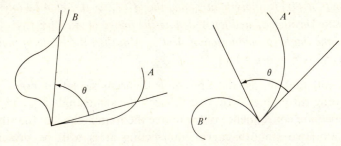

FIGURE 3.22 CONFORMAL MAPPING

EXERCISES—SET 2

1. Use the function $w = z^2$, an appropriate point z_0, and two conveniently chosen curves to illustrate the fact that if the condition $f'(z_0) \neq 0$ is omitted, then conformality is destroyed.
2. Use the properties of conformal transformations to argue that, as we saw in Section 11, the linear mapping $w = az + b$, $a \neq 0$, rotates every curve in the plane through an angle equal to arg a.
3. Discuss the stretching property of the linear function (Section 11), connecting it to the scale-preserving property of conformal transformations.
4. (a) Verify that, under the reciprocal transformation, the line $x = \frac{1}{2}$ maps onto the circle $(u - 1)^2 + v^2 = 1$, the line $y = \frac{1}{2}$ maps onto the circle $u^2 + (v + 1)^2 = 1$, and the line $y = x + \frac{1}{2}$ maps onto the circle $(u + 1)^2 + (v + 1)^2 = 2$ (Section 13).
   (b) By actually calculating the angles of intersection of the circles in (a), demonstrate the conformality of the reciprocal map.

ON THE INVERSE OF AN ANALYTIC FUNCTION

The question of whether a given function has an inverse that is itself a function is of crucial importance in many problems in the theory as well as in the applications of complex variables.

In Section 9 we noted that if a function is one-to-one, then its inverse is also a function.* This is a rather general result which does not apply to many-to-one functions. It turns out, however, that if a function is analytic, then not only is its inverse a *function* but an analytic function. This is one more illustration of the strong inner structure and exceptional properties possessed by analytic functions. The following theorem describes the entire concept in precise terms.

Theorem

Suppose that $w = f(z)$ is analytic at z_0 and that $f'(z_0) \neq 0$.

Then f has an inverse $z = f^{-1}(w)$ which is a continuous (single-valued) function throughout some neighborhood N of the point $w_0 = f(z_0)$ and such that $f^{-1}(w_0) = z_0$. Moreover, f^{-1} is analytic throughout N and, for every w in N,

$$\frac{dz}{dw} = \frac{1}{dw/dz};$$

i.e., the derivative of the inverse function is the reciprocal of the derivative of the function.

---

* The proof of this fact is quite easy once one grasps the basic definitions of the concepts involved. See Halmos, *Naïve Set Theory* (New York: Van Nostrand Reinhold, 1967), Sec. 10.

A complete proof of this theorem utilizes a number of results that ordinarily belong to a course in advanced calculus, and we cannot rightly assume knowledge of them on the part of the reader to whom this book addresses itself. On the other hand, development of the necessary prerequisites is beyond the scope of this book. We therefore choose to omit the proof of the above theorem. The interested reader will find the proof of this fundamental result in the theory of transformations given in Courant, *Differential and Integral Calculus*, Vol. II (New York: Wiley, 1968), p. 152.

Instead of a proof we discuss a number of examples that illustrate some of the aspects of the theorem above.

EXAMPLE 1

The power function $w = z^n$ is, of course, analytic with a nonzero derivative at every nonzero z_0. Therefore, according to the above theorem, in some neighborhood of the point $w_0 = z_0^n$, its inverse

$$z = w^{1/n}$$

is an analytic function, and for every point in that neighborhood

$$\frac{dz}{dw} = \frac{1}{dw/dz} = \frac{1}{nz^{n-1}} = \frac{1}{nw^{(n-1)/n}} = \frac{1}{n}\frac{1}{w^{1-(1/n)}} = \frac{1}{n}w^{(1/n)-1}.$$

The reader who has studied Part A of this appendix knows already that the analyticity of the inverse function is guaranteed as long as we avoid $z = 0$.

EXAMPLE 2

The logarithmic function $w = \text{Log } z$, $z \neq 0$, has a nonzero derivative at every nonzero point z_0. Hence, according to the above theorem, it possesses an inverse function in some neighborhood of the point $w_0 = \text{Log } z_0$, which, we know, is the exponential function

$$z = e^w$$

and whose derivative is given by the formula of the theorem,

$$\frac{dz}{dw} = \frac{1}{dw/dz} = \frac{1}{1/z} = z = e^w,$$

to nobody's surprise.

EXAMPLE 3

We recall that the function $w = \cos z$ is entire and that its derivative $w' = -\sin z$ is nonzero for all $z \neq k\pi$. The theorem then guarantees that the cosine function has an inverse in the neighborhood of every point w except $w = \cos(k\pi) = \pm 1$, and the inverse in question is analytic there. In Exercise 12, of Part A we described the steps of the derivation, which yields*

$$w = -i \log [z + (z^2 - 1)^{1/2}] \tag{10}$$

as the defining relation of the **inverse cosine function**, which is customarily denoted

$$w = \cos^{-1} z \qquad \text{or} \qquad w = \text{arc } \cos z.$$

Obviously, the function in (10) is multivalued, but when a choice of branches is made for the logarithm and the square root we will have a (single-valued) analytic function. Its derivative, given by the formula of the theorem, is calculated as follows: We have

$$\frac{dw}{dz} = \frac{1}{dz/dw} = \frac{1}{-\sin w}.$$

However, $z = \cos w$ implies that $\cos^2 w = z^2$, which implies that $1 - \cos^2 w = 1 - z^2$, and hence $\sin^2 w = 1 - z^2$. But then $\sin w = (1 - z^2)^{1/2}$ and by substitution

$$\frac{dw}{dz} = -\frac{1}{(1 - z^2)^{1/2}},$$

where the sign of the square root is that which we chose earlier in order to make Equation (10) single-valued.

EXERCISES—SET 3

1. (a) Find the inverse of the linear function function $w = az + b$, $a \neq 0$. Is it analytic? If so, where?
   (b) Find the derivative of the inverse: first, by direct differentiation and then by the formula of the theorem.
2. Find the derivative of the inverse sine function $w = \sin^{-1} z$; see Exercise 13 of Part A and Example 3.

---

* In that derivation, we interchange the roles of z and w so that the dependent and independent variables of the inverse cosine will be w and z, respectively. Note, however, that because of that interchange the given function here should be taken as $z = \cos w$.

ELEMENTARY APPLICATIONS

At this point of our development, any discussion of the uses of conformal mapping in applied problems can only serve as an indication of some of the methods used in connecting the theory with the physical problem. Consequently, we are not about to *solve* any problems facing the applied scientist; we merely intend to illustrate the *passage* from the physical to the mathematical concepts, and vice versa.

We begin with an application from *electrostatics*. It is known that if $\phi(x, y)$ is a two-dimensional potential function in a region free of charges, then ϕ satisfies Laplace's equation,

$$\phi_{xx} + \phi_{yy} = 0,$$

and the field intensity corresponding to this potential is given by the vector

$$E = -\phi_x - i\phi_y.$$

Now, since ϕ is a harmonic function, its conjugate harmonic $\psi(x, y)$ can be derived to within an additive constant (see Example 5, Section 8), so we can form an analytic function

$$f(z) = \phi(x, y) + i\psi(x, y);$$

the latter is called the **complex potential** corresponding to the (real) potential $\phi(x, y)$. Now we examine some basic attributes of f.

We note, first, that if we set

$$\phi(x, y) = c, \qquad c = \text{real}, \tag{1}$$

and we let c take on all possible values, we obtain a family of curves each of which has the property that the potential is the same at each of its points; for obvious reasons, each curve of the family is called an **equipotential line**. If we also consider the family of curves

$$\psi(x, y) = k, \qquad k = \text{real}, \tag{2}$$

then it is easy to show that (1) and (2) constitute an *orthogonal system*; i.e., every curve of (1) intersects every curve of (2) at right angles. See Exercise 3 at the end of this section. The family of curves in (2) represents what is often called the **force** or **lines of force** of the field.

Another basic property that we can derive from the equation

$$f = u + iv$$

of a complex potential is that the magnitude of the intensity of the field is given by the magnitude of f':

$$|E| = |f'|;$$

see Exercise 4 at the end of this discussion.

EXAMPLE 1

Find the potential, the force, and the complex potential between two vertical conducting plates that are kept at the constant potentials

$$\phi = P_1 \qquad \text{and} \qquad \phi = P_2,$$

making the simplifying assumption that the plates are infinitely long. (Of course, this assumption allows one to ignore fringe effects.)

Since the two plates are parallel to the y-axis, then, clearly, ϕ is a function of x *only* and hence Laplace's equation reduces to the ordinary differential equation

$$\frac{d^2\phi}{dx^2} = 0.$$

Antidifferentiating twice, we obtain the (real) potential

$$\phi = c_1 x + c_2. \tag{3}$$

The values of c_1 and c_2 depend on the boundary conditions of the problem. For instance, if we coordinate the xy-plane so that the plates will intersect the plane along the lines $x = -d$ and $x = d$, then appropriate substitutions into (3) yield

$$c_1 = \frac{1}{2d}(P_2 - P_1) \qquad \text{and} \qquad c_2 = \frac{1}{2}(P_1 + P_2).$$

Next we determine the lines of force by finding the conjugate harmonic of (3). Imitating the method described in Example 5, Section 8, we find that the electrical force is

$$\psi = c_1 y + c.$$

Thus the complex potential is

$$f(z) = (c_1 x + c_2) + i(c_1 y + c) = c_1 z + k.$$

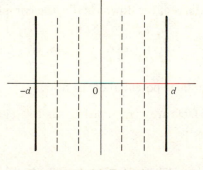

FIGURE 3.23 EXAMPLE 1

The next area from which we will draw a simple illustration is *fluid mechanics.* In particular, we will look at the case of a steady, two-dimensional, irrotational flow of an incompressible fluid* so that, essentially, we will examine a planar cross section of the moving fluid, parallel to the direction of the flow.

In the theory of fluid mechanics it is shown that, associated with each such flow, there is a function

$$\phi(x, y)$$

with the property that its partial derivatives ϕ_x and ϕ_y constitute the x- and y-components of the velocity vector of the moving fluid. The function ϕ is called the **velocity potential** of the flow and the level curves of the family

$$\phi(x, y) = c$$

are called the **equipotential lines** of the flow. Since any complex quantity can be thought of as a two-dimensional vector, we can write the velocity vector as

$$V = \phi_x + i\phi_y. \tag{4}$$

Going a step further, one may show that the function ϕ satisfies Laplace's equation

$$\phi_{xx} + \phi_{yy} = 0$$

in every region where there is neither a *sink* nor a *source,* i.e., in every region where fluid is neither lost nor introduced. Thus the velocity potential is a harmonic function, and hence its conjugate harmonic $\psi(x, y)$ can be determined; the latter is called the **stream function** and the level curves $\psi(x, y) = k$ are called the **streamlines** of the flow. Finally, the complex function

$$f(x) = \phi(x, y) + i\psi(x, y)$$

is again called the **complex potential** of the flow.

The following two facts are easy to prove; see Exercise 5 at the end of this section.

---

* These are simplifying assumptions about the flow that are valid for most fluids moving at low speeds.

(A) The velocity V of the flow is given by the complex conjugate of the derivative of f : $V = \overline{f'}$.
(B) The speed (magnitude of V) is $|f'|$.

EXAMPLE 2

1. In the context of the preceding development, let us think of the function

$$f(z) = \alpha z, \qquad \alpha = \text{real and positive,}$$

as representing a flow. Then its velocity potential is $\phi = \alpha x$ and its stream function is $\psi = \alpha y$. The streamlines, which are the curves of the family $\psi = c$, are the horizontal lines $y = c/\alpha$, from which we see that the flow described by f is parallel to the real axis. Its velocity, calculated either from (4) or from (A), is found to be

$$V = \alpha;$$

see Figure 3.24(a).

2. Next we examine a flow around a 90° turn and for that we appeal to the square root function

$$w = z^{1/2}.$$

From our discussion of multivalued functions, we know that, under this function, the upper half of the z-plane is mapped onto the first quadrant of the w-plane. On the other hand, its inverse function

$$z = x + iy = (u^2 - v^2) + (2uv)i$$

transforms the first quadrant of the w-plane [as in Figure 3.24(b)] into the upper half of the z-plane [as in Figure 3.24(a)]. Then, in view of our development in the first part of this example, we see that the velocity potential in this case is given by

$$\phi = \alpha x = \alpha(u^2 - v^2)$$

and the stream function becomes

$$\psi = \alpha y = 2\alpha uv.$$

Thus the streamlines $\psi = c$ are the rectangular hyperbolas

$$uv = \frac{c}{2\alpha}$$

[see Figure 3.24(b)], and the complex potential is

$$f = w^2.$$

Finally, the velocity of the flow at any point (u, v) of the first quadrant is

$$V = \overline{f'} = 2\alpha(u - iv)$$

and hence its speed is

$$|V| = 2\alpha(u^2 + v^2)^{1/2}.$$

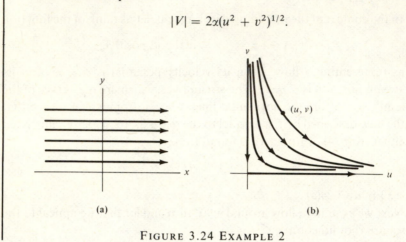

FIGURE 3.24 EXAMPLE 2

As we remarked at the beginning of this discussion, our intention was to give only an indication of the *passage* from the mathematical to the physical concepts. Consequently, we have treated only a very small part of the problems we examined. Beyond these, there are many theoretical and "practical" questions that we are not prepared to tackle at this point of our development.

EXERCISES—SET 4

1. Given the complex potential $f(z) = z + 1$, find and plot the equipotential lines.
2. Repeat the preceding exercise for $f(z) = 1/z$.
3. Given the complex potential $f(z) = \phi(x, y) + i\psi(x, y)$, consider the two families of curves

$$\phi(x, y) = c \qquad \text{and} \qquad \psi(x, y) = k.$$

(a) Find the total differentials $d\phi = 0$ and $d\psi = 0$. [The total differential of a function $g(x, y)$ is given by $dg = g_x\, dx + g_y\, dy$.]
(b) From your answers in (a), verify that

$$\frac{dy}{dx} = -\frac{\phi_x}{\phi_y} \qquad \text{and} \qquad \frac{dy}{dx} = -\frac{\psi_x}{\psi_y}.$$

(c) Use the Cauchy–Riemann equations to show that the two families of curves above constitute an orthogonal system (p. 140).

4. Show that the magnitude of the intensity of a field represented by the complex potential $f(z) = u + iv$ is given by $|E| = |f'|$.

5. Verify the two assertions immediately preceding Example 2.

Part C
Riemann Surfaces

The concept of a Riemann surface, first introduced by the German mathematician Bernhard Riemann in his 1851 dissertation, is an ingenious solution to the problem of attaching a geometric meaning to a multivalued function in such a way that the function appears to be single-valued.

In this part of Appendix 3 we propose to introduce the reader to the general idea of a Riemann surface by examining two specific cases: the cube root function and the logarithm function. To avoid repetition, we shall use some of the developments and the notation from Part A.

THE CUBE ROOT SURFACE

In our earlier discussion of the cube root function we decomposed the multivalued function

$$w = z^{1/3}$$

into three *single-valued* functions defined by

$$w_0 = r^{1/3}e^{it/3}, \qquad 0 \le t < 2\pi,$$

$$w_1 = r^{1/3}e^{i(t+2\pi)/3}, \qquad 0 \le t < 2\pi,$$

$$w_2 = r^{1/3}e^{i(t+4\pi)/3}, \qquad 0 \le t < 2\pi.$$

A careful examination of the behavior of these three functions as t varies from 0 to 2π, in each case, will show that the same results will be obtained if we write them in the alternative form

$$w_0 = r^{1/3}e^{it/3}, \qquad 0 \le t < 2\pi,$$

$$w_1 = r^{1/3}e^{it/3}, \qquad 2\pi \le t < 4\pi,$$

$$w_2 = r^{1/3}e^{it/3}, \qquad 4\pi \le t < 6\pi.$$

Going one step further, it is easy to see that the last three equations can be collapsed into one:

$$w = r^{1/3}e^{it/3}, \qquad 0 \le t < 6\pi. \tag{1}$$

Again, a careful examination of these three alternatives will reveal that one fact remains unaltered in all of them: The z-plane must be traversed three times before the w-plane can be covered entirely. However, Equation (1) lends itself to a "geometrical construction," which leads to a Riemann surface for the cube root function.

Instead of traversing the z-plane three times, we take three copies of the z-plane, S_1, S_2, and S_3, forming a strange configuration that we may describe as follows; see Figure 3.25(a). As t varies from 0 to 6π, let us trace the path of a point

$$z = re^{it}$$

as it describes a continuous curve around the origin, say, a circle. The point z starts from the positive real axis and travels on S_1 until its argument t reaches 2π. At that instant, z ascends onto S_2 and it continues its trip around the origin while it remains on this second level. When its argument reaches 4π, z ascends onto S_3 and goes around the origin once again until t approaches 6π. By this time, the function given by Equation (1) has covered the w-plane completely, and the point z descends onto the first level S_1. The actual configuration looks more like that of Figure 3.25(b), in that the three sheets of the surface are joined along the nonnegative real axis.

This ingenious scheme has enabled us to effect a one-to-one mapping *from the three-sheet Riemann surface* onto the w-plane by means of the function of Equation (1).

The reader should have no difficulty in visualizing the Riemann surfaces of $w = z^{1/2}$ (having two sheets), $w = z^{1/4}$ (with four sheets), and, in general, of $w = z^{1/n}$ (with n sheets.)

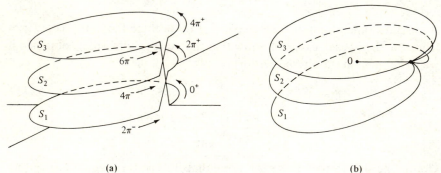

(a)   (b)

FIGURE 3.25 RIEMANN SURFACE OF $w = z^{1/3}$

THE LOGARITHM SURFACE

We recall from our discussion of the multivalued function

$$w = \log z$$

that an equivalent way to write this function is

$$w = \ln |z| + i(\text{Arg } z + 2k\pi), \qquad k = \text{integer}. \qquad (2)$$

Then, as k takes on all integral values, we obtain the single-valued functions

$$\ldots w_{-2}, w_{-1}, w_0, w_1, w_2, \ldots, \qquad (3)$$

each of which maps the z-plane, except $z = 0$, onto a horizontal strip of the w-plane, having width 2π. Therefore, in order to cover the entire w-plane we must map the z-plane an infinite number of times, by using all the functions given in (3). Instead, using a scheme similar to that used for the cube root function, we employ an infinite number of copies of the z-plane,

$$\ldots S_{-2}, S_{-1}, S_0, S_{-1}, S_{-2}, \ldots,$$

each with its origin deleted and with a slit starting at the origin and running along the nonpositive real axis. Consecutive sheets are then glued together along the slits so that they form a continuous sheet resembling a circular staircase of infinite height and depth. A point moving around the origin in a counterclockwise direction will be going up the staircase, and every time it completes a circle with center at the origin, its position will be on the next sheet up, directly above its starting point.

Again, this geometrical scheme enables us to think of the multivalued function $w = \log z$ as single-valued, mapping its Riemann surface in a one-to-one fashion onto the entire w-plane.

NOTE: The above Riemann surfaces are the simplest types of surfaces that result from multivalued functions. Some less simple surfaces appear in the exercises that follow.

EXERCISES—SET 5

1. Sketch the Riemann surfaces of $w = z^{1/2}$ and $w = z^{1/4}$.
2. Discuss and sketch the Riemann surface of the function $w = (z - a)^{1/2}$, where a is any nonzero complex number.
3. Generalize your answer in the preceding exercise to sketch the surfaces corresponding to the functions $w = (z - a)^{1/3}$ and $w = (z - a)^{1/4}$.
4. Sketch the Riemann surface of $w = \log (z - a), a \neq 0$.
5. Sketch the Riemann surface of the function $w = (z^2 - 1)^{1/2}$. Your work will be greatly facilitated if you first discuss carefully all the parts of Exercise 11 of Part A.

PART II

The Foundations of Complex Function Theory

CHAPTER 4
Complex Integration

Section 17 Introduction to elementary topological concepts: Smooth curve; path (open, closed, simple, multiple); the interior and exterior of a simple closed path; orientation of a path; connected, simply connected, and multiply connected regions.

Section 18 Definition and methods of evaluation of a real line integral.

Section 19 Definition of a complex integral and its evaluation by means of line integrals.

Section 17
Paths. Connectedness

The notion of expressing curves in parametric form is familiar to the reader from calculus. For instance, the parabola

$$y = x^2$$

can be represented, in parametric form, by

$$x = t, \qquad y = t^2, \qquad \text{with} \qquad -\infty < t < \infty.$$

If the parameter t were allowed to vary in the interval $-1 \le t \le 2$ only, one would obtain that part of the parabola which extends from the point $(-1, 1)$ to the point $(2, 4)$; see Figure 4.1.

The concept of planar curves capable of being expressed parametrically will be a very important tool in our work on complex integration. However, certain forms of such curves are of no use to us at this point and hence certain restrictions must be imposed by which the undesirable types of curves will be avoided. In particular, we wish to exclude curves or parts of curves whose length over a finite range of the parameter is infinite. We have the following definition.

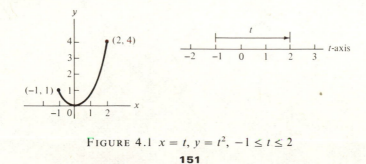

FIGURE 4.1 $x = t, y = t^2, -1 \le t \le 2$

Let t be a real variable. Then a curve in the plane will be called a **smooth curve** if and only if it is representable by two real-valued functions

$$x = \phi(t), \qquad y = \psi(t), \qquad \text{with} \qquad \alpha \le t \le \beta, \tag{1}$$

such that their derivatives

$$\frac{dx}{dt} = \phi'(t) \qquad \text{and} \qquad \frac{dy}{dt} = \psi'(t)$$

exist and are continuous functions of t on the same interval.

REMARK 1

The above definition imposes three conditions, each with a specific purpose. First, it requires that the functions ϕ and ψ be continuous on the given interval by requiring existence of their derivatives. This, in turn, guarantees that the curve is continuous. Second, the condition that the derivatives exist at every point of the interval ensures that the curve is "smooth," i.e., that it has a tangent at every point of the interval. Finally, the requirement that the derivatives also be continuous on the given interval guarantees that the curve has an assignable length. Indeed, one proves in calculus that, under the above conditions, the length of the curve is finite and is given by

$$L = \int_{\alpha}^{\beta} \left[\left(\frac{dx}{dt} \right)^2 + \left(\frac{dy}{dt} \right)^2 \right]^{1/2} dt.$$

EXAMPLE 1

The curve C described by

$$x = 2 \cos t, \qquad y = 2 \sin t, \qquad \text{with} \qquad 0 \le t \le \frac{3\pi}{2},$$

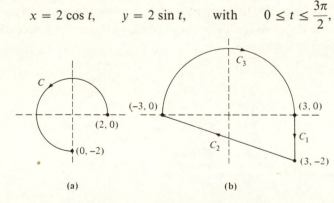

(a)   (b)

FIGURE 4.2 EXAMPLES 1 AND 2

is a smooth curve, since the conditions of the definition given above are satisfied. It is not difficult to recognize that C is the circular arc of Figure 4.2(a).

As the parameter t varies from 0 to $3\pi/2$, the curve is traced in the counterclockwise direction beginning with the point $(2, 0)$. Thus an "orientation" is induced on C, which is indicated in the figure by an arrowhead.

EXAMPLE 2

The reader may easily verify that each of the following sets of parametric equations represents a smooth curve:

$$C_1 : x = 3, \qquad y = -t, \qquad 0 \le t \le 2;$$
$$C_2 : x = -6t + 3, \qquad y = 2t - 2, \qquad 0 \le t \le 1;$$
$$C_3 : x = -3 \cos t, \qquad y = 3 \sin t, \qquad 0 \le t \le \pi.$$

The resulting configuration is drawn in Figure 4.2(b), where we also indicate the "orientation" of each curve induced by the variation of the parameter over the allotted interval in each case. Thus, for instance, as t varies from 0 to 2, C_1 is traced from $(3, 0)$ to $(3, -2)$.

As we shall see shortly, such combinations of smooth curves will be useful in our work on complex integration; we shall use the symbolic notation

$$C_1 + C_2 + C_3$$

to denote the entire curve.

If C is a smooth curve expressed parametrically as in (1), then the point of C corresponding to the value $t = \alpha$ is called the **initial point** of C and the point obtained from the value $t = \beta$ is called the **terminal point** of C. A curve C (not necessarily smooth) will be called a **path** if it consists of a

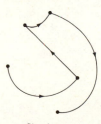
Simple open

Simple closed

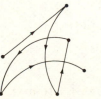
Multiple open

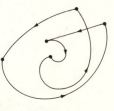
Multiple closed

FIGURE 4.3 TYPES OF PATHS

finite number of smooth curves $C_1, C_2, \ldots, C_n$, joined in such a way that the terminal point of C_k coincides with the initial point of C_{k+1}, for $k = 1, 2, \ldots, n$.* Clearly, one may now designate the initial point of C_1 as the *initial point of the path* C and the terminal point of C_n as the *terminal point of* C. We use the symbolic notation

$$C = C_1 + C_2 + \cdots + C_n.$$

If the terminal point of a path C coincides with its initial point, then C is called a **closed path;** otherwise, it is called an **open path.** If a path does not intersect itself (except that its initial and terminal points may coincide), it is called a **simple path;** otherwise, we call it a **multiple path.** With the help of Figure 4.3, the meaning of the terms **simple open, simple closed, multiple open,** and **multiple closed** paths should be obvious.

The reader is cautioned not to confuse the concepts of open and closed *paths* just introduced with the concepts of open and closed *sets*, which we discussed in Section 3.

Paths will be employed extensively in our discussion of complex integration, where they will take the place of the *interval of integration* familiar to the reader from elementary calculus. In the process, paths will be traversed from their initial to their terminal point, or vice versa. Consequently, a sense of direction—more precisely, an *orientation*—will have to be associated with each path and some conventions must be agreed upon concerning the orientation of an arbitrary path. But before we do this we state, without proof, a celebrated theorem by the French mathematician Camille Jordan, which will provide us with a most convenient terminology needed for our purposes.

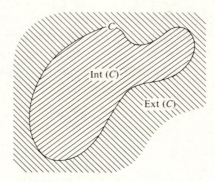

FIGURE 4.4 JORDAN CURVE THEOREM

---

* Other names are used for this concept. They include the terms "piecewise smooth" and "sectionally smooth."

Theorem 4.1 (*Jordan Curve Theorem*)

Suppose that C is a simple closed path in the plane.

Then the plane is divided by C into three mutually disjoint sets as follows:

1. *The curve C itself.*
2. *The interior of C, denoted* Int (*C*), *which is an open and bounded set.*
3. *The exterior of C, denoted* Ext (*C*), *which is an open and unbounded set.*

Furthermore, C is the boundary of both Int (*C*) *and* Ext (*C*).

The truth of the above theorem is certainly intuitively obvious and its simple geometric meaning is illustrated in Figure 4.4. However, the proof of Jordan's theorem is by no means a simple matter and is definitely beyond the scope and the level of this book.

We proceed now with the concept of the orientation of a smooth curve and define it for the cases of a simple open and a simple closed path only. The reason for this is that we will deal almost exclusively with these two types of curves. If and when we deal with multiple paths, we will take care of any problems that may arise in connection with their orientation. A simple open path will be said to be **positively oriented** if it is traversed from its initial to its terminal point. A simple closed path C will be said to be positively oriented if it is traversed in such a way that its interior, Int (*C*), lies to one's immediate left. In both cases, a path traversed in the sense opposite to its positive orientation will be said to be **negatively oriented.** If C is an oriented path, then the symbol

$$- C$$

will be used to denote the same path but with an orientation opposite to that of C.

EXAMPLE 3

The path

$$C : x = t, \qquad y = t^2, \qquad 0 \le t \le 2,$$

is the part of the parabola $y = x^2$ between the points $(0, 0)$ and $(2, 4)$ and is (positively) oriented from $(0, 0)$ to $(2, 4)$. On the other hand, note that the path given by

$$x = -t, \qquad y = t^2, \qquad -2 \le t \le 0,$$

is the same curve but this time oriented from $(2, 4)$ to $(0, 0)$. So, actually, the latter set of equations represents $- C$.

EXAMPLE 4

The curve

$$C : x = \cos t, \qquad y = -\sin t, \qquad 0 \le t \le 2\pi,$$

is a simple closed path, namely, the unit circle traversed in the clockwise direction. Clearly, C is negatively oriented for, as t goes from 0 to 2π, the curve is traversed in such a way that Int (C) is to our immediate right. On the other hand, the same curve can be expressed by

$$x = \cos t, \qquad y = \sin t, \qquad 0 \le t \le 2\pi,$$

and be positively oriented.

We close this section with the introduction of two types of sets that play a prominent role in the study of complex integration, which will be used often enough to deserve special names. A set R is called **connected** if and only if every two points in R can be joined by a path that lies entirely in R. On the other hand, a set S is called **simply connected** if and only if, given any simple closed path C lying entirely in S, its interior Int (C) also lies entirely in S. We will often refer to a set that is not simply connected as **multiply connected.**

REMARK 2

The term "simply connected" is very widely used and we adopt it here mainly because of respect for established terminology. However, the reader should bear in mind that the presence of the word "connected" in it is slightly misleading. For, as we will see in the following example, *a set may be simply connected but not connected.*

EXAMPLE 5

1. The interior of any circle or triangle or polygon or, in general, the interior of any simple closed path is a simply connected region that is also connected. Justify this statement.
2. The right half-plane, $R(z) > 0$, is a region that is connected and simply connected. Justify this.
3. The set of all z such that $R(z) \ne 0$, i.e., the plane with the imaginary axis removed, is a simply connected region which, however, is not connected. Justify both parts of this assertion.
4. The set of all z such that $1 < |z| < 2$ is a connected region that is not simply connected. Justify this.

The washer-type region contained between two concentric circles (such as in part 4 of this example) is called a **circular annulus**. In general, the region bounded by two simple closed paths, one of which is contained entirely in the interior of the other, is called an **annulus** or **annular region**; see Figure 4.5.

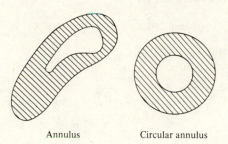

Annulus   Circular annulus

FIGURE 4.5 EXAMPLE 5, PART 4

EXERCISE 17

A

17.1. In each of the following cases, draw the given smooth curve, mark its initial and terminal point, and determine its orientation. Also, in each case, eliminate the parameter and express the curve in its rectangular form.
(a) $x = t^2 - 1, y = t, -1 \le t \le 1$.
(b) $x = 3 \cos t, y = 2 \sin t, 0 \le t \le \pi$.
(c) $z = -i + e^{it}, -\pi \le t \le \pi$ (see Exercise 17.5).
(d) $x = e^{-t}, y = t + 1, 0 \le t \le 1$.
(e) $z = z_0 + 2e^{-it}, -\pi/2 \le t \le \pi$.

17.2. Sketch the path $C = C_1 + C_2 + C_3 + C_4$, where
$C_1 : x = -\sin t, y = \cos t, 0 \le t \le \pi/2$;
$C_2 : y = -x - 1$, from $(-1, 0)$ to $(0, -1)$;
$C_3 : x = 2t + 2, y = t, -1 \le t \le 0$;
$C_4 : z = 1 + e^{it}, 0 \le t \le \pi$.

17.3. Find a parametric representation of the line segment joining the points $(1, 1)$ and $(-3, -1)$ in that order.

17.4. Give one example (other than those given in the text) for each of the following:
(a) A set that is connected and simply connected. •
(b) A set that is connected but not simply connected.
(c) A set that is simply connected but not connected.
(d) A set that is neither connected nor simply connected.

B

17.5. Verify that a circle of radius r and center at $z_0 = (a, b)$ can be represented

(a) In rectangular form by $(x - a)^2 + (y - b)^2 = r^2$.

(b) In parametric form by

$$x = a + r \cos t \qquad y = b + r \sin t \qquad 0 \le t \le 2\pi.$$

(c) In parametric form by

$$x = a + \frac{mr}{(m^2 + 1)^{1/2}} \qquad y = b + \frac{r}{(m^2 + 1)^{1/2}}.$$

$$-\infty < m < \infty.$$

(d) In complex form by $z = z_0 + re^{it}$, $0 \le t \le 2\pi$.

17.6. Verify that (b) and (d) in the preceding exercise are essentially one and the same representation.

17.7. Suppose that a smooth curve C is represented by

$$x = \phi(t), \qquad y = \psi(t), \qquad \alpha \le t \le \beta.$$

Exercise 17.8 essentially claims that one may find a different representation for C over any other interval $\gamma \le \tau \le \delta$ by replacing, in the functions ϕ and ψ, the variable t by the expression

$$\alpha + \frac{\beta - \alpha}{\delta - \gamma}(\tau - \gamma).$$

Use this transformation to find a different representation for each of the following paths by changing from the first to the second of the intervals given in each case.

(a) $x = t + 3$, $y = t^2 - 1$, $-1 \le t \le 1$, $0 \le \tau \le 1$.

(b) $x = -\sin t$, $y = 2 \cos t$, $0 \le t \le \pi$, $\pi \le \tau \le 2\pi$.

(c) $z = -i + e^{it}$, $-\pi \le t \le \pi$, $0 \le \tau \le \pi$.

C

17.8. Show that any smooth curve $C : x = \phi(t)$, $y = \psi(t)$, $\alpha \le t \le \beta$, can also be represented by $x = \phi(t)$, $y = \psi(t)$, where

$$t = \alpha + \frac{\beta - \alpha}{\delta - \gamma}(\tau - \gamma)$$

over any interval $\gamma \le \tau \le \delta$.

Section 18
Line Integrals

In our discussion of the complex integral in the next section, we shall see that the integral of a complex function can be decomposed into the sum of four "line integrals" (see Theorem 4.4). As a consequence, properties of line integrals are inherited by complex integrals, and methods of evaluation of line integrals can be effectively utilized quite often in evaluating complex integrals. With these facts in mind, we devote this section to the study of line integrals and thus prepare the way for the study of the first phase of complex integration.

We begin by setting the stage for the definition of a line integral. Let C be a smooth curve in the plane defined by

$$x = \phi(t), \qquad y = \psi(t), \qquad \alpha \le t \le \beta.$$

Subdivide the interval $\alpha \le t \le \beta$, arbitrarily, into n subintervals by use of a partition

$$\alpha = t_0 < t_1 < t_2 < \cdots < t_{n-1} < t_n = \beta$$

and on each subinterval

$$t_{k-1} \le t \le t_k$$

choose, arbitrarily, a point τ_k for all $k = 1, 2, \ldots, n$; see Figure 4.6. Denote by $(\Delta t)_k$ the length $t_k - t_{k-1}$ of the kth subinterval and by μ the maximum of these lengths. Now, any partition of the interval $\alpha \le t \le \beta$ induces a similar configuration on C (see Figure 4.6), since each t_k of the partition yields a point on C with coordinates $x_k = \phi(t_k)$ and $y_k = \psi(t_k)$. Thus C is subdivided into n arcs by means of the points

$$(x_0, y_0), (x_1, y_1), \ldots, (x_k, y_k), \ldots, (x_n, y_n).$$

Furthermore, on each of these arcs we obtain a point that corresponds to the point τ_k of the t-axis, i.e., a point (σ_k, ω_k) where $\sigma_k = \phi(\tau_k)$ and $\omega_k = \psi(\tau_k)$, for all $k = 1, 2, \ldots, n$.

Suppose now that a *real-valued* function

$$M(x, y)$$

of the real variables x and y is well defined at least on C. Then, certainly, the function is defined at the points

$$(\sigma_1, \omega_1), (\sigma_2, \omega_2), \ldots, (\sigma_n, \omega_n).$$

For each $k = 1, 2, \ldots, n$, form the product

$$M(\sigma_k, \omega_k)(\Delta x)_k,$$

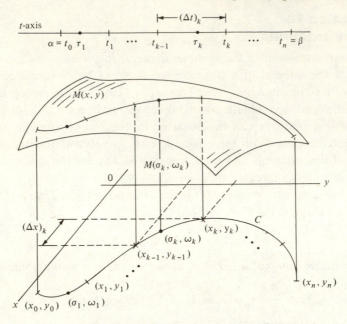

FIGURE 4.6 DEFINITION OF LINE INTEGRAL

where $(\Delta x)_k$ is the projection of the arc determined by (x_{k-1}, y_{k-1}) and (x_k, y_k) onto the x-axis, and then form the sum

$$\sum_{k=1}^{n} M(\sigma_k, \omega_k)(\Delta x)_k.$$

In the context of the above configuration, we define the **line integral** of $M(x, y)$ along C with respect to x to be the limit of the above sum as $\mu \to 0$, provided that this limit exists, and we denote this line integral by

$$\int_C M(x, y)\, dx.$$

In short, by definition,

$$\int_C M(x, y)\, dx = \lim_{\mu \to 0} \sum_{k=1}^{n} M(\sigma_k, \omega_k)(\Delta x)_k,$$

provided that this limit exists independently of the choice of the partition and the points τ_k. The curve C will be referred to as the **path of integration.**

REMARK 1

The line integral of $M(x, y)$ along C *with respect to y* is defined analogously. To that end, the above definition may be used word for word, except that the arcs into which C is subdivided must be projected onto the y-axis, and the corresponding $(\Delta y)_k$'s must replace the $(\Delta x)_k$'s. The line integral thus defined is denoted

$$\int_C M(x, y)\, dy.$$

REMARK 2

The manner in which the interval $\alpha \le t \le \beta$ was subdivided implies that the integration along C is in the direction of increasing t, i.e., from the initial to the terminal point of C. This is true also when C is a closed curve, even though the orientation induced on C as t increases from α to β may not be the positive orientation of C.

We now proceed to show that when some rather mild conditions are satisfied, the existence of a line integral is guaranteed.

Theorem 4.2 *(Existence of Line Integral)*
Suppose that
1. C is a smooth curve expressed parametrically by $x = \phi(t)$, $y = \psi(t)$, $\alpha \le t \le \beta$.
2. The function $M(x, y)$ is continuous on C.
Then the line integral of M along C with respect to x exists and its value is given by the formula

$$\int_C M(x, y)\, dx = \int_\alpha^\beta M(\phi(t), \psi(t)) \cdot \phi'(t)\, dt.$$

Proof:
See Appendix 4.

REMARK 3

As we shall see during our discussion of how a line integral is evaluated, later in this section, the preceding theorem not only establishes the existence of a line integral, but it supplies the fundamental formula for evaluating such an integral. An analogous theorem is also true when the integration is with respect to y, in which case its value is given by

$$\int_C M(x, y)\, dy = \int_\alpha^\beta M(\phi(t), \psi(t)) \cdot \psi'(t)\, dt.$$

REMARK 4

The line integral has a rather simple geometrical interpretation, although not as meaningful as some interpretations attached to the ordinary (Riemann)

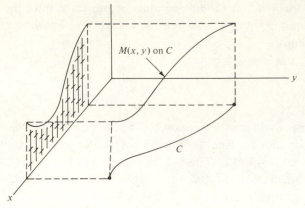

$M(x, y)$ on C

y

C

x

FIGURE 4.7 GEOMETRICAL REALIZATION OF A LINE INTEGRAL.
REMARK 4

integral $\int f(x)\, dx$. We arrive at this interpretation as follows. In the familiar three-dimensional cartesian space, consider the function $M(x, y)$ defined on a set of points of the xy-plane including the smooth curve C; see Figure 4.7. The graph of $M(x, y)$ over C is, of course, a space curve. Project this curve on the xz-plane. Then the area on the xz-plane between this projection and the x-axis is a geometrical realization of

$$\int_C M(x, y)\, dx.$$

Before tackling the problem of the evaluation of line integrals, we take a look at some of their basic properties that the reader should find familiar, at least in form. We have the following theorem.

Theorem 4.3

Suppose that ζ is an arbitrary complex constant and that $C + K$ is a path consisting of the smooth curves C and K. Suppose further that the line integrals of the functions $M(x, y)$ and $N(x, y)$ along each of the above paths exists.
Then,

1. $\int_C \zeta \cdot M(x, y)\, dx = \zeta \int_C M(x, y)\, dx.$
2. $\int_C (M(x, y) + N(x, y))\, dx = \int_C M(x, y)\, dx + \int_C N(x, y)\, dx.$
3. $\int_{C+K} M(x, y)\, dx = \int_C M(x, y)\, dx + \int_K M(x, y)\, dx.$
4. $\int_{-C} M(x, y)\, dx = -\int_C M(x, y)\, dx.$

Proof:
See Exercise 18.16 for parts 1, 2, and 4, and Appendix 4 for part 3.

REMARK 5

Part 3 of Theorem 4.3 can be extended, by induction, to the case of a path consisting of n smooth curves. This, in turn, would constitute a proof of the fact that Theorem 4.2, which establishes the existence of a line integral along a smooth curve, is true also when C is any path.

EVALUATION OF LINE INTEGRALS

The key step in the process of evaluating line integrals is an appropriate *substitution from the equations of the path of integration* into the integrand. A careful examination of the formula of Theorem 4.2 will make this fact abundantly clear. For, in it the right-hand side is obtained from the left by use of the familiar equations

$$x = \phi(t), \qquad y = \psi(t), \qquad dx = \phi'(t) \, dt.$$

These substitutions yield an ordinary (Riemann) integral in the variable t whose limits α and β determine the interval of integration. Analogous substitutions are obvious in the formula of Remark 3. Then, the evaluation of the right-hand side is a matter of ordinary integration. See Examples 1, 2, and 3 later in this section.

Quite often, the path of integration is given in nonparametric form:

$$y = f(x) \qquad \text{or} \qquad x = g(y) \tag{1}$$

with initial point (a, b) and terminal point (c, d). In such cases, as above, straight substitution yields the formulas

$$\int_C M(x, y) \, dx = \int_{x=a}^{c} M(x, f(x)) \, dx$$

and

$$\int_C M(x, y) \, dy = \int_{y=b}^{d} M(g(y), y) \, dy.$$

Note, once again, that the integrals on the right-hand side of each formula are ordinary (Riemann) integrals and can be evaluated as usual. See Example 4 of this section.

A variation of the preceding case makes use of the formulas

$$dy = f'(x) \, dx \qquad \text{and} \qquad dx = g'(y) \, dy,$$

which result from Equations (1) to yield the formulas

$$\int_C M(x, y) \, dx = \int_{y=b}^{d} M(g(y), y)g'(y) \, dy$$

and

$$\int_C M(x, y)\, dy = \int_{x=a}^c M(x, f(x)) f'(x)\, dx.$$

Examples 5 and 6 of this section point to another problem associated with the evaluation of line integrals. We are referring to the case where the path is given either verbally or via a drawing. This leaves the reader with the task of choosing an appropriate representation of the path so that, after the ensuing substitution, the resulting integration will be as simple as possible. It should be pointed out that the "wrong" choice of path representation may lead to an integral nearly impossible to evaluate, although the path representation may be correct. No panacea is known for this problem, but one method is known to have the best results: practice.

EXAMPLE 1
Evaluate the line integrals with respect to both variables of the function $M(x, y) = xy$ along $C : x = 4t,\ y = t^2,\ 0 \le t \le 2$.

Substituting in the formula of Theorem 4.2, we find that

$$\int_C xy\, dx = \int_0^2 4t \cdot t^2 \cdot 4\, dt = 16 \int_0^2 t^3\, dt = 64.$$

Similarly, using the formula of Remark 3, we find that

$$\int_C xy\, dy = \int_0^2 4t \cdot t^2 \cdot 2t\, dt = 8 \int_0^2 t^4\, dt = \tfrac{256}{5}.$$

EXAMPLE 2
Evaluate both line integrals of $M(x, y) = x + y$ along the path

$$C : x = e^t, \qquad y = \sin t, \qquad 0 \le t \le \pi.$$

Using the same formulas as in the preceding example, we find

$$\int_C (x + y)\, dx = \int_0^\pi (e^t + \sin t)e^t\, dt = \tfrac{1}{2}(e^{2\pi} + e^\pi)$$

and

$$\int_C (x + y)\, dy = \int_0^\pi (e^t + \sin t) \cos t\, dt = -\tfrac{1}{2}(e^\pi + 1).$$

EXAMPLE 3

Evaluate both line integrals of the function $M(x, y) = x + y$ along the path $C + K$ of Figure 4.8.

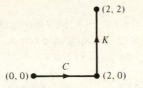

FIGURE 4.8 EXAMPLE 3

The given path consists of the smooth curves

$$C : y = 0 \qquad \text{with } 0 \leq x \leq 2$$

and

$$K : x = 2 \qquad \text{with } 0 \leq y \leq 2.$$

It is important to realize that along C, $dy = 0$, and along K, $dx = 0$. Then, by direct substitution, we find that

$$\int_{C+K} (x + y)\, dx = \int_C (x + y)\, dx + \int_K (x + y)\, dx = \int_0^2 x\, dx = 2.$$

On the other hand,

$$\int_{C+K} (x + y)\, dy = \int_C (x + y)\, dy + \int_K (x + y)\, dy = \int_0^2 (2 + y)\, dy = 6.$$

EXAMPLE 4
Evaluate both line integrals of $M(x, y) = x + 2y$ along $C + K$, where

$$C : y = x^2 \qquad \text{from } (-2, 4) \text{ to } (0, 0)$$

and

$$K : x = y^2 \qquad \text{from } (0, 0) \text{ to } (4, 2).$$

Note that on C, $dy = 2x\, dx$ and on K, $dx = 2y\, dy$. Then

$$\int_{C+K} (x + 2y)\, dx = \int_C (x + 2y)\, dx + \int_K (x + 2y)\, dx$$

$$= \int_{-2}^0 (x + 2x^2)\, dx + \int_0^2 (y^2 + 2y)2y\, dy = \tfrac{66}{3} = 22.$$

Similarly, we have

$$\int_{C+K} (x + 2y)\, dy = \int_C (x + 2y)\, dy + \int_K (x + 2y)\, dy$$

$$= \int_{-2}^0 (x + 2x^2)2x\, dx + \int_0^2 (y^2 + 2y)\, dy = -4.$$

EXAMPLE 5

Evaluate the line integral $\int (x^2 + y^2)\, dx$ along the upper half of the circle $|z| = 2$ traversed in the counterclockwise sense.

We can express the given path C by

$$y = (4 - x^2)^{1/2} \qquad \text{with } x : 2 \to -2.$$

Then, substituting in the given integral, we find that

$$\int_C (x^2 + y^2)\, dx = \int_2^{-2} [x^2 + (4 - x^2)]\, dx = -16.$$

EXAMPLE 6

Evaluate the line integrals (a) $\int_C y^2\, dx$ and (b) $\int_C x\, dx$, where C is the semicircle in Figure 4.9.

There are at least two representations for C:
1. $y = (1 - x^2)^{1/2}$ with $x : 1 \to -1$.
2. $x = \cos t,\ y = \sin t,\ 0 \le t \le \pi$.

In the exercises at the end of this section, the reader is asked to verify that the problem of evaluating the given integrals becomes very easy if we use 1 in the first integral and 2 (or no substitution at all) to evaluate the second.

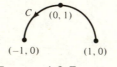

(−1, 0)   (1, 0)

FIGURE 4.9 EXAMPLE 6

EXERCISE 18

A

18.1. Evaluate both line integrals of the function $M(x, y) = xy - y^2$ along each of the following paths.
  (a) $x = t^2,\ y = t^3,\ 0 \le t \le 1$.
  (b) $x = t^2,\ y = t,\ 1 \le t \le 3$.
  (c) $y = x^3$ from $(-1, -1)$ to $(1, 1)$.
  (d) $y = 3 - x$ from $(3, 0)$ to $(5, -2)$.

18.2. If C consists of the line segments joining $(0, 0)$ to $(1, 1)$ and $(1, 1)$ to $(1, 0)$ show that

$$\int_C (x^2 + y^2)\, dy = -\tfrac{2}{3}.$$

18.3. If C is the semicircle $x = \cos t$, $y = \sin t$, $0 \le t \le \pi$, show that

$$\int_C \left(\frac{y}{x^2 + y^2} \, dx - \frac{x}{x^2 + y^2} \, dy \right) = -\pi.$$

18.4. If C is the line segment from $(0, 0)$ to $(2, 1)$, show that

$$\int_C [x^2 y \, dx - (x + y) \, dy] = \tfrac{1}{2}.$$

18.5. Let C be the square with vertices $(1, 1), (-1, 1), (-1, -1)$, and $(1, -1)$. Traversing C in the positive sense, evaluate
(a) $\int_C (x \, dy + y \, dx)$.
(b) $\int_C (x + y)(dx + dy)$.

18.6. Carry out the integrations in Example 6, this section.

B

18.7. Review the definition of a line integral with all the preliminaries, and then argue that, whereas $(\Delta t)_k \to 0$ implies that $(\Delta x)_k \to 0$, the converse is not always true.

18.8. Evaluate the integral of Exercise 18.5(b) around the unit circle centered at the origin and traversed positively.

18.9. Evaluate both line integrals of $M(x, y) = y^3$ along the upper half of the circle $|z| = 1$ traversed in the clockwise sense.

18.10. Evaluate the line integral of $M(x, y) = (x + y)^{-1}$ along $y = x^2$ from $(1, 1)$ to $(3, 9)$, with respect to x.

18.11. Repeat the integration of the preceding exercise with respect to y.

18.12. Evaluate $\int x^2 y \, dy$ along $|z| = 1$ in the negative orientation.

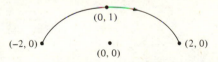

(0, 1)

(−2, 0)   (0, 0)   (2, 0)

18.13. Evaluate $\int (y + x) \, dx$ along the semiellipse shown in the adjoining figure.

18.14. Evaluate $\int (x^2 + 1) \, dy$ along the quarter-circle shown in the adjoining figure.

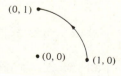

(0, 1)

• (0, 0)   (1, 0)

18.15. Evaluate $\int_{(0,0)}^{(1,1)} x^2 e^y \, dx$ along $y = x^3$.

C

18.16. Prove parts 1, 2, and 4 of Theorem 4.3 by use of the formula of Theorem 4.2 and known properties of ordinary (Riemann) integrals, noting that the right-hand side of the formula in question is an integral of that type.

18.17. Green's Theorem: *Let C be a positively oriented simple closed path C in the plane and suppose that $u(x, y)$ and $v(x, y)$ as well as their partial derivatives are continuous functions of x and y on C and in Int (C). Then the following formula connecting the line integral along C with the ordinary double integral over the area Int (C) is true:*

$$\int_C u(x, y)\, dx + v(x, y)\, dy = \int \int_{\text{Int}(C)} \left(\frac{\partial v}{\partial x} - \frac{\partial u}{\partial y} \right) dy\, dx.$$

Use this formula to evaluate the line integral $\int u\, dx + v\, dy$ along the respective path in each of the following problems. In each case, verify first that the hypotheses of the theorem are satisfied.

(a) The integral of Exercise 18.5(a).

(b) The integral of Exercise 18.5(b).

(c) The integral of Exercise 18.12.

18.18. Use Green's theorem to show that if

$$f(z) = u(x, y) + iv(x, y)$$

is analytic on a simple closed path C and at every point of Int (C), then

$$\int_C u\, dx - \int_C v\, dy + i \int_C u\, dy + i \int_C v\, dx = 0.$$

Section 19
The Complex Integral

The definition of the integral of a complex function is formally identical with the definition of the integral of a real function, where the interval of integration is replaced by a path. We recall that the integral of a real function $f(x)$ over the interval $a \le x \le b$ is defined by

$$\int_a^b f(x)\, dx = \lim_{\mu \to 0} \sum_{k=1}^n f(x_k^*)(\Delta x)_k,$$

where μ is the largest of all the $(\Delta x)_k$'s representing the lengths of the n subintervals into which the interval $a \le x \le b$ has been partitioned, and where x_k^* is a random point in the kth subinterval. We also recall that a sufficient condition for the existence of the defining limit is that f be continuous on the interval of integration. It turns out that the same condition will guarantee integrability of a complex function.

The definite complex integral is defined as follows.

Let C be a smooth curve with the usual parametrization [see Equation (1) on p. 152] and note that we can also express C in complex form as

$$C : z(t) = x(t) + iy(t), \qquad \alpha \le t \le \beta.$$

Suppose now that a function $w = f(z)$ is defined (and single-valued) at each point of C. As in the definition of a line integral, a partition of the interval $\alpha \le t \le \beta$ will induce a partition

$$z_0, z_1, z_2, \ldots, z_{k-1}, z_k, \ldots, z_n$$

on C, where z_0 and z_n coincide with the initial and terminal points of C, respectively; see Figure 4.10. We denote the difference $z_k - z_{k-1}$ by $(\Delta z)_k$, for $k = 1, 2, \ldots, n$, and in each arc $\overparen{z_{k-1}z_k}$ we choose at random a point ζ_k (either endpoint being an allowable choice). Then we form the sum

$$\sum_{k=1}^{n} f(\zeta_k)(\Delta z)_k.$$

Denoting by μ the maximum of the $(\Delta t)_k$'s (see Figure 4.6) and letting $\mu \to 0$, we see that, for all $k = 1, 2, \ldots, n$, $(\Delta t)_k \to 0$ and hence $(\Delta z)_k \to 0$. In this context, we define the **complex integral** of $f(z)$ along C by

$$\int_C f(z)\, dz = \lim_{\mu \to 0} \sum_{k=1}^{n} f(\zeta_k)(\Delta z)_k.$$

If the limit exists *independently of the partition of the interval $\alpha \le t \le \beta$ and independently of the choice of the ζ_k's*, then f is said to be **integrable** along C. As in the case of the line integral, C is called the **path of integration.**

As we remarked earlier, continuity of f along C guarantees the existence of the above integral. Specifically, we have the following theorem.

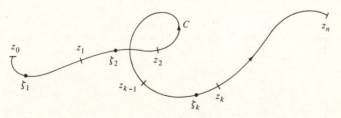

FIGURE 4.10 PATH OF INTEGRATION

Theorem 4.4 (*Existence of Complex Integral*)

Suppose that $f(z) = u(x, y) + iv(x, y)$ is continuous at every point of a smooth curve C.

Then the integral of f along C exists and

$$\int_C f(z)\, dz = \int_C u\, dx - \int_C v\, dy + i \int_C u\, dy + i \int_C v\, dx.$$

Proof:

See Appendix 4.

The use of the formula in Theorem 4.4 will be illustrated by means of examples following the listing of some basic properties of the complex integral in Theorem 4.5.

It is interesting to note that if C is actually an interval of the real axis, then f, at the points of C, will reduce to a function of x alone, and so the existence of the integral of a continuous function of x will be but a special case of the above theorem. In that sense, a complex integral can be thought of as a natural extension of a real integral. This intimacy between the two integrals, however, does not last through many more developments, and this fact will be most pronounced when we attempt to evaluate complex integrals.

The following three examples will be of use to us in a number of later developments. At this point, however, the first two will help us see how the definition of the complex integral can be used to evaluate some simple cases.

EXAMPLE 1

Show that if C is any smooth curve from z_0 to ζ, then $\int_C dz = \zeta - z_0$.

Here, $f(z) = 1$ for all z. In particular, $f(\zeta_k) = 1$ for every ζ_k on C. Then, using the definition, we find that

$$\int_C dz = \lim_{\mu \to 0} \sum_{k=1}^{n} f(\zeta_k)(\Delta z)_k$$

$$= \lim_{\mu \to 0} \sum_{k=1}^{n} (z_k - z_{k-1})$$

$$= \lim_{\mu \to 0} [(z_1 - z_0) + (z_2 - z_1) + \cdots + (\zeta - z_{n-1})]$$

$$= \lim_{\mu \to 0} (\zeta - z_0)$$

$$= \zeta - z_0,$$

as claimed.

Note that, in particular, if C is any closed smooth curve, then $\zeta = z_0$, and hence

$$\int_C dz = 0.$$

EXAMPLE 2

With C as in Example 1, show that $\int_C z\, dz = \frac{1}{2}(\zeta^2 - z_0^2)$.

Since the function $f(z) = z$ is everywhere continuous, the above integral exists by Theorem 4.4. Therefore, its value is independent of any partition of C and of the choice of the ζ_k's. So, choosing $\zeta_k = z_{k-1}$ for all

$$k = 1, 2, \ldots, n,$$

we have

$$\int_C z\, dz = \lim_{\mu \to 0} \sum_{k=1}^{n} f(\zeta_k)(\Delta z)_k$$

$$= \lim_{\mu \to 0} \sum_{k=1}^{n} z_{k-1}(z_k - z_{k-1})$$

$$= \lim_{\mu \to 0}[(z_0 z_1 - z_0^2) + \cdots + (z_{n-1}z_n - z_{n-1}^2)].$$

On the other hand, choosing $\zeta_k = z_k$, we have

$$\int_C z\, dz = \lim_{\mu \to 0} \sum_{k=1}^{n} f(\zeta_k)(\Delta z)_k$$

$$= \lim_{\mu \to 0} \sum_{k=1}^{n} z_k(z_k - z_{k-1})$$

$$= \lim_{\mu \to 0} [(z_1^2 - z_1 z_0) + \cdots + (z_n^2 - z_n z_{n-1})].$$

Adding and simplifying, one then obtains

$$2 \int_C z\, dz = z_n^2 - z_0^2,$$

and since $z_n = \zeta$, the assertion made at the beginning of this example follows. Again, as in the preceding example, if C is closed, then

$$\int_C z\, dz = 0.$$

EXAMPLE 3

We show that if C is the circle $z = z_0 + re^{it}, 0 \le t \le 2\pi, r > 0$, then

$$\int_C \frac{dz}{z - z_0} = 2\pi i.$$

From the equation of C we find that $dz = ire^{it}\, dt$; hence,

$$\int_C \frac{dz}{z - z_0} = \int_0^{2\pi} \frac{ire^{it}\, dt}{re^{it}} = i \int_0^{2\pi} dt = i(2\pi - 0) = 2\pi i,$$

where in the next-to-last step we used part 1 of Theorem 4.5.

Complex integrals possess five basic properties, listed in Theorem 4.5, the first four of which correspond to the basic properties of ordinary integrals and are actually identical with the four properties of line integrals of Theorem 4.3. We establish them in the following.

Theorem 4.5

Suppose that ζ is an arbitrary complex constant and that $C + K$ is a path consisting of the two smooth curves C and K. Suppose further that $f(z)$ and $g(z)$ are integrable along C and along K.

Then,

1. $\int_C \zeta f(z)\, dz = \zeta \int_C f(z)\, dz.$
2. $\int_C (f(z) + g(z))\, dz = \int_C f(z)\, dz + \int_C g(z)\, dz.$
3. $\int_{C+K} f(z)\, dz = \int_C f(z)\, dz + \int_K f(z)\, dz.$
4. $\int_{-C} f(z)\, dz = -\int_C f(z)\, dz.$
5. *If, in addition, for some positive M the function f satisfies the relation $|f(z)| \le M$ for every z on C and if the length of C is L, then*

$$\left| \int_C f(z)\, dz \right| \le ML.$$

Proof:

See Appendix 4 for part 5 and Exercise 19.13 for parts 1–4.

REMARK

Part 3 of Theorem 4.5 shows that whatever was developed so far for integrals along a smooth curve holds, also, for any *path*. Review Remark 5, p. 163.

As we noted in the opening remarks of Section 18, the evaluation of a complex integral often reduces to the problem of evaluating a sum of line integrals, and this is certainly obvious from the formula of Theorem 4.4. The formula in question is fundamental in the evaluation of complex integrals; some typical cases are illustrated in the examples that follow.

E X A M P L E 4

Integrate the function $f(z) = x$ along $C_1 + C_2 + C_3$ as in Figure 4.11.

The path of integration can be represented as follows:

$$C_1 : y = 0, 0 \le x \le 1; \text{ hence } dy = 0.$$

$$C_2 : x = 1, 0 \le y \le 1; \text{ hence } dx = 0.$$

$$C_3 : y = x, \text{ from } (1, 1) \text{ to } (0, 0); \text{ hence } dy = dx.$$

Therefore, substituting appropriately, we find

$$\int_{C_1 + C_2 + C_3} f(z)\, dz = \int_{C_1} f(z)\, dz + \int_{C_2} f(z)\, dz + \int_{C_3} f(z)\, dz$$

$$= \int_{C_1} x(dx + i\, dy) + \int_{C_2} x(dx + i\, dy)$$

$$+ \int_{C_3} x(dx + i\, dy)$$

$$= \int_0^1 x\, dx + \int_0^1 i\, dy + \int_1^0 (1 + i)x\, dx$$

$$= \frac{i}{2}.$$

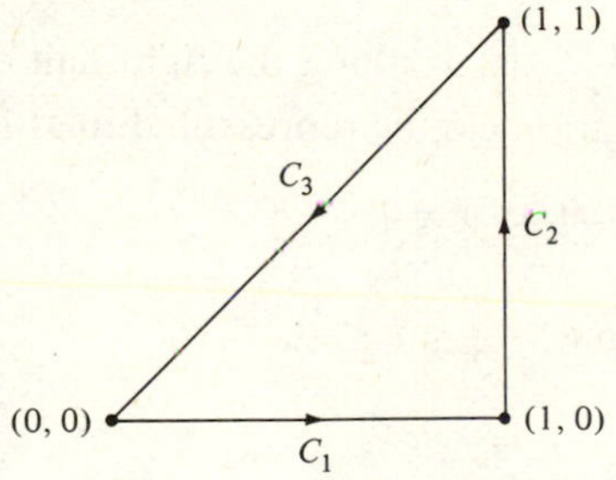

FIGURE 4.11 EXAMPLE 4

E X A M P L E 5

Evaluate $\int_C y\, dz$ along $C : x = t - 1, y = e^{t-1}, 2 \le t \le 3$.

From the equation of C we first find that $dx = dt$ and that $dy = e^{t-1}\, dt$. Then, appropriate substitutions yield the following:

$$\int_C y\, dz = \int_C y(dx + i\, dy)$$

$$= \int_2^3 e^{t-1}(dt + ie^{t-1}\, dt)$$

$$= \int_2^3 (e^{t-1} + ie^{2t-2})\, dt$$

$$.= e^2 - e + \frac{i}{2}(e^4 - e^2).$$

EXAMPLE 6

Evaluate the integral of $f(z) = 2z + 3i$ along the path of Example 4; see Figure 4.11.

The reader will note that in this problem there is no need for calculations, since, by using the results of Examples 1 and 2 and the properties of integrals listed in Theorem 4.5, we have

$$\int_C (2z + 3i)\, dz = \int_C 2z\, dz + \int_C 3i\, dz$$

$$= 2 \int_C z\, dz + 3i \int_C dz$$

$$= 0 + 0$$

$$= 0,$$

since C is a simple closed path. See Remark after Theorem 4.5.

EXAMPLE 7

Evaluate the integral $\int_{-i}^{i} |z|\, dz$ along the right half of the unit circle.

The path of integration can be represented in at least three ways:

(a) $x = (1 - y^2)^{1/2},\ -1 \le y \le 1$.

(b) $x = \cos t,\ y = \sin t,\ -\dfrac{\pi}{2} \le t \le \dfrac{\pi}{2}$.

(c) $z = e^{it},\ -\dfrac{\pi}{2} \le t \le \dfrac{\pi}{2}$.

Of course, (b) and (c) are essentially the same representation; see Exercise 17.6. For this particular problem the most convenient choice would be (b), by use of which we find

$$\int_{-i}^{i} |z|\, dz = \int_{-i}^{i} 1(dx + i\, dy)$$

$$= \int_{-\pi/2}^{\pi/2} (-\sin t + i \cos t)\, dt$$

$$= 2i.$$

EXAMPLE 8

By virtue of Example 3 of this section, the value of the integral

$$\int_C \frac{dz}{z - i},$$

where C is the circle $|z - i| = 1$ positively oriented, is $2\pi i$.

EXAMPLE 9

The formula in part 5 of Theorem 4.5 can be used to find an upper bound to the value of an integral whose actual evaluation may not be possible. We illustrate this use by finding an upper bound to the value of the integral

$$\int_C (z^4 + 1)\, dz,$$

where C is the line segment from 0 to $1 + i$.

First, we must find a number M such that for every z on C,

$$|z^4 + 1| \leq M.$$

Clearly, for any z on C, $|z| \leq \sqrt{2}$. Therefore,

$$|z^4 + 1| \leq |z^4| + |1| = |z|^4 + 1 \leq 5.$$

Thus $M = 5$ has the desired property.

Next, we find that the length of C is

$$L = \sqrt{2}.$$

Therefore, by use of the formula referred to above, we find that

$$\left| \int_C (z^4 + 1)\, dz \right| \leq 5\sqrt{2},$$

which tells us that, whatever the actual value of the given integral, the modulus of that value will not exceed $5\sqrt{2}$.

EXERCISE 19

A

19.1. Evaluate $\int (x^2 + y^2)\, dz$ along the path $C : x = t^2,\ y = 1/t,\ 1 \leq t \leq 3$.

19.2. Evaluate $\int \bar{z}\, dz$ along $C = C_1 + C_2 + C_3$ as in the adjoining figure.

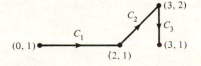

19.3. Evaluate $\int z \, dz$ along each of the two paths in the adjoining figure by direct calculation. Then check your answers by use of Example 2 of this section.

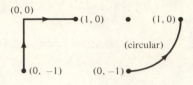

19.4. Evaluate the integral of $f(z) = e^z$ along $y = 2x$ from $(-1, -2)$ to $(1, 2)$.

19.5. Evaluate $\displaystyle\int_C \frac{dz}{z}$ if $C : x^2 + y^2 = 16$, positively oriented.

19.6. Integrate the function $f(z) = (\bar{z})^2$ along the path $y = x^2$ from $(0, 0)$ to $(1, 1)$.

19.7. Use the result stated in Exercise 19.10 to carry out the integration of Example 7, this section, using the exponential representation of C.

19.8. Evaluate $\int x \, dz$ along $C : |z| = 1$, positively oriented.

19.9. Evaluate $\int y \, dz$ from $(-2, 0)$ to $(2, 0)$ for each of the following paths of integration:

(a) The line segments from $(-2, 0)$ to $(-2, -1)$ to $(2, -1)$ to $(2, 0)$.
(b) The lower half of a circle.
(c) The upper half of a circle.

B

19.10. In a later section we will show that, for any nonzero constant k and for any path C,

$$\int_C e^{kz} \, dz = \frac{1}{k} e^{kz},$$

evaluated appropriately. Use this fact, whenever necessary, to establish the following formulas; in each case, C is the circle

$$z = z_0 + re^{it}, \qquad 0 \le t \le 2\pi, \quad r > 0.$$

(a) $\int_C \alpha \, dz = 0$, for any constant α.
(b) $\int_C (z - z_0) \, dz = 0$.
(c) $\int_C (z - z_0)^2 \, dz = 0$.
(d) $\int_C (z - z_0)^n \, dz = 0$, for any integer $n \ne -1$.
(e) From the preceding and Example 3, this section, conclude that if C is the circle given above, then

$$\int_C (z - z_0)^n \, dz = \begin{cases} 2\pi i, & \text{if } n = -1, \\ 0, & \text{for all other } n. \end{cases}$$

19.11. Use Example 9 of this section as a guide to find an upper bound for each of the following without actually evaluating the integrals.

(a) $|\int z \, dz|$ along the line segment from 0 to i.

(b) $|\int e^z \, dz|$ along the circle $|z| = 2$, positively oriented.

(c) $|\int (2z + 1) \, dz|$ along the line segment from i to $2 + i$.

(d) $\left| \int \dfrac{dz}{z^2} \right|$ along the line segment from i to $2 + i$.

(e) $\left| \int_{-2}^{2} \dfrac{dz}{z^3} \right|$ along the upper half of the circle in part (b).

19.12. Evaluate the integral of the function $\dfrac{2z - 3}{z}$ from -2 to 2 along the three paths of Exercise 19.9.

C

19.13. Use the results of Theorems 4.3 and 4.4 to prove parts 1, 2, 3, and 4 of Theorem 4.5.

19.14. Consider the formula

$$\int_C e^{kz} \, dz = \frac{1}{k} e^{kz}$$

given in Exercise 19.10. Let $k = a + ib$ and suppose that C is an interval of the real axis. In this context, use the above formula to derive the following two integration formulas:

(a) $\displaystyle \int e^{ax} \cos bx \, dx = \frac{e^{ax}(a \cos bx + b \sin bx)}{a^2 + b^2}.$

(b) $\displaystyle \int e^{ax} \sin bx \, dx = \frac{e^{ax}(a \sin bx - b \cos bx)}{a^2 + b^2}.$

19.15. Prove the following variation of Theorem 4.5 (5): If $f(z)$ is integrable along a path C, then

$$\left| \int_C f(z) \, dz \right| \leq \int_C |f(z)| \, |dz|.$$

(Imitate part of the proof of Theorem 4.5, given in Appendix 4.)

REVIEW EXERCISES — CHAPTER 4

1. Evaluate $\int (\bar{z})^2 \, dz$ along the upper half of $|z| = 2$ from -2 to 2.
2. Evaluate the integral of $f(z) = i \sin z$ from $-i$ to i along a straight line.

3. Evaluate the line integral $\int (x^2 + xy)\, dy$ from $(0, 0)$ to $(1, 2)$ along a straight line.

4. Find an upper bound for $\left| \int \dfrac{dz}{z^2 + 1} \right|$ along the path of Exercise 1.

5. Why is the evaluation of the integral $\int_{(0,0)}^{(1,1)} y\, dx$ along $y = e^x$ impossible?

6. Represent the following curves in parametric form:
   (a) The circle of radius 2 and center at $1 + i$.
   (b) The parabola $y = x^2$ from $(-2, 4)$ to $(0, 0)$.
   (c) The parabola $y = x^2$ from $(0, 0)$ to $(-2, 4)$.
   (d) The hyperbola $y^2 - x^2 = 4$ from $(0, 2)$ to $(2, 2\sqrt{2})$.
   (e) The line segment from $(-1, 2)$ to $(2, -1)$.

7. Evaluate $\int_0^{1+i} (x - y)\, dz$ along $y = 0$ to the point $z = 1$ and then along $x = 1$.

8. Repeat the integration in Exercise 7 along $x = 0$ to the point $z = i$ and then along $y = 1$.

9. Evaluate $\int_{-1}^{2} |\bar{z}|^2\, dz$ along the upper half of the unit circle from -1 to 1 and then along the real axis to the point $z = 2$.

10. In the preceding exercise replace the upper half of the unit circle by its lower half, leaving everything else the same.

11. Repeat the integration of Exercise 9 using the straight line segment between -1 and 2 as your path of integration.

APPENDIX 4

Proofs of Theorems

Theorem 4.2   (*Existence of Line Integral*)
Suppose that
1. C is a smooth curve expressed parametrically by $x = \phi(t)$, $y = \psi(t)$, $\alpha \le t \le \beta$.
2. The function $M(x, y)$ is continuous on C.
Then the line integral of M along C with respect to x exists and its value is given by the formula

$$\int_C M(x, y)\, dx = \int_\alpha^\beta M(\phi(t), \psi(t)) \cdot \phi'(t)\, dt.$$

Proof:
In the context of the preliminary discussion setting the stage for the definition of a line integral, consider the n subintervals into which the entire

interval $J : \alpha \leq t \leq \beta$ was subdivided. Since $\phi(t)$ is continuous and differentiable on J (why?), then it is continuous and differentiable on each of the above subintervals. Hence, by the Mean Value Theorem for derivatives,* a point η_k exists in each $t_{k-1} < t < t_k$ with the property that

$$\frac{\phi(t_k) - \phi(t_{k-1})}{t_k - t_{k-1}} = \phi'(\eta_k),$$

which, by use of different notation, we can write in the form

$$\frac{(\Delta x)_k}{(\Delta t)_k} = \phi'(\eta_k).$$

Therefore,

$$(\Delta x)_k = \phi'(\eta_k)(\Delta t)_k.$$

But then, along the points of C, we have

$$\sum_{k=1}^{n} M(\sigma_k, \omega_k)(\Delta x)_k = \sum_{k=1}^{n} M[\phi(\tau_k), \psi(\tau_k)]\phi'(\eta_k)(\Delta t)_k.$$

Next, taking limits of both sides in the preceding equation, as $\mu \to 0$, we note that the limit of the left-hand side is the line integral

$$\int_C M(x, y)\, dx,$$

whereas the limit of the right-hand side is the Riemann integral

$$\int_\alpha^\beta M[\phi(t), \psi(t)] \cdot \phi'(t)\, dt,$$

whose existence is guaranteed by the hypothesis that the functions M, ϕ, ψ, and ϕ' are continuous on the interval J.

The assertion of the theorem then follows.

Theorem 4.3†

Suppose that $C + K$ is a path consisting of the smooth curves

$$C : x = \phi(t), \qquad y = \psi(t), \qquad \alpha \leq t \leq \beta,$$

and

$$K : x = \kappa(t), \qquad y = \lambda(t), \qquad \gamma \leq t \leq \delta.$$

---

*The Mean Value Theorem for Derivatives: If $f(x)$ is continuous in the closed interval $a \leq x \leq b$ and differentiable in the open interval $a < x < b$, then there is at least one point χ in $a < x < b$ such that

$$\frac{f(b) - f(a)}{b - a} = f'(\chi).$$

† This is actually only part 3 of Theorem 4.3.

Suppose further that the line integrals of a function M(x, y) exist along each of the above smooth curves.

Then,

$$\int_{C+K} M(x, y)\, dx = \int_{C} M(x, y)\, dx + \int_{K} M(x, y)\, dx.$$

Proof:

By use of Exercise 17.8, we know that the smooth curve K can be represented parametrically by two functions

$$x = \mu(t), \qquad y = v(t), \qquad \text{over the interval } \beta \leq t \leq \delta.$$

These two functions are found from the given $\kappa(t)$ and $\lambda(t)$ by use of the transformation given in the above exercise; specifically,

$$\mu(t) \text{ on } \beta \leq t \leq \delta \text{ agrees with } \kappa(t) \text{ on } \gamma \leq t \leq \delta$$

and

$$v(t) \text{ on } \beta \leq t \leq \delta \text{ agrees with } \lambda(t) \text{ on } \gamma \leq t \leq \delta.$$

As a result, the path $C + K$ can now be represented by a single set of equations over a single interval:

$$x = f(t), \qquad y = g(t), \qquad \alpha \leq t \leq \delta,$$

where

$$f(t) = \phi(t) \qquad \text{and} \qquad g(t) = \psi(t) \qquad \text{when } \alpha \leq t \leq \beta$$

and

$$f(t) = \mu(t) \qquad \text{and} \qquad g(t) = v(t) \qquad \text{when } \beta \leq t \leq \delta.$$

With the above preliminaries at our disposal we now complete the proof as follows.

$$\int_{C+K} M(x, y)\, dx = \int_{\alpha}^{\delta} M[f(t), g(t)] f'(t)\, dt$$

$$= \int_{\alpha}^{\beta} M[f(t), g(t)] f'(t)\, dt + \int_{\beta}^{\delta} M[f(t), g(t)] f'(t)\, dt$$

$$= \int_{\alpha}^{\beta} M[\phi(t), \psi(t)] \phi'(t)\, dt + \int_{\beta}^{\delta} M[\mu(t), v(t)] \mu'(t)\, dt$$

$$= \int_{\alpha}^{\beta} M[\phi(t), \psi(t)] \phi'(t)\, dt + \int_{\gamma}^{\delta} M[\kappa(t), \lambda(t)] \kappa'(t)\, dt$$

$$= \int_{C} M(x, y)\, dx + \int_{K} M(x, y)\, dx$$

and the proof is complete since, by hypothesis, the last integrals exist.

Note that all the substitutions in the preceding chain of equalities will be found in the preliminary work done earlier in the proof.

Theorem 4.4 *(Existence of Complex Integral)*
Suppose that $f(z) = u(x, y) + iv(x, y)$ is continuous at every point of a smooth curve C.
Then the integral of f along C exists and

$$\int_C f(z)\, dz = \int_C u\, dx - \int_C v\, dy + i \int_C u\, dy + i \int_C v\, dx.$$

Proof:
Since $f(z) = u(x, y) + iv(x, y)$, $\zeta_k = (\sigma_k, \omega_k)$, and $(\Delta z)_k = (\Delta x)_k + i(\Delta y)_k$ (see the definition of complex integral at the beginning of Section 19), we have the following decomposition, where all summations are taken from $k = 1$ to $k = n$:

$$\sum f(\zeta_k)(\Delta z)_k = \sum [u(\sigma_k, \omega_k) + iv(\sigma_k, \omega_k)][(\Delta x)_k + i(\Delta y)_k]$$

$$= \sum u(\sigma_k, \omega_k)(\Delta x)_k - \sum v(\sigma_k, \omega_k)(\Delta y)_k$$

$$+ i \sum u(\sigma_k, \omega_k)(\Delta y)_k + i \sum v(\sigma_k, \omega_k)(\Delta x)_k.$$

Then, substituting in the definition of complex integral and using Theorem 2.3(1), as well as the definition of line integral, we find

$$\int_C f(z)\, dz = \lim_{\mu \to 0} \sum_{k=1}^{n} f(\zeta_k)(\Delta z)_k$$

$$= \lim_{\mu \to 0} \sum_{k=1}^{n} u(\sigma_k, \omega_k)(\Delta x)_k - \lim_{\mu \to 0} \sum_{k=1}^{n} v(\sigma_k, \omega_k)(\Delta y)_k$$

$$+ i \lim_{\mu \to 0} \sum_{k=1}^{n} u(\sigma_k, \omega_k)(\Delta y)_k + i \lim_{\mu \to 0} \sum_{k=1}^{n} v(\sigma_k, \omega_k)(\Delta x)_k$$

$$= \int_C u\, dx - \int_C v\, dy + i \int_C u\, dy + i \int_C v\, dx.$$

But the last four integrals exist according to Theorem 4.2, since the continuity of f on C implies continuity of u and v on C. The theorem then follows.

Theorem 4.5.*
Suppose that $f(z)$ is integrable along a smooth curve C and that, for some positive number M, f satisfies the relation $|f(z)| \le M$ for every z on C. Suppose

---

* This is only part 5 of Theorem 4.5.

further that the length of C is L.
  Then

$$\left| \int_C f(z)\, dz \right| \le ML.$$

Proof:
First, we establish the following fact (see Remark 1, p. 152):

$$L = \int_\alpha^\beta \left[\left(\frac{dx}{dt}\right)^2 + \left(\frac{dy}{dt}\right)^2 \right]^{1/2} dt = \int_C [(dx)^2 + (dy)^2]^{1/2} = \int_C |dz|.$$

Then, by use of the triangle inequality, the properties of absolute values and the hypothesis $|f(z)| \le M$ for every point of C, we have the following:

$$\left| \sum_{k=1}^n f(\zeta_k)(\Delta z)_k \right| \le \sum_{k=1}^n |f(\zeta_k)(\Delta z)_k| = \sum_{k=1}^n |f(\zeta_k)| \, |(\Delta z)_k| \le M \sum_{k=1}^n |(\Delta z)_k|.$$

Finally, taking limits of the first and last expressions, as $\mu \to 0$, we get

$$\left| \int_C f(z)\, dz \right| \le M \int_C |dz|,$$

which, in view of the first development in this proof, yields

$$\left| \int_C f(z)\, dz \right| \le ML,$$

and the proof is complete.

CHAPTER 5
Cauchy Theory of Integration

Section 20   The Cauchy integral theorem. Independence of path. Fundamental theorem of complex integration.

Section 21   The annulus theorem and its extension.

Section 22   The Cauchy integral formulas. Cauchy inequalities. Morera's theorem.

Section 20
Integrals of Analytic Functions. Cauchy's Theorem

In Chapter 4 we concerned ourselves with the integral of an arbitrary complex function. The three sections of the present chapter are devoted to the study of integration of analytic functions. The family of such functions is, of course, a more restricted category of functions, but it is precisely their restricted nature which allows the development of a powerful theory of exceptional mathematical beauty and, at the same time, of inestimable value in the hands of the applied scientist. It is in the theory of integration of analytic functions that the extremely strong properties and inner structure possessed by such functions are manifested in the form of the results that follow and, eventually, in the applications of these results.

The basis for the development of nearly the remainder of this book consists of what may be called *Cauchy's theory of analytic functions,* and the cornerstone of the theory is the celebrated *Cauchy integral theorem.* Concerning this theorem, it is interesting to note that, actually, Cauchy first proved a weaker form of the theorem that now bears his name, and it was Goursat who later proved that one of the hypotheses in the original form of the theorem was not only unnecessary but redundant; see the Note following Theorem 5.1.

Theorem 5.1   (*Cauchy Integral Theorem*)
Suppose that
1. $f(z)$ is analytic on a simply connected region R.
2. C is a closed path lying entirely in R.
Then

$$\int_C f(z)\, dz = 0.$$

183

Proof:
See Appendix 5(B).

NOTE: Cauchy's original theorem included the additional hypothesis that the derivative f' is continuous on C and in Int (C). However, it will be shown as a consequence of Theorem 5.1 that if f is analytic in a region R, then f' is also analytic in R and hence continuous there. Therefore, as we remarked earlier, the continuity of f' would be a redundant hypothesis, since it is implied by the analyticity of f.

REMARK 1

The converse of Cauchy's theorem does not hold; i.e., the following statement is, in general, false:

> If $\int_C f(z)\, dz = 0$ for every closed path C in a simply connected region R, then f is analytic in R.

An example that illustrates this fact is provided by the function $f(z) = 1/z^2$, whose integral can be shown to be zero along any closed path for which the integral is defined. And yet, f clearly fails to be analytic at $z = 0$. A partial converse of Cauchy's theorem, known as *Morera's theorem*, is proved in Section 22.

The proof of the Cauchy integral theorem is rather involved and it presupposes familiarity with a number of analytical concepts that are developed in Appendix 5(B) prior to the proof proper. A brief description of the proof of this theorem is given here for the benefit of those who may be interested in an outline of the long proof.

BRIEF OUTLINE OF THE PROOF OF CAUCHY'S THEOREM

The proof is divided into three major parts:

1. The theorem is first proved for the case in which C, the path of integration, is any triangle. This part of the proof is a straightforward analytical proof involving some geometrical considerations of very elementary nature. The remainder of the proof rests heavily on this part (see Figure 5.1).

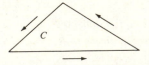

FIGURE 5.1 PROOF OF CAUCHY'S THEOREM, CASE 1

2. The second stage of the proof is devoted to establishing a result of technical nature: Any closed polygon (with a finite number of sides) can be decomposed into finitely many triangles. Then, since (by part 1 of the proof) the integral is zero along each triangle, one easily argues that the integral is zero when C is any polygon. Essential in this part of the proof is the fact that every side of the decomposition triangles, which does not fall on a side of the polygon is traversed twice and in opposite directions (see Figure 5.2). Therefore, the value of the integral along such parts of the path is zero.

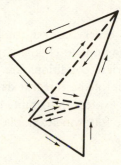

FIGURE 5.2 PROOF OF CAUCHY'S THEOREM, CASE 2

3. In the final stage it is proved that any closed path C can be approximated to any desired degree by a closed polygon (see Figure 5.3). As a consequence of this, one proves that the integral along C differs from the integral along the approximating polygon by as small a quantity as desired, and hence its value is zero, according to part 2 of the proof.

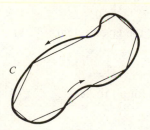

FIGURE 5.3 PROOF OF CAUCHY'S THEOREM, CASE 3

The Cauchy integral theorem, important as it is in its own right, is almost overshadowed by the immensity of its consequences, both in depth and extent. We begin their study with an equivalent form of the Cauchy theorem. See Exercise 20.17.

Theorem 5.2   (*Independence of Path*)
Suppose that
1. R is a simply connected region.
2. z_1 and z_2 are points in R.
3. $f(z)$ is analytic throughout R.
Then the value of the integral $\int_{z_1}^{z_2} f(z)\, dz$ is the same along any C joining
z_1 and z_2, in that order, provided that C is a path which lies entirely within R.

Proof:
See Appendix 5(A).

The preceding theorem is a very convenient tool in the evaluation of
certain integrals, since it allows one to choose convenient paths of integration, as long as the appropriate conditions are met. More precisely, the
theorem states that the value of the integral depends only on the initial
and terminal points of C; we express this simply by saying that the integral
is **independent of the path.** The following example illustrates this concept.

EXAMPLE 1

Evaluate the integral $\int_C (3z^2 - 2z)\, dz$, where $C = C_1 + C_2$ as in Figure 5.4.
   Without Theorem 5.2 at our disposal, we would have to find equations
for the quarter-circle C_1 and the line segment C_2 and then proceed using
methods developed in Sections 18 and 19. However, since the integrand
$f(z) = 3z^2 - 2z$ is analytic in any simply connected region containing the
points $z_1 = -1$ and $z_2 = 1$, Theorem 5.2 allows one to choose any path
from z_1 to z_2. Clearly, the most convenient path in this case is the line
segment

$$K : y = 0, \qquad -1 \le x \le 1.$$

Then, substitution in the given integral yields the real integral

$$\int_{-1}^{1} (3x^2 - 2x)\, dx,$$

whose value is easily found to be 2.

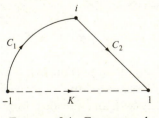

FIGURE 5.4   EXAMPLE 1

The reader may recall from calculus the intimate connection between the definite integral and an antiderivative of a function. This close relationship is described in precise terms by the Fundamental Theorem of Integral Calculus, which states that if $G(x)$ is an antiderivative of a real function $g(x)$ continuous on the interval $a \le x \le b$, then

$$\int_a^b g(x)\, dx = G(b) - G(a).$$

This relationship has its analog in the case of analytic functions, and we begin its study by means of a sequence of observations which find their culmination in Theorem 5.3 and, subsequently, in Theorem 5.4, which is the analog of the Fundamental Theorem of Integration for complex functions. We argue deductively as follows:

1. In the context of Theorem 5.2, we saw that the integral

$$\int_C f(z)\, dz$$

has a value that is independent of the path C joining the points z_1 and z_2. Therefore, the only points that "count" in evaluating the integral are the points z_1 and z_2.

2. It follows that one then would be justified in writing

$$\int_{z_1}^{z_2} f(z)\, dz$$

in place of the above integral, with the understanding that the path from z_1 to z_2 will remain within a simply connected region R throughout which f is analytic.

3. Now, suppose that the upper limit of the integral in part 2 is allowed to vary, *always within R*. Then the equation

$$F(\zeta) = \int_{z_1}^{\zeta} f(z)\, dz$$

defines a single-valued function F of the variable ζ, since the integral on the right depends only on the limits z_1 and ζ, of which z_1 is constant.

4. Finally, we can claim that the function $F(\zeta)$ is an antiderivative of f:

$$F'(\zeta) = f(\zeta), \qquad \text{for all } \zeta \text{ in } R.$$

This, of course, also means that F is analytic throughout R. We formalize our claim in the following theorem.

Theorem 5.3

Suppose that

1. R is a simply connected region.

2. z_1 is a point in R.

3. $f(z)$ is analytic at every point of R.

Then, for any ζ in R,

$$\frac{d}{d\zeta} \int_{z_1}^{\zeta} f(z)\, dz = f(\zeta).$$

Proof:

See Appendix 5(A).

With the above theorem at our disposal, the extension of the fundamental theorem of integration to the complex case is now a simple matter.

Theorem 5.4   *(Fundamental Theorem of Integration)*

Suppose that

1. R is a simply connected region.

2. z_1 and z_2 are points in R.

3. $f(z)$ is analytic on R.

4. $\Phi(z)$ is an antiderivative of $f(z)$ in R.

Then

$$\int_{z_1}^{z_2} f(z)\, dz = \Phi(z_2) - \Phi(z_1).$$

Proof:

See Appendix 5(A).

The reader will certainly find Theorem 5.4 familiar, at least in form. The examples that follow illustrate the use of the theorem as well as a typical case in which the theorem cannot be used because at least one of its hypotheses is not satisfied. But before we work out some examples, the following remark is in order.

REMARK 2

It is clear from Theorem 5.4 that the integral of an entire function along *any* path connecting *any* two points in the plane can be readily evaluated, provided, of course, that an antiderivative of the function can be found. The same is true for the integral of an analytic function, provided that the initial and terminal points of the path of integration as well as the path itself are contained in some simply connected region on which the function is analytic.

EXAMPLE 2

In view of the preceding remark, the following evaluations of integrals require no explanatory remarks.

1. $\int_{-i}^{1+i} 2z\, dz = z^2\big|_{-i}^{1+i} = (1+i)^2 - (-i)^2 = 1 + 2i.$
2. $\int_0^{i\pi} e^{z+1}\, dz = e^{z+1}\big|_0^{i\pi} = e^{i\pi+1} - e^1 = -2e.$
3. $\int_\pi^i \sin z\, dz = -\cos z\big|_\pi^i = -\cos i + \cos \pi = -1 - \cos i.$

EXAMPLE 3

Evaluate the integral $\displaystyle\int_{-i}^1 \frac{dz}{z}$ along the quarter-circle described by $z = e^{it}$, $-\pi/2 \le t \le 0$.

Since the only point of nonanalyticity of the integrand is at $z = 0$, one can easily find a simply connected region R containing the path of integration and such that the integrand $f(z) = 1/z$ is analytic throughout R; make a simple drawing of the configuration. Hence, by Theorem 5.4, we have

$$\int_{-i}^1 \frac{1}{z}\, dz = \text{Log } z\,\bigg|_{-i}^1 = \text{Log }(1) - \text{Log }(-i) = \frac{\pi i}{2}.$$

EXAMPLE 4

Evaluate the integral $\displaystyle\int \frac{dz}{z-i}$ along the path $C = C_1 + C_2 + C_3$ of Figure 5.5.

In this case the fundamental theorem of integration does not apply directly because any simply connected region R containing C must also contain the point $z = i$, which is a singularity of the integrand. However, by use of Theorem 4.5(3), we can write

$$\int_C \frac{dz}{z-i} = \int_{C_1+C_3} \frac{dz}{z-i} + \int_{C_2} \frac{dz}{z-i}.$$

The first integral on the right may now be evaluated by use of Theorem 5.4 to give

$$\text{Log }(-2-i) - \text{Log }(2-i).$$

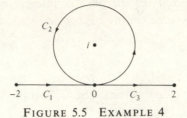

FIGURE 5.5   EXAMPLE 4

On the other hand, according to Example 8 of Section 19, the value of the integral along C_2 is $2\pi i$. Hence

$$\int_C \frac{dz}{z - i} = \text{Log}\,\frac{-2 - i}{2 - i} + 2\pi i.$$

EXERCISE 20

A

In Exercises 20.1–20.8 evaluate the integral of the given function along the respective path. In the case of a simple closed path assume the positive orientation.

20.1. $f(z) = z^3 - 1, |z - 1| = 1$.

20.2. $f(z) = z^3 - iz + 3i, |z + i| = 2$.

20.3. $f(z) = z/(z^2 - 1), |z - \pi| = 1$.

20.4. $f(z) = \dfrac{3}{z} - \dfrac{2}{z - 2i}, |z - 2i| = 1$.

20.5. $f(z) = 6z^5 - 1$, along the straight line segments from $z = i$ to $z = 1 + i$ and then to $z = 1$.

20.6. $f(z) = z^2/(z - 2)$ along the triangle with vertices at $-1, 0$, and $2i$.

20.7. $f(z) = e^z - 1/z^2$ along the lower half of the unit circle with center at the origin traversed in the clockwise direction.

20.8. $f(z) = \cos z/z^3, |z + 2i| = 1$.

In Exercises 20.9–20.14, evaluate the given definite integral

20.9. $\displaystyle\int_{-i}^{2i} (z^2 + z + 1)\, dz.$ **20.10.** $\displaystyle\int_{0}^{\pi} (e^z - \sin z)\, dz.$

20.11. $\displaystyle\int_{\pi i}^{2\pi i} \cosh z\, dz.$ **20.12.** $\displaystyle\int_{0}^{\pi i} z \cos z^2\, dz.$

20.13. $\displaystyle\int_0^i z e^{z^2}\, dz.$ **20.14.** $\displaystyle\int_0^\pi \cos^2 z\, dz.$

B

20.15. Use decomposition into partial fractions to evaluate the integral

$$\int_C \frac{2}{z^2 - 1}\, dz,$$

where C is the circle $|z - 1| = \frac{1}{2}$ positively oriented.

20.16. Suppose that $H(z)$ and $G(z)$ are antiderivatives of a function $h(z)$ at every point of a region R throughout which $h(z)$ is analytic. Prove that, at every point of R, $H(z) = G(z) +$ constant. The following two steps, if completed correctly, amount to a proof.
(a) Define the function $f(z) = H(z) - G(z)$ and show that $f'(z) = 0$ for every z in R.
(b) Complete the proof by use of Exercise 8.22. Note that you must prove analyticity of f in R before you can use that exercise.

C

20.17. In Appendix 5(A) we prove that Theorem 5.2 (independence of path) is a consequence of the Cauchy integral theorem. Study that proof and then "reverse the process" to prove that the Cauchy integral theorem is a consequence of Theorem 5.2, thus proving that the two theorems are equivalent.

20.18. Prove that Exercise 18.18 amounts to a proof of the Cauchy integral theorem as a consequence of Green's theorem (see Exercise 18.17).

Section 21
The Annulus Theorem and Its Extension

In this section, as well as in the next one, we continue with the exploration and exploitation of some of the more immediate consequences of Cauchy's theorem. As we proceed, the reader should bear in mind that, beside their beauty from a mathematical standpoint, nearly all the developments in these two sections provide us with extremely convenient tools which make the evaluation of a very large class of integrals extremely simple. We have already seen that this was the case with the results of the preceding section. But before we continue, *a word of advice:* Do not ignore the statements of the theorems because the details contained in them tell you under what conditions you may use each tool; remember that using the wrong tool to solve a given problem will almost always result in frustrating and often ridiculous situations.

We recall the definition of an *annulus* or *annular region*, which is the region "between" two simple closed paths, one of which is in the interior of the other. More precisely, if C and K are two simple closed paths such that K is in Int (C), then the set of points common to Int (C) and Ext (K) is the **annulus determined** by C and K. Occasionally, we shall be interested in the set consisting of the annulus determined by C and K along with the paths C and K themselves; we shall refer to such a set as the **closed annulus** determined by C and K.

The following theorem will prove to be very useful in further theoretical developments, as well as in the evaluation of certain types of integrals whose path of integration is a simple closed path. We will see that, under appropriate conditions, the theorem allows one to choose alternate, more convenient paths of integration; in a sense, it is a theorem of independence of path for the case of a simple closed path.

Theorem 5.5   (*Annulus Theorem*)
Suppose that $f(z)$ is analytic on the closed annulus determined by two simple closed paths C and K.

Then

$$\int_C f(z)\, dz = \int_K f(z)\, dz,$$

provided that C and K are traversed with the same orientation. See Figure 5.6.

Proof:
See Appendix 5(A).

It is very important to note that the above theorem does not require analyticity of the function in Int (K). On the other hand, it is easy to see

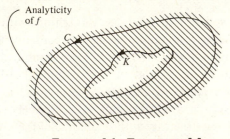

Analyticity of f

FIGURE 5.6   THEOREM 5.5

that if f were analytic in Int (K), then the theorem would reduce to a triviality, since, if such were the case, then according to Cauchy's theorem we would have

$$\int_C f(z)\, dz = \int_K f(z)\, dz = 0.$$

Thus one may conclude that if Theorem 5.5 is to be of any essential value, then f must fail to be analytic somewhere in Int (K), and this indeed can be the case. We shall have numerous occasions to use the above theorem, and we begin with the following.

EXAMPLE 1

Evaluate the integral $\displaystyle\int_C \frac{dz}{z - i}$, where C is the path of Figure 5.7.

An attempt to evaluate this integral by the methods of Chapter 4 would involve an excessive amount of calculation. On the other hand, Theorem 5.5 provides us with an extremely easy alternative.

First, we note that the only singularity of the integrand is $z = i$. So, if we choose a positively oriented circle K, centered at $z = i$ and having a radius small enough so that K will lie in Int (C), we put ourselves in the context of the annulus theorem; see Figure 5.7. Specifically, we see that the integrand $f(z) = 1/(z - i)$ is analytic on the closed annulus determined by C and K. Hence, by Theorem 5.5 and making use of Example 8, Section 19, we find that

$$\int_C \frac{dz}{z - i} = \int_K \frac{dz}{z - i} = 2\pi i,$$

and the given integral has been evaluated.

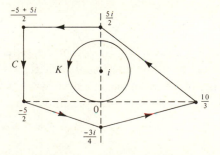

FIGURE 5.7   EXAMPLE 1

The next theorem is an extension of Theorem 5.5 to the case in which more than two simple closed paths are involved in the following sense.

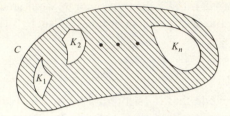

FIGURE 5.8   MULTIPLE ANNULUS

Suppose that $C, K_1, K_2, \ldots, K_n$ are $n + 1$ simple closed paths such that
1. All the K_j's are in Int (C).
2. Each K_j lies in Ext (K_s), for all $s \neq j$.
See Figure 5.8. Then, the region common to Int (C) and Ext (K_j) for all
$j = 1, 2, \ldots, n$ is called the **multiple annulus** determined by $C, K_1, \ldots, K_n$.
A multiple annulus along with the paths that determine it is called a **closed
multiple annulus.** In this context, we have the following.

Theorem 5.6   (*Multiple Annulus Theorem*)
*Suppose that $f(z)$ is analytic on the closed multiple annulus determined by
the simple closed paths $C, K_1, K_2, \ldots, K_n$.*
Then

$$\int_C f(z)\,dz = \int_{K_1} f(z)\,dz + \cdots + \int_{K_n} f(z)\,dz,$$

provided that all $n + 1$ paths are traversed with the same orientation.

Proof:
See Appendix 5(A).

As in the case of Theorem 5.5, Theorem 5.6 reduces to a trivially true
statement if f is analytic in the interior of each K_j. (Why? Justify this assertion.) Hence its usefulness becomes essential when f fails to be analytic at
points in Int (C), as the following examples illustrate.

EXAMPLE 2

Evaluate the integral $\int_C \left(\dfrac{4}{z + 1} + \dfrac{3}{z + 2i} \right) dz$, where C is the circle $|z| = 4$
positively oriented; see Figure 5.9.

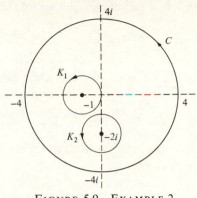

FIGURE 5.9   EXAMPLE 2

First, we note that the integrand fails to be analytic at $z = -1$ and $z = -2i$. Then, we "isolate each singularity" by choosing the paths

$$K_1 : z = -1 + e^{it}, \qquad 0 \le t \le 2\pi$$

and

$$K_2 : z = -2i + e^{it}, \qquad 0 \le t \le 2\pi.$$

Clearly, the integrand is analytic on the closed multiple annulus determined by C, K_1, and K_2. Hence

$$\int_C \left(\frac{4}{z+1} + \frac{3}{z+2i} \right) dz$$

$$= \int_{K_1} \left(\frac{4}{z+1} + \frac{3}{z+2i} \right) dz + \int_{K_2} \left(\frac{4}{z+1} + \frac{3}{z+2i} \right) dz$$

$$= 4 \int_{K_1} \frac{dz}{z+1} + 3 \int_{K_1} \frac{dz}{z+2i} + 4 \int_{K_2} \frac{dz}{z+1} + 3 \int_{K_2} \frac{dz}{z+2i}.$$

Now, with the help of Figure 5.9, we see that the function $1/(z + 2i)$ is analytic on K_1 and its interior and, similarly, the function $1/(z + 1)$ is analytic on K_2 and its interior. Consequently, the second and third of the above four integrals are each equal to zero according to Cauchy's theorem. On the other hand, by using Example 3, Section 19, we find that the first integral above has a value of $8\pi i$, whereas the last one has a value of $6\pi i$. Therefore, the given integral is equal to $14\pi i$.

EXAMPLE 3
Evaluate the integral

$$\int_C \frac{2i}{z^2 + 1} \, dz$$

along $C : |z - 1| = 6$, positively oriented.

By use of partial fraction decomposition we find that

$$\frac{2i}{z^2 + 1} = \frac{1}{z - i} - \frac{1}{z + i}.$$

Clearly, the singularities of the integrand are i and $-i$. Choosing the paths

$$K_1 : |z - i| = \tfrac{1}{2} \quad \text{and} \quad K_2 : |z + i| = \tfrac{1}{2},$$

oriented positively, we place the problem in the context of Theorem 5.6. Then, as in Example 2, we find

$$\int_C \frac{2i}{z^2 + 1}\, dz = \int_C \left(\frac{1}{z - i} - \frac{1}{z + i} \right) dz$$

$$= \int_{K_1} \frac{dz}{z - i} - \int_{K_1} \frac{dz}{z + i} + \int_{K_2} \frac{dz}{z - i} - \int_{K_2} \frac{dz}{z + i} = 0.$$

The reader should justify the final answer by examining the value of each of the last four integrals as in Example 2.

EXERCISE 21

A

In Exercises 21.1–21.10 evaluate the integral of the given function along the respective path, taking all simple closed paths in their positive orientation, unless otherwise instructed. In each case, locate the singularities of the function and draw the path of integration.

21.1. $f(z) = z^2 + 3 + 4/z$, $|z| = 4$.

21.2. $f(z) = 6/(z + a)$, $|z + a| = 3$.

21.3. $f(z) = 1/(z^2 - 1)$, $z = -i + 5e^{it}$, $-\pi \leq t \leq \pi$.

21.4. $f(z) = 1/(z^2 - 2z)$, $|z| = 1$.

21.5. $f(z) = 1/(z^2 - \pi^2)$, $|z| = 3$.

21.6. $f(z) = 2z + ize^{iz} - \dfrac{i}{z + i}$ along $C = C_1 + C_2 + C_3$, where C_1 is the line segment from $(-1, 0)$ to $(0, 0)$; C_2 is the negatively oriented circle of radius 2 centered at $-2i$; and C_3 is the line segment from $(0, 0)$ to $(1, 0)$.

21.7. $f(z) = \sin z/(z - 1)$, $|z + i| = 1$.

21.8. $f(z) = (z^3 - z^2 - 1)/z$, along the path of Exercise 21.5.

21.9. $f(z) = \text{Log}\left(\dfrac{z}{2}\right) + \dfrac{i}{z-3}$, $|z-2| = \dfrac{3}{2}$.

21.10. $f(z) = 3i$, along the path of Exercise 21.9.

B

21.11. Evaluate the integral of $f(z) = \cos z / z^3$ along $C = C_1 + C_2$, where $C_1 : |z| = 2$, positively oriented, and $C_2 : |z| = 3$, negatively oriented.

21.12. Evaluate the integral of $f(z) = e^z/(z-i)$, along the path of Exercise 21.11.

21.13. Evaluate the integral of $f(z) = 1/(z^3 - z)$, along $|z| = 2$, positively oriented.

21.14. Let C be a positively oriented simple closed path and let z_0 be a point in Int (C). Review Exercise 19.10(e) and prove that, for any integer n,

$$\int_C (z - z_0)^n \, dz = \begin{cases} 2\pi i, & \text{for } n = -1, \\ 0, & \text{for } n \ne -1. \end{cases}$$

21.15. Let R be the circular annulus $2 < |z| < 3$ and let C be a simple closed path contained in R. Prove that, for any such C,

$$\int_C \frac{dz}{z^2 + 1} = 0.$$

Section 22
The Cauchy Integral Formulas. Morera's Theorem

The Cauchy integral theorem is generally believed to be the single most important result in the theory of analytic functions. The **Cauchy integral formula,** which we introduce in the next theorem, is probably the next most important result. The theorem in question demonstrates the intimate connection that exists among the values attained by an analytic function within its region of analyticity and, in particular, in the interior of a given simple closed path throughout which the function is analytic. However, the importance of this theorem reaches considerably farther than that; as one of its consequences we may mention the fact that it forms the basis for the development of the theory of complex power series, which, in turn, leads to the theory of residues with its diverse uses in a multitude of applied fields. On the more elementary level, the Cauchy integral formula and its extensions provide convenient tools for the evaluation of a large class of complex integrals.

Theorem 5.7   (*Cauchy Integral Formula*)

Suppose that

1. A function $f(z)$ is analytic on a positively oriented simple closed path C and on $\text{Int }(C)$.

2. z_0 is any point in $\text{Int }(C)$.

Then

$$f(z_0) = \frac{1}{2\pi i} \int_C \frac{f(z)}{z - z_0}\, dz.$$

Proof:

See Appendix 5(A).

REMARK

In using the formula of Theorem 5.7, the reader should be careful to distinguish between the function $f(z)$, which is, by hypothesis, analytic in $\text{Int }(C)$, and the function $f(z)/(z - z_0)$, which has a singularity in $\text{Int }(C)$, namely, at $z = z_0$. Also, note should be made of the fact that the formula is valid as given under the hypothesis that C is positively oriented. Of course, if C were negatively oriented, then the left-hand side of the formula would be $-f(z_0)$.

Some of the more profound implications of the above theorem will manifest themselves in subsequent developments. In the examples that follow, we illustrate some of its more practical uses.

EXAMPLE 1

Evaluate the integral $\int_C \frac{z^2}{z - i}\, dz$, where $C : |z| = 2$, positively oriented.

In the notation of Theorem 5.7, $f(z) = z^2$, which is an entire function, and $z_0 = i$, which is in $\text{Int }(C)$. It follows by the Cauchy integral formula that

$$\int_C \frac{z^2}{z - i}\, dz = 2\pi i[f(i)] = -2\pi i.$$

EXAMPLE 2

Evaluate the integral $\int_C \frac{dz}{z(z + \pi i)}$, where $C : z = -3i + e^{it}, 0 \le t \le 2\pi$.

The integrand is analytic except at $z = 0$, which is in $\text{Ext }(C)$ and at $z = -\pi i$, which is in $\text{Int }(C)$. Thus, writing the given integral in the form

$$\int_C \frac{1/z}{z + \pi i}\, dz,$$

we can employ the Cauchy integral formula with $f(z) = 1/z$ and $z_0 = -\pi i$:

$$\int_C \frac{dz}{z(z + \pi i)} = \int_C \frac{1/z}{z + \pi i}\, dz = 2\pi i[f(-\pi i)] = -2.$$

We proceed next with an extension of Theorem 5.7 that will yield the generalized form of the Cauchy integral formula. From the practical standpoint, the result in question is a much stronger tool than the formula of Theorem 5.7 when it comes to the evaluation of certain types of complex integrals. More importantly, the generalized formula establishes the truth of a very strong and far-reaching attribute of complex functions, namely, that an analytic function possesses derivatives *of all orders* at every point at which it is analytic. This, in turn, shows that if a function is analytic, then not only does it possess derivatives of all orders, but the derivatives themselves are analytic functions; see Corollary 1, below.

Theorem 5.8   (*Generalized Cauchy Integral Formula*)

Suppose that

1. A function $f(z)$ is analytic on a positively oriented simple closed path C and on Int (*C*),

2. z_0 is any point in Int (*C*).

Then, for any integer $n = 0, 1, 2, \ldots$, the derivative $f^{(n)}(z_0)$ exists and is given by the formula

$$f^{(n)}(z_0) = \frac{n!}{2\pi i} \int_C \frac{f(z)}{(z - z_0)^{n+1}}\, dz.$$

Proof:

See Appendix 5(A).

N O T E: Theorem 5.7 is a special case of Theorem 5.8 for $n = 0$, since by $f^{(0)}(z)$ we simply mean $f(z)$.

If a function f is analytic at z_0, then, by definition, it is also analytic in some $N(z_0, \varepsilon)$. Now, if a simple closed path C is drawn within N with z_0 in Int (*C*), then the hypotheses of Theorem 5.8 are satisfied. Repeating the same argument for each point of a region R on which f is analytic, one may then establish the following.

Corollary 1

Suppose that $f(z)$ is analytic in a region R.

Then f possesses derivatives of all orders each of which is analytic in R and is given by the formula of Theorem 5.8.

Another consequence of Theorem 5.8 consists of the **Cauchy inequalities**, which place an upper bound on the values attained by the derivatives of an analytic function at points within its region of analyticity. We formalize this concept in Corollary 2, which follows; an outline of its proof appears in Exercise 22.15.

Corollary 2   (*Cauchy Inequalities*)
Suppose that
1. $f(z)$ *is analytic on* $C : |z - z_0| = \rho$ *and in* Int (C).
2. $f(z)$ *is bounded on* C; *i.e., there is a positive real number* M *such that*
   $|f(z)| \leq M$, *for all z on C.*
Then, for all $n = 0, 1, \ldots$,

$$|f^{(n)}(z_0)| \leq \frac{n!M}{\rho^n}.$$

Closely associated with the above result is the concept of "maximum modulus," which we discuss in Part III.

EXAMPLE 3

Evaluate the integral $\int_C \frac{z^3 + e^z}{(z + \pi i)^3}\, dz$, where $C : z = 7e^{it}, 0 \leq t \leq 2\pi$.

In the context of Theorem 5.8, we have $f(z) = z^3 + e^z$, $z_0 = -\pi i$, and $n = 2$. Now, since $f(z)$ is an entire function and z_0 is in Int (C), the hypotheses of the theorem are met. Therefore, using the formula in the conclusion of the theorem for $n = 2$, we find that

$$\int_C \frac{z^3 + e^z}{(z + \pi i)^3}\, dz = \frac{2\pi i}{2!} \cdot f''(-\pi i) = 6\pi^2 - \pi i.$$

EXAMPLE 4

Evaluate the integral $\int_C \frac{dz}{(z - 2)^2 z^3}$ for $C : |z - 3| = 2$, positively oriented.

First, we note that the only singularity of the integrand which lies in Int (C) is $z = 2$. Thus, in the context of Theorem 5.8, we can take $f(z) = 1/z^3$, $z_0 = 2$, and $n = 1$. Then the formula of the theorem yields the following:

$$\int_C \frac{dz}{(z - 2)^2 z^3} = \int_C \frac{1/z^3}{(z - 2)^2}\, dz = 2\pi i f'(2) = -\frac{3\pi i}{8}.$$

EXAMPLE 5

Evaluate the integral $\displaystyle\int_C \frac{dz}{(z-2)^2 z^3}$ for $C : |z-1| = 3$, positively oriented.

In handling this problem, we will use first the multiple annulus theorem (Theorem 5.6) to isolate the singularities $z = 0$ and $z = 2$, both of which are in Int (C). Then, we will use Theorem 5.8 in the form employed in Example 4 to finish the problem.

Let C_1 be a circle centered at 0 and let C_2 be a circle centered at 2, both positively oriented and with radii small enough so that they do not intersect each other and so that both lie in Int (C). Sketch the configuration. Then

$$\int_C \frac{dz}{(z-2)^2 z^3} = \int_{C_1} \frac{dz}{(z-2)^2 z^3} + \int_{C_2} \frac{dz}{(z-2)^2 z^3}$$

$$= \int_{C_1} \frac{(z-2)^{-2}}{z^3}\, dz + \int_{C_2} \frac{z^{-3}}{(z-2)^2}\, dz$$

$$= \frac{2\pi i}{2!}\frac{3}{8} + \frac{2\pi i}{1!}\left(-\frac{3}{16}\right)$$

$$= 0.$$

We close this section with a partial converse of the Cauchy integral theorem due to Morera; see Remark 1, p. 184.

Theorem 5.9   (*Morera's Theorem*)
Suppose that
1. $f(z)$ is continuous in a simply connected region R.
2. $\int_C f(z)\, dz = 0$ for every simple closed path C contained in R.
Then $f(z)$ is analytic in R.

Proof:
See Appendix 5(A).

EXERCISE 22

A

In Exercises 22.1–22.8 evaluate the integral of the given function along the given path traversed positively. In each case, locate the singularities of the function and draw the path of integration.

22.1. $f(z) = 3z^4/(z - 6i)$, $|z| = 10$.

22.2. $f(z) = 1/(z + i)z^4$, $|z - i| = \frac{3}{2}$.

22.3. $f(z) = (e^{2z} - z^2)/(z - 2)^3$, $|z - 1| = 3$.

22.4. $f(z) = \sin z/(z - 1)^2$, $|z| = 2$.

22.5. $f(z) = \cos z/(z + 3i)^6$, $|z + i| = 1$.

22.6. $f(z) = 3/z^2(z + i)^2$, $|z| = 5$.

22.7. $f(z) = 1/(z + 1)^3 - z^5/(z - \pi i)^4$, $|z| = 5$.

22.8. $f(z) = (z + 1 - e^z)/z(z + 3)$, $|z - i| = 2$.

B

22.9. If a is a positive real number, evaluate the integral

$$\int_C \frac{e^z}{z^2 + a^2}\, dz$$

for each of the following two paths, both positively oriented:

(a) $C : |z| = 2a$.   (b) $C : |z - ai| = a$.

22.10. Evaluate $\displaystyle\int_C \frac{e^z - z^3}{(z - a)^3}\, dz$, where C is a positively oriented simple closed path and a is in Int (C).

22.11. Evaluate $\displaystyle\int_C \frac{dz}{z^4 - 1}$ along $C : |z - 1| = 5$, positively oriented.

22.12. Evaluate $\int_C e^{-z}z^{-2} \sin z \, dz$ along C of the preceding exercise.

22.13. Verify the Cauchy inequality for $f(z) = z^4$, $n = 2$, for the circle $|z - i| = r > 0$.

C

22.14. Show that if $f(z)$ is analytic in a simply connected region R and the circle $|z - z_0| = r$ is in R, then

$$f(z_0) = \frac{1}{2\pi} \int_0^{2\pi} f(z_0 + re^{it})\, dt.$$

22.15. The various parts of this exercise constitute the main steps of a proof of Corollary 2 of this section. Supply the details for each step and then combine them to give a complete proof. In the notation of the statement of the corollary:

(a) Find the length L of C.

(b) Show that

$$\left| \frac{f(z)}{(z - z_0)^{n+1}} \right| \le \frac{M}{\rho^{n+1}}.$$

(c) Use the above two results, the generalized Cauchy integral formula (Theorem 5.8), and, finally, part 5 of Theorem 4.5 to establish
the truth of the Cauchy inequality.

REVIEW EXERCISES — CHAPTER 5

In Exercises 1–7 integrate each of the given functions along the respective
path. All simple closed paths are to be traversed positively.

1. $f(z) = 3z^2 - \sin z$, from $z_1 = 0$ to $z_2 = 2i$.
2. $f(z) = z(z^4 - 1)^{-1}$, $C: |z - 1 - i| = 2$.
3. $f(z) = z^2(z - i)^{-1}(z + 2)^{-3}$, $C: |z - 1| = 2$.
4. $f(z) = (e^z - z^6)(z - 2i)^{-6}$, $C: |z| = 3$.
5. $f(z) = [\text{Log}\,(z - i)](z + i)^{-1}$, $C: |z + 2i| = 2$.
6. $f(z) = 1/(z^2 + 4)$, $C: |z - 1| = 6$.
7. $f(z) = (z^3 - 8)(z^2 - 4z + 4)^{-1}$, $C: |z - 1| = 8$.

8. Evaluate $\displaystyle\int_{-1}^{1+i} (z - 3i)\,dz$:

    (a) by use of line integration along a straight line segment.
    (b) by any other method.

9. If α is any complex number with $|\alpha| \neq 1$, find all possible values of

$$\int_C \frac{e^z\,dz}{(z - \alpha)^3},$$

where $C: |z| = 1$, positively oriented.

10. Find the value of $\displaystyle\frac{d}{d\zeta}\int_0^\zeta (z^2 + 3z - i)\,dz$ in two different ways.

11. Suppose that a function $f(z)$ is analytic on the set $0 < |z| < 1$ and that
the integral of f is zero along *every* $C: |z| = r$, where $0 < r < 1$. Is f
necessarily analytic at $z = 0$? Justify your answer.

12. Consider the integral $\displaystyle\int_C \frac{z\,dz}{z^2 - \alpha^2}$, where α is any fixed nonzero complex
number and C is a simple closed path not passing through α or $-\alpha$.
Find the value of this integral for each of the four essentially different
positions of C with respect to α and $-\alpha$.

13. If C is the circle $|z| = r > 0$, find sufficient conditions on r so that
$\int_C \tan z\,dz = 0$. Verify that your choice of r works.

14. (a) Use the Cauchy integral formula to show that if C is the positively
    oriented circle $|z| = 1$, then $\displaystyle\int_C \frac{dz}{z} = 2\pi i.$

(b) Decompose C in the form $C = C_1 + C_2$, where C_1 is the upper half of C from 1 to -1 and C_2 is the lower half of C from -1 to 1. Then consider the following argument and find the fallacy which leads one to conclude that $2\pi i = 0$.

$$\int_C \frac{dz}{z} = \int_{C_1} \frac{dz}{z} + \int_{C_2} \frac{dz}{z} = \int_1^{-1} \frac{dz}{z} + \int_{-1}^1 \frac{dz}{z}$$

$$= \operatorname{Log} z|_1^{-1} + \operatorname{Log} z|_{-1}^1$$

$$= \operatorname{Log}(-1) - \operatorname{Log}(1) + \operatorname{Log}(1) - \operatorname{Log}(-1) = 0.$$

15. Supply the missing details in the proof of Theorem 5.8, Appendix 5(A).

APPENDIX 5

Part A
Proofs of Theorems

Theorem 5.2   *(Independence of Path)*
Suppose that
1. R is a simply connected region.
2. z_1 and z_2 are points in R.
3. $f(z)$ is analytic throughout R.
Then the value of the integral

$$\int_C f(z)\,dz$$

is the same along any path C joining z_1 and z_2, provided that C lies entirely within R.

Proof:
Let C and K be any two paths joining z_1 and z_2 and lying entirely in R; see Figure 5.10. Then $C - K$ is a closed path that lies in the simply connected region R throughout which f is analytic. Hence, by the Cauchy integral theorem,

$$\int_{C-K} f(z)\,dz = 0.$$

Then, using the properties of integrals (Theorem 4.5), we have

$$\int_C f(z)\,dz + \int_{-K} f(z)\,dz = 0$$

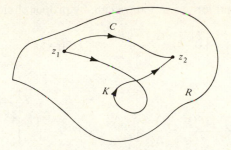

FIGURE 5.10   THEOREM 5.2

from which

$$\int_C f_\cdot(z)\, dz - \int_K f(z)\, dz = 0.$$

Therefore,

$$\int_C f(z)\, dz = \int_K f(z)\, dz,$$

and since C and K are arbitrary paths from z_1 to z_2, the theorem is proved.

Theorem 5.3

Suppose that

1. R is a simply connected region.

2. z_1 is a point in R.

3. f(z) is analytic at every point of R.

Then, for every ζ in R,

$$\frac{d}{d\zeta} \int_{z_1}^{\zeta} f(z)\, dz = f(\zeta).$$

Proof:

In the context and the notation of our discussion on p. 187, we restrict our considerations to points of R *only* and we denote the integral

$$\int_{z_1}^{\zeta} f(z)\, dz$$

by $F(\zeta)$. Using this notation and the definition of derivative, in order to prove the theorem we must establish that

$$\lim_{\Delta\zeta \to 0} \frac{F(\zeta + \Delta\zeta) - F(\zeta)}{\Delta\zeta} = f(\zeta).$$

or, equivalently, that for any $\varepsilon > 0$ we can, by proper choice of $\Delta\zeta$, make

$$\left| \frac{F(\zeta + \Delta\zeta) - F(\zeta)}{\Delta\zeta} - f(\zeta) \right| < \varepsilon. \tag{1}$$

We have

$$\frac{F(\zeta + \Delta\zeta) - F(\zeta)}{\Delta\zeta} = \frac{1}{\Delta\zeta} \int_{z_1}^{\zeta + \Delta\zeta} f(z)\, dz - \frac{1}{\Delta\zeta} \int_{z_1}^{\zeta} f(z)\, dz$$

$$= \frac{1}{\Delta\zeta} \int_{\zeta}^{\zeta + \Delta\zeta} f(z)\, dz$$

$$= \frac{1}{\Delta\zeta} \int_{\zeta}^{\zeta + \Delta\zeta} [f(z) - f(\zeta) + f(\zeta)]\, dz$$

$$= \frac{1}{\Delta\zeta} \int_{\zeta}^{\zeta + \Delta\zeta} [f(z) - f(\zeta)]\, dz + \frac{1}{\Delta\zeta} \int_{\zeta}^{\zeta + \Delta\zeta} f(\zeta)\, dz$$

$$= \frac{1}{\Delta\zeta} \int_{\zeta}^{\zeta + \Delta\zeta} [f(z) - f(\zeta)]\, dz + f(\zeta),$$

where $f(\zeta)$ has been obtained by direct evaluation of the integral

$$\frac{1}{\Delta\zeta} \int_{\zeta}^{\zeta + \Delta\zeta} f(\zeta)\, dz,$$

whose integrand, $f(\zeta)$, is constant, since the variable of integration is z. Thus, from the first and last steps of the above sequence of equalities, we have

$$\frac{\Delta F}{\Delta\zeta} - f(\zeta) = \frac{1}{\Delta\zeta} \int_{\zeta}^{\zeta + \Delta\zeta} [f(z) - f(\zeta)]\, dz. \tag{2}$$

Now, by hypothesis, f is analytic at every ζ in R and hence continuous there. Therefore, given any $\varepsilon > 0$, there exists $\delta > 0$ such that for any z satisfying $|z - \zeta| < \delta$, we have $|f(z) - f(\zeta)| < \varepsilon$; in other words,

$$|\Delta\zeta| < \delta \qquad \text{implies that} \qquad |f(z) - f(\zeta)| < \varepsilon.$$

So, provided that we stay within $N(\zeta, \delta)$, the integrand $[f(z) - f(\zeta)]$ never exceeds ε in magnitude. Now, the integral in (2) is independent of the path (why?); in particular, within $N(\zeta, \delta)$ we can choose the straight line segment joining ζ to $\zeta + \Delta\zeta$ as our path of integration. But then, the length of the path is $L = |\zeta + \Delta\zeta - \zeta| = |\Delta\zeta|$. Finally, applying Theorem 4.5(5), with $M = \varepsilon$

and L as above, we have

$$\left| \frac{F(\zeta + \Delta\zeta) - F(\zeta)}{\Delta\zeta} - f(\zeta) \right| = \left| \frac{1}{\Delta\zeta} \right| \left| \int_{\zeta}^{\zeta + \Delta\zeta} [f(z) - f(\zeta)]\, dz \right|$$

$$\leq \frac{1}{|\Delta\zeta|} \cdot \varepsilon \cdot |\Delta\zeta|$$

$$= \varepsilon.$$

In view of Equation (1), the proof is complete.

Theorem 5.4 (*Fundamental Theorem of Integration*)
Suppose that
1. *R is a simply connected region.*
2. *z_1 and z_2 are points in R.*
3. *$f(z)$ is analytic throughout R.*
4. *$\Phi(z)$ is an antiderivative of f.*

Then

$$\int_{z_1}^{z_2} f(z)\, dz = \Phi(z_2) - \Phi(z_1).$$

Proof:
We know that any two antiderivatives of a function differ by a constant; see Exercise 20.16. Then, in view of hypothesis 4 and Theorem 5.3, we have

$$\Phi(\zeta) = \int_{z_1}^{\zeta} f(z)\, dz + k \tag{1}$$

for some constant k and any ζ in R. In particular, Equation (1) is true for $\zeta = z_1$ and $\zeta = z_2$. If we let $\zeta = z_1$, (1) yields

$$\Phi(z_1) = k.$$

On the other hand, $\zeta = z_2$ yields

$$\Phi(z_2) = \int_{z_1}^{z_2} f(z)\, dz + \Phi(z_1),$$

and therefore

$$\int_{z_1}^{z_2} f(z)\, dz = \Phi(z_2) - \Phi(z_1),$$

as asserted by the theorem.

Theorem 5.5   (*Annulus Theorem*)

Suppose that $f(z)$ is analytic on the closed annulus determined by two simple closed paths C and K.

Then

$$\int_C f(z)\,dz = \int_K f(z)\,dz,$$

provided that C and K are traversed with the same orientation.

Proof:

In the context of Figure 5.11, draw two nonintersecting paths Γ and Π from C to K,* assigning their orientation to be that from C to K. The points at which Γ and Π intersect C and K induce a decomposition of C and K:

$$C = C_1 + C_2 \qquad \text{and} \qquad K = K_1 + K_2.$$

Consider now the two simple closed paths

$$C_1 + \Gamma - K_1 - \Pi \qquad \text{and} \qquad C_2 + \Pi - K_2 - \Gamma.$$

Clearly, f is analytic on and in the interior of each of these paths, by hypothesis. Therefore, by Cauchy's theorem the integral of f along each of these paths is zero. Hence

$$\int_{C_1+\Gamma-K_1-\Pi} f(z)\,dz + \int_{C_2+\Pi-K_2-\Gamma} f(z)\,dz = 0.$$

Since Γ and Π are traversed in both directions, they contribute nothing to the above integration. Therefore,

$$\int_{C_1-K_1} f(z)\,dz + \int_{C_2-K_2} f(z)\,dz = 0.$$

But then

$$\int_{C_1+C_2} f(z)\,dz - \int_{K_1+K_2} f(z)\,dz = 0,$$

and finally

$$\int_C f(z)\,dz - \int_K f(z)\,dz = 0,$$

from which the theorem follows.

* This is a possible process; for a proof, see Knopp, *Theory of Functions*, Part 1 (New York: Dover, 1945), p. 57.

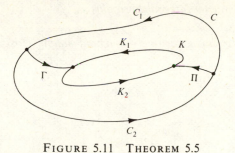

FIGURE 5.11 THEOREM 5.5

Theorem 5.6 (*Multiple Annulus Theorem*)
Suppose that $f(z)$ is analytic on the closed multiple annulus determined by the simple closed paths $C, K_1, K_2, \ldots, K_n$.
Then

$$\int_C f(z)\, dz = \int_{K_1} f(z)\, dz + \cdots + \int_{K_n} f(z)\, dz,$$

provided that all $n + 1$ paths are traversed with the same orientation.

Proof:
Draw $n + 1$ simple open nonintersecting auxiliary paths $\Gamma_0, \Gamma_1, \ldots, \Gamma_n$, all within R, in such a way that Γ_0 joins C to some K_i, Γ_1 joins K_i to some K_j with $j \neq i$, and so forth until Γ_n joins the last of the K_m's with C again; see Figure 5.12. Moreover, assign to each Γ_s the orientation implied by the above process.

The argument now parallels that of the proof of Theorem 5.5, since each K_i has been decomposed into two subpaths and so has C. Consequently, the multiple annulus has been subdivided into two simply connected regions (shaded differently in Figure 5.12) around each of which the integral of f is zero. The reader may now duplicate the process used in the preceding proof to establish the truth of this theorem.

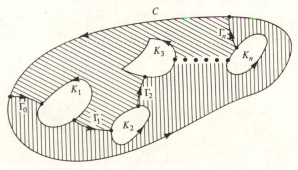

FIGURE 5.12 THEOREM 5.6

Theorem 5.7   *(Cauchy Integral Formula)*

Suppose that

1. $f(z)$ is analytic on a positively oriented simple closed path C and on
   Int (C).

2. z_0 is any point in Int (C).

Then

$$f(z_0) = \frac{1}{2\pi i} \int_C \frac{f(z)}{z - z_0} \, dz.$$

Proof:

Elementary algebraic manipulations yield the following:

$$\int_C \frac{f(z)}{z - z_0} \, dz = \int_C \frac{f(z) - f(z_0) + f(z_0)}{z - z_0} \, dz$$

$$= \int_C \frac{f(z) - f(z_0)}{z - z_0} \, dz + \int_C \frac{f(z_0)}{z - z_0} \, dz. \tag{1}$$

Now consider separately each of the integrals in Equation (1).

First, by use of Exercise 21.14, we have

$$\int_C \frac{f(z_0)}{z - z_0} \, dz = f(z_0) \int_C \frac{dz}{z - z_0} = f(z_0) \cdot 2\pi i.$$

Next, concerning the first integral in (1), we note the following. Since $f(z)$ is continuous at z_0 (why?), given any $\varepsilon > 0$ there exists $\delta > 0$ such that

$$|z - z_0| < \delta \qquad \text{implies that} \qquad |f(z) - f(z_0)| < \varepsilon.$$

Take $\lambda \leq \delta/2$ and consider the circle

$$K : |z - z_0| = \lambda,$$

positively oriented, with the understanding that, if need be, we will take λ as small as necessary so that K will be in Int (C); this is possible, since z_0 is a point of the open set Int (C). Then, for every z on K we have

$$\left| \frac{f(z) - f(z_0)}{z - z_0} \right| = \frac{|f(z) - f(z_0)|}{|z - z_0|} < \frac{\varepsilon}{\lambda}.$$

We also note that the length of K is $2\pi\lambda$. Now, using the Annulus Theorem (Theorem 5.5) and Theorem 4.5(5), with $M = \varepsilon/\lambda$ and $L = 2\pi\lambda$, we have

$$\left| \int_C \frac{f(z) - f(z_0)}{z - z_0} \, dz \right| = \left| \int_K \frac{f(z) - f(z_0)}{z - z_0} \, dz \right| \leq \frac{\varepsilon}{\lambda} 2\pi\lambda = 2\pi\varepsilon.$$

Since the above relation is true for any $\varepsilon > 0$, it follows that the first integral in Equation (1) is zero. Hence

$$\int_C \frac{f(z)}{z - z_0}\, dz = f(z_0) \cdot 2\pi i,$$

and the assertion of the theorem follows.

Theorem 5.8   (*Generalized Cauchy Integral Formula*)
Suppose that
1. $f(z)$ is analytic on a positively oriented simple closed path C and on Int (C).
2. z_0 is any point in Int (C).
Then, for any $n = 0, 1, 2, \ldots$, the derivative $f''(z_0)$ exists and is given by the formula

$$f^{(n)}(z_0) = \frac{n!}{2\pi i} \int_C \frac{f(z)}{(z - z_0)^{n+1}}\, dz.$$

Proof:
We give the main steps of the proof, leaving a good part of the details to be supplied by the reader.
The proof is by induction on n.
(A) $n = 0$: In this case the formula of the theorem reduces to the formula of Theorem 5.7; hence it is true (see Note on p. 199).
(B) *Inductive step:* Here, we assume that the formula is true for $n - 1$ and we prove that it holds for n. In other words, we assume that

$$f^{(n-1)}(z_0) = \frac{(n-1)!}{2\pi i} \int_C \frac{f(z)}{(z - z_0)^n}\, dz,$$

and we proceed to prove that

$$f^{(n)}(z_0) = \frac{n!}{2\pi i} \int_C \frac{f(z)}{(z - z_0)^{n+1}}\, dz. \tag{1}$$

According to the principle of mathematical induction, if this is accomplished, then the theorem will have been proved for all nonnegative integers n.
Before we outline the proof proper, we note the following:

(a) By definition,

$$f^{(n)}(z_0) = \lim_{\Delta z \to 0} \frac{f^{(n-1)}(z_0 + \Delta z) - f^{(n-1)}(z_0)}{\Delta z}.$$

(b) Since z_0 belongs to the open set Int (C), one may choose a circle $K : |z - z_0| = \lambda$ sufficiently small so that K will lie entirely in Int (C). We then observe that, by the Annulus Theorem, all integrals involved in this proof can be evaluated along K instead of along C.

(c) Since $\Delta z \to 0$ [see (a)], we may assume that $|\Delta z| < \lambda$.

(d) For the purpose of keeping our notation simple, we let $\xi = z - z_0$.

(e) All integrations involved here will be along K, positively oriented. Then since the variable of integration z must traverse K, it must satisfy the relation $|z - z_0| = \lambda$ and, equivalently, $|\xi| = \lambda$.

(f) By Theorem E, in part B of this appendix, there is a positive real number M such that, for every z on the circle K, $|f(z)| \le M$.

(g) In view of (1) and (a), above, and in terms of the notation introduced in (d), the theorem will have been proved if we show that

$$\left| \frac{f^{(n-1)}(z_0 + \Delta z) - f^{(n-1)}(z_0)}{\Delta z} - \frac{n!}{2\pi i} \int_K \frac{f(z)}{\xi^{n+1}} \, dz \right| \to 0 \qquad (2)$$

as $\Delta z \to 0$.

With the above preliminaries at our disposal, we now outline the proof proper.

By substitution from the inductive step and algebraic manipulations, the first fraction in (2) yields

$$\frac{f^{(n-1)}(z_0 + \Delta z) - f^{(n-1)}(z_0)}{\Delta z} = \frac{(n-1)!}{2\pi i \, \Delta z} \int_K f(z) \frac{\xi^n - (\xi - \Delta z)^n}{(\xi - \Delta z)^n \xi^n} \, dz.$$

By use of the identity

$$v^n - w^n = (v - w)(v^{n-1} + v^{n-2}w + \cdots + w^{n-1}),$$

the right-hand side of the last equality may be written in the form

$$\frac{(n-1)!}{2\pi i} \int_K f(z) \frac{\xi^{n-1} + \xi^{n-2}(\xi - \Delta z) + \cdots + (\xi - \Delta z)^{n-1}}{(\xi - \Delta z)^n \xi^n} \, dz.$$

Now, substitute this expression for the left-hand quantity in (2), combine the resulting two integrals into one integral, and complete the proof by showing that the resulting one integral tends to 0 as $\Delta z \to 0$. In this last step, use Theorem 4.5(5), along with (e) and (f) and the fact that the length of K is $2\pi\lambda$. This will complete the proof.

Theorem 5.9 *(Morera's Theorem)*

Suppose that

1. $f(z)$ is continuous in a simply connected region R.

2. $\int_C f(z) \, dz = 0$ for every simple closed path C contained in R.

Then $f(z)$ is analytic in R.

Proof:

Fix a point z_0 in R and take any other point ζ also in R. By hypothesis 2, the value of the integral

$$\int_{z_0}^{\zeta} f(z)\, dz$$

is independent of the path taken from z_0 to ζ and depends only on the values of z_0 and ζ. But z_0 is constant; hence the value of the integral depends on ζ alone. It follows, then, that the function $F(\zeta)$ defined by

$$F(\zeta) = \int_{z_0}^{\zeta} f(z)\, dz$$

is a single-valued function; see the discussion on p. 187.

Now, following the same argument as in the proof of Theorem 5.3, in which we only made use of the continuity of $f(z)$, we conclude that $F(\zeta)$ is analytic at every ζ in R and, in fact,

$$F'(z) = f(z).$$

But by Corollary 1, p. 199, the derivatives of analytic functions are themselves analytic functions. Hence $f(z)$, being the derivative of the analytic function $F(z)$, is itself an analytic function and Morera's theorem is proved.

Part B
Proof of the Cauchy Integral Theorem

The main goal of this section is a relatively detailed proof of the Cauchy theorem. The proof proper is preceded by a number of results of preliminary nature, including the Bolzano–Weierstrass theorem, the Heine–Borel theorem, and the theorem of nested sets, all of which are cornerstones in the field of mathematical analysis.

Since Goursat published his proof of the Cauchy theorem (see *Transactions of the American Mathematical Society*, Vol. 1, 1900, and *Acta Mathematica*, Vol. 4, 1884), there have been other approaches to the proof of this theorem. A complete discussion of technical and historical details is found in G. N. Watson, *Complex Integration and Cauchy's Theorem* (New York: Cambridge University Press, 1914). Alfred Pringsheim improved on Goursat's proof; see Vol. 2, 1901, of the *Transactions of the American Mathematical Society*. In this section we discuss a form of Pringsheim's proof in a development similar to one found in Knopp, *Theory of Functions*, Vol. 2 (New York: Dover, 1945).

PRELIMINARIES

We begin with a number of definitions.

A *real number b* is said to be an **upper bound** for a set A of real numbers provided $x \leq b$ for every x in A; a *real number c* is said to be a **lower bound** for A if and only if $c \leq x$ for every x in A. If A has an upper bound, then A is said to be **bounded above**, and if it has a lower bound, it is said to be **bounded below**. If A has both of these attributes, then we say that A is **bounded**; otherwise, A is said to be **unbounded**. For example,

1. The set A of all x such that $x < 3$ is bounded above,
2. The set B of all x such that $-4 < x$ is bounded below,
3. The set C of all x such that $-6 < x < 1533$ is bounded.

Clearly, A and B are unbounded sets. The reader is very strongly urged to verify that the above definition of a bounded set can be easily adapted to the form of the definition of a bounded set of *complex numbers* on p. 35.

It is evident from the above definitions that if b is an upper bound for a set A, then every number $d > b$ is also an upper bound for A. Thus, if A is bounded above, then it has no largest upper bound. The natural question then arises: If A is bounded above, does it have a smallest upper bound? The answer to this question is in the affirmative and is given by the *completeness axiom*, which postulates that

> every nonempty set A of real numbers that is bounded above has a smallest upper bound.

The latter is called a **supremum** of A and is denoted sup A; in precise terms, it may be defined as follows: A real number s is said to be a supremum of a set A (of real numbers), provided that (1) s is an upper bound for A and (2) if $x < s$, then x cannot be an upper bound for A. For example, if A is the set of all real numbers y such that $0 \leq y < 1$, then sup $A = 1$. A similar development concerning a set B bounded below will result in the concept of the **infimum** of B, which is defined analogously and denoted inf B.

We return now to the field of complex numbers, and we consider a nonempty set S of points in the plane. We define the **diameter** of S, denoted diam(S), to be the supremum of the distances of *all* pairs of points in S:

$$\text{diam}(S) = \sup |z - \zeta|, \qquad \text{for all } z, \zeta \text{ in } S.$$

If S is bounded (see p. 35), then it can be argued, by use of the completeness axiom, that diam(S) exists as a unique and finite number. If S is unbounded, then we agree to say that the set has an infinite diameter. For example, the set of all points on a circular disk of radius 3 has a diameter equal to 6, the set of points in a square of side 2 has a diameter of $\sqrt{8}$, and the set of points on the parabola $y = x^2$ has an infinite diameter.

Again, consider a nonempty set S of points in the plane. A point z (which may or may not belong to S) will be called a **cluster point** or an **accumulation point** of S, provided that *every deleted* neighborhood of z contains infinitely many points of S. Clearly, a set that consists of only a finite number of points cannot have a cluster point. On the other hand, a set may contain infinitely many points and have no cluster point; for instance, consider the set of all positive integers that is infinite but has no cluster point. By way of some more simple examples, consider the following: The set of points

$$1, \frac{1}{2}, \frac{1}{3}, \ldots, \frac{1}{n}, \ldots,$$

has exactly one cluster point, $z = 0$, which does not belong to the set. The open unit disk S consisting of all z such that $|z| < 1$ has infinitely many cluster points, namely, all points z on the closed unit disk $|z| \leq 1$. Note that S contains some but not all of its cluster points.

We now employ the concept of a cluster point to give an alternative definition of a closed set which is more conveniently adaptable to the current development. A set S will be called **closed** if and only if it contains all its cluster points.

We continue with a sequence of theorems that, directly or indirectly, will be needed in the proof of the Cauchy theorem, some of which, as we mentioned earlier, are cornerstones in the field of mathematical analysis. The first one is the so called Bolzano–Weierstrass theorem, whose proof we omit. The reader who is interested in its proof is referred to Chapter 6 of Knopp, *Elements of the Theory of Functions* (New York: Dover, 1952).

Theorem A (*Bolzano–Weierstrass*)
Suppose that S is a set in the plane such that
1. S contains infinitely many points.
2. S is bounded.
Then S has at least one cluster point,

Put simply, the above theorem asserts that if a set has an infinite number of elements and is a bounded set, then there is at least one point z_0 (which may or may not belong to the set) such that the points belonging to the set "pile up" arbitrarily close to z_0. In general, the theorem fails to be true if either of the two hypotheses is deleted. For, if S contains only finitely many points, then, as we remarked earlier, it cannot have a cluster point, and if S is not bounded, then the example of the set of positive integers shows that the theorem fails.

We now make use of Theorem A to prove the next result.

Theorem B   (*Theorem of Nested Sets*)

Suppose that $S_1, S_2, \ldots, S_n, \ldots$ is a sequence of sets in the plane having the following properties:

1. Every S_n is a closed set and contains at least one point.

2. S_{n+1} is contained in S_n for all $n = 1, 2, 3, \ldots$.

3. $\mathrm{diam}(S_n) \to 0$, as $n \to \infty$.

Then there is one and only one point that belongs to all S_n.

Proof:

(a) We prove that there is *at most one* such point.

The proof is by contradiction. Suppose that ζ and ξ both have the property of belonging to all S_n and that $\zeta \neq \xi$. Then there is a positive distance between them: $|\zeta - \xi| = d > 0$. But this says that the diameter of every S_n cannot be smaller than the positive number d, and this is a contradiction to the third hypothesis of the theorem. It follows then that if such point ζ exists, then it is unique.

(b) We now prove that there is *at least one* point which belongs to all the S_n.

It is clear from hypothesis 3 that, with the possible exception of the first M sets, the S_n's are bounded. From each of these bounded sets take one point: z_n from S_n. Concerning the set of these z_n's, there are two possibilities:

Either they are one and the same number, say, ξ from some point on, or they form an infinite and bounded set of distinct points.

If the first is true, then ξ belongs to all the sets and the proof is complete. In the second case, the set of these points has a cluster point ζ by virtue of Theorem A. We now prove that ζ belongs to all S_n.

Take any S_n. By hypothesis 2, all of the subsequent sets $S_{n+1}, S_{n+2}, \ldots$ belong to S_n, and hence the points $z_{n+1}, z_{n+2}, \ldots$, which we chose in these sets, all belong to S_n. But S_n is a closed set and hence, by definition, it contains all its cluster points. In particular, it contains ζ. But S_n was arbitrarily chosen. It follows that ζ is in every S_n.

Therefore, it has been proved that there is at least one point that belongs to all the S_n's. This, along with part (a) of the proof, establishes the truth of the theorem.

The next theorem is a simplified version of the celebrated Heine-Borel theorem. In its more general form, the theorem involves open sets of arbitrary nature, whereas the form in which we discuss it here deals with sets each of which is an open circular disk. We choose this version of the theorem because it satisfies our needs and, also, because the proof in this form is much more simple.

Theorem C (*Heine–Borel*)

Suppose that

1. S is a closed and bounded set in the plane.

2. Φ is an arbitrary family of open circular disks that cover S; i.e., every point of S belongs to at least one disk from Φ.

Then there is a finite subfamily (i.e., a finite number of open disks) of Φ that still covers S.

Proof:

The proof is by contradiction. We thus assume that infinitely many disks are always needed to cover S and we arrive at an absurdity.

Since S is bounded, we can draw a rectangle R having its sides parallel to the coordinate axes and such that S is in Int (R); see Figure 5.13. Next, we subdivide R into four congruent subrectangles by means of one vertical and one horizontal line, and we consider each of these subrectangles as containing its perimeter, so that each is a closed set. Thus the set S (which is contained in R) has been subdivided into at most four subsets, each of which belongs to one of the subrectangles. According to our assumption at the beginning of the proof, *at least one* of these subsets of S needs infinitely many disks from Φ to cover it; call R_1 the subrectangle that contains this particular subset of S.

Next, we subdivide R_1 as we did R and by a similar argument we obtain a subrectangle R_2, of R_1, containing a part of S that needs an infinity of disks from Φ to cover it. If we continued this process ad infinitum, we would obtain a sequence

$$R_1, R_2, R_3, \ldots, R_n, \ldots \tag{1}$$

of rectangles, each of which contains a part of S that requires infinitely many disks from Φ to cover it. Moreover, this sequence of rectangles would satisfy the following properties:

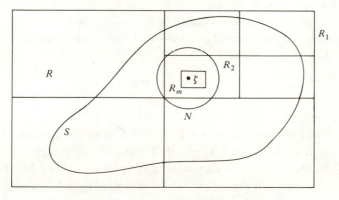

FIGURE 5.13 HEINE–BOREL THEOREM

1. Each R_k is closed and contains at least one point.
2. Each R_k is contained in R_{k-1}, for $k = 2, 3, \ldots$.
3. $\mathrm{diam}(R_k) \to 0$ as $k \to \infty$.

According to Theorem B, there is one and only one point ζ common to all R_k and since $\mathrm{diam}(R_k)$ tends to zero, any neighborhood N of ζ contains some R_k from expression (1) entirely in its interior. Hence N contains that part of S that is in R_k, and therefore N contains infinitely many points of S. Then, by definition, ζ is a cluster point of S and, since S is closed, it follows that S contains ζ.

Now, since every point of S is in some disk from Φ, there is at least one such disk, say D, that contains ζ. But, again, since the diameters of the R_k's tend to zero and since ζ belongs to all the R_k's, there is some R_m with diameter small enough so that R_m is in D. Thus the one disk D from Φ covers the part of S that is in R_m. This is a contradiction to the basic property of (every) R_m described immediately following expression (1), earlier in the proof.

This contradiction establishes the theorem.

In the next theorem we prove a result of technical nature but rather interesting in its own right. It asserts that if a path C lies inside a region R, then C cannot get arbitrarily close to the boundary of R. When we utilize this result in the proof of the Cauchy theorem, we will see that its practical aspect is that if we have a path C inside a region R, we can draw another path K between C and the boundary of R, regardless of how close C is to that boundary. A more general version of this theorem is also true with C replaced by any closed and bounded set contained in a region.

Theorem D

Suppose that

1. R is a region in the plane.

2. C is an arbitrary path contained in R.

Then there is a positive real number λ such that, for any point z on C and any point w on the boundary of R,

$$|z - w| > \lambda.$$

Proof:

Since every point z of C is in R and since R is an open set, there is a sufficiently small neighborhood N_z of z that is entirely contained in R. Find one such N_z for each z on C and then take another neighborhood M_z of each such z with radius one-half that of N_z. The family Φ of the M_z's clearly covers C. But C is a closed and bounded set and hence, according to Theorem C, a finite number of these M_z's suffice to cover C; call them

$$M_1, M_2, \ldots, M_k.$$

The radii of these open disks from Φ are k positive real numbers and hence by a finite process of comparison we may locate the smallest one, say, λ. We proceed now to verify the fact that the number λ satisfies the conclusion of the theorem.

Let z be any point of C and w be any point on the boundary of R; z must belong to one of the M_n's, whose center we may denote ξ. Then $|\xi - z| < \lambda$ and $|\xi - w| > 2\lambda$. Finally,

$$2\lambda < |\xi - w| \leq |\xi - z| + |z - w| < \lambda + |z - w|.$$

from which the inequality of the theorem follows. The reader should find it instructive to supply justifications for the last steps of the proof. This will complete the proof of the theorem.

The last in this sequence of theorems is an immediate consequence of the Heine–Borel theorem and is listed here for reference. In informal terms the theorem asserts that if a function $f(z)$ is continuous on a closed and bounded set D, then, in magnitude, the values attained by f over points of D have a finite "ceiling." Associated with this concept, we have the following definition: A function $f(z)$ is said to be **bounded** on a set D if and only if a positive number M exists such that

$$|f(z)| \leq M, \qquad \text{for every } z \text{ in } D.$$

The reader who, at this point, is only interested in what is preliminary to the proof of the Cauchy theorem may omit the next result.

Theorem E
Suppose that
1. S is a closed and bounded set.
2. $f(z)$ is continuous at every point of S.
Then f is bounded on S.

Proof:
Let z be an arbitrary point in S. Since f is continuous at z, given any $\varepsilon > 0$, there is a $\delta > 0$ such that

$$|f(z) - f(\xi)| < \varepsilon$$

for all ξ satisfying

$$|z - \xi| < \delta;$$

in other words, whenever ξ from S is in $N(z, \delta)$, $f(\xi)$ is in $N(f(z), \varepsilon)$. The totality of all such δ-neighborhoods, one for each z of S, covers S. Therefore, according to Theorem C, there are finitely many such δ-neighborhoods

$$N_1, N_2, \ldots, N_k,$$

corresponding to points

$$z_1, z_2, \ldots, z_k$$

in S, which also cover S. Considering now the corresponding ε-neighborhoods

$$M_1, M_2, \ldots, M_k$$

of $f(z_1), f(z_2), \ldots, f(z_k)$, we see that given any ξ in S, ξ belongs to some N_i, and hence $f(\xi)$ belongs to the corresponding M_i. Thus the M_i's cover every $f(\xi)$ for every ξ in S. Since there are only k such M_i's, each of which is a bounded set, there is a circle $|w| = M$ that contains all the M_i's in its interior. [For instance, one may take M to be the largest of the k finite numbers $|f(z_1)| + \varepsilon, |f(z_2)| + \varepsilon, \ldots, |f(z_k)| + \varepsilon$.] It follows that

$$|f(\xi)| \leq M,$$

for every ξ in S and the proof is complete.

Since analyticity of a function at a point implies continuity of the function at that point, the truth of the following corollary is an immediate consequence of Theorem E.

Corollary

If $f(z)$ is analytic at every point of a closed and bounded set S, then f is bounded on S.

We turn now to the concept of uniform continuity, and we begin with the following definition.

Suppose that $f(z)$ is a function defined (and single-valued) at every point of a set D. Then, f is said to be **uniformly continuous** on D, provided that for any $\varepsilon > 0$, there is a $\delta > 0$ such that whenever any two points z and ζ in D satisfy $|z - \zeta| < \delta$, then $|f(z) - f(\zeta)| < \varepsilon$. At this point of the development, our interest in the concept just defined is restricted to the result put forth by Theorem F. However, before we state and prove the theorem in question, a number of elementary remarks may prove helpful to the reader who has had very little or no experience with the notion of uniform continuity until now. Recalling the definition of continuity (p. 47), we note the following.

1. Continuity of a function is defined *at a point*, whereas uniform continuity is defined *on a set*.

2. In defining continuity, we begin with a point z_0. We are then given a positive number ε, and we find a $\delta > 0$ that depends on both z_0 and ε. In the case of uniform continuity, we are given an $\varepsilon > 0$ and we find a $\delta > 0$ that depends only on ε and is good (in terms of the definition) for any two points of the set under consideration.

3. If a function is uniformly continuous on a set D, then it is continuous at every point of D. For, given any w in D and any $\varepsilon > 0$, there is a $\delta > 0$ (by the definition of uniform continuity) such that for any z in D, $|w - z| < \delta$ implies that $|f(w) - f(z)| < \varepsilon$, which is precisely the definition of continuity at the point w.

4. The converse of the preceding property is not true, in general; i.e., continuity of a function f at every point of a set D does not imply that f is uniformly continuous on D. However, not all is lost. The following theorem gives us sufficient conditions under which the converse is true.

Theorem F

Suppose that $f(z)$ is continuous at every point of a closed and bounded set B. Then f is uniformly continuous on B.

Proof:

In order to prove uniform continuity of $f(z)$ on B, for any given $\varepsilon > 0$ we must produce a $\delta > 0$ such that, for any z and ζ in B,

$$|z - \zeta| < \delta \qquad \text{will imply that} \qquad |f(z) - f(\zeta)| < \varepsilon. \tag{1}$$

Since f is continuous on B, then for every ξ in B there is $\lambda_\xi > 0$ such that

$$|f(\xi) - f(w)| < \frac{\varepsilon}{2}$$

for every w in B satisfying

$$|\xi - w| < \lambda_\xi.$$

Repeating the same process for every ξ in B, we generate a family of neighborhoods N_ξ (one for each ξ in B), each of which is centered at the respective ξ and has radius the corresponding λ_ξ. Now, for each such N_ξ, consider a new open disk M_ξ concentric with N_ξ and having radius $\lambda_\xi/2$. The family of the M_ξ's covers B and hence, by Theorem C, there are finitely many of them which still cover B. Their radii, which we may denote

$$\frac{\lambda_1}{2}, \frac{\lambda_2}{2}, \dots, \frac{\lambda_k}{2},$$

are k positive numbers and therefore there is a smallest one among them; call it δ.

We prove that δ satisfies expression (1). To that end, suppose that for z and ζ in B, it is true that

$$|z - \zeta| < \delta.$$

Then, since z belongs to one of the k disks that still cover B, say M_v, and since v is the center of M_v, we have

$$|z - v| < \frac{\lambda_v}{2} < \lambda_v$$

and hence

$$|f(z) - f(v)| < \frac{\varepsilon}{2}.$$

On the other hand, since

$$|z - \zeta| < \delta \leq \frac{\lambda_v}{2},$$

then, by use of the triangle inequality, we have

$$|\zeta - v| \leq |z - \zeta| + |z - v| < \frac{\lambda_v}{2} + \frac{\lambda_v}{2} = \lambda_v,$$

and hence

$$|f(\zeta) - f(v)| < \frac{\varepsilon}{2}.$$

Finally, using the triangle inequality once again, we get

$$|f(z) - f(\zeta)| \leq |f(z) - f(v)| + |f(\zeta) - f(v)|$$

$$< \frac{\varepsilon}{2} + \frac{\varepsilon}{2}$$

$$= \varepsilon,$$

and the theorem is proved.

We conclude discussion of the preliminaries with two definitions and one theorem of plane-geometrical nature, whose proof we shall omit.

A **polygon** is a closed path that consists of (a finite number of) smooth curves, each of which is a straight line segment. A **simple polygon** is a polygon that is a simple closed path.

Theorem G

Suppose that P is a polygon.

Then

1. *P can be decomposed into a finite number of simple polygons and a finite number of straight line segments, the latter traversed twice, once in each direction.*
2. *Each of the simple polygons in part 1 is traversed either entirely in the positive or entirely in the negative orientation.*
3. *Every simple polygon in part 1 can be decomposed into a finite number of triangles by means of diagonals, each of which lies entirely within the simple polygon.*

For a proof of the above theorem, the reader is referred to Knopp, *Theory of Functions*, Part 1 (New York: Dover, 1945), p. 15.

THE PROOF PROPER

Theorem 5.1   (*Cauchy Integral Theorem*)

Suppose that

1. *$f(z)$ is analytic on a simply connected region R.*
2. *C is a closed path lying entirely in R.*

Then

$$\int_C f(z)\, dz = 0.$$

Proof:

C A S E 1. *The path C is a triangle.* Divide C into four congruent triangles C_1, C_2, C_3, C_4 as in Figure 5.14. Traversing all the C_i's in the positive sense, we observe that the sides of C_3 are traversed twice in opposite directions. Hence the relation

$$\int_{C_1} + \int_{C_2} + \int_{C_3} + \int_{C_4} = \int_C$$

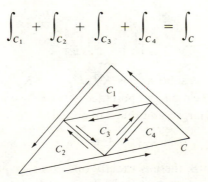

FIGURE 5.14   CAUCHY INTEGRAL THEOREM

is clearly true. From this relation it follows that, for at least one C_i, it is true that

$$\left| \int_{C_i} \right| \geq \tfrac{1}{4} \left| \int_C \right|; \tag{1}$$

for, if

$$\left| \int_{C_i} \right| < \tfrac{1}{4} \left| \int_C \right|$$

were true for all four C_i, then by use of the triangle inequality we could easily arrive at the absurdity

$$\left| \int_C \right| < \left| \int_C \right|.$$

Denoting by K_1 any one of the triangles for which inequality (1) holds, we have

$$\left| \int_C \right| \leq 4 \left| \int_{K_i} \right|. \tag{2}$$

Next, focusing our attention on K_1, we repeat the argument used on C to single out a subtriangle K_2 of K_1 such that

$$\left| \int_{K_1} \right| \leq 4 \left| \int_{K_2} \right|. \tag{3}$$

From (2) and (3) we then have

$$\left| \int_C \right| \leq 4^2 \left| \int_{K_2} \right|,$$

and by induction

$$\left| \int_C \right| \leq 4^n \left| \int_{K_n} \right|. \tag{4}$$

Now, note that the subtriangles $K_1, K_2, \ldots, K_n, \ldots$ (each taken along with its interior) are closed and bounded sets whose diameters tend to zero, as n gets larger and larger, and they are nested:

$$K_{n+1} \text{ is in } K_n, \qquad \text{for all } n = 1, 2, 3, \ldots.$$

Hence, by Theorem B, there is exactly one point ζ common to all K_n and hence ζ is either on C or in Int (C). By hypothesis, $f'(\zeta)$ exists. Thus, by definition of the derivative, given $\varepsilon > 0$, there is a $\delta > 0$ such that

$$|z - \zeta| < \delta \qquad \text{implies that} \qquad \left| \frac{f(z) - f(\zeta)}{z - \zeta} - f'(\zeta) \right| < \varepsilon$$

or, equivalently,

$$|z - \zeta| < \delta \qquad \text{implies that} \qquad |f(z) - f(\zeta) - f'(\zeta)(z - \zeta)| < \varepsilon |z - \zeta|.$$

Then, there exists a complex number ξ (which, in general, depends on the value of z) with the property that $|\xi| < \varepsilon$ and such that

$$f(z) = f(\zeta) + f'(\zeta)(z - \zeta) + \xi(z - \zeta).^* \tag{5}$$

Now, since the diameters of the K_n's tend to zero, and since ζ lies in all the K_n's, we can certainly find a sufficiently large n so that K_n will be contained in the δ-neighborhood of ζ; see Figure 5.15. Consequently, every z on or inside this K_n has the property that $|z - \zeta| < \delta$ and hence Equation (5) holds for every such z. But then

$$\int_{K_n} f(z) \, dz = \int_{K_n} f(\zeta) \, dz + \int_{K_n} z f'(\zeta) \, dz - \int_{K_n} \zeta f'(\zeta) \, dz$$

$$+ \int_{K_n} \xi(z - \zeta) \, dz$$

$$= f(\zeta) \int_{K_n} dz + f'(\zeta) \int_{K_n} z \, dz - \zeta f'(\zeta) \int_{K_n} dz$$

$$+ \int_{K_n} \xi(z - \zeta) \, dz$$

$$= 0 + 0 - 0 + \int_{K_n} \xi(z - \zeta) \, dz,$$

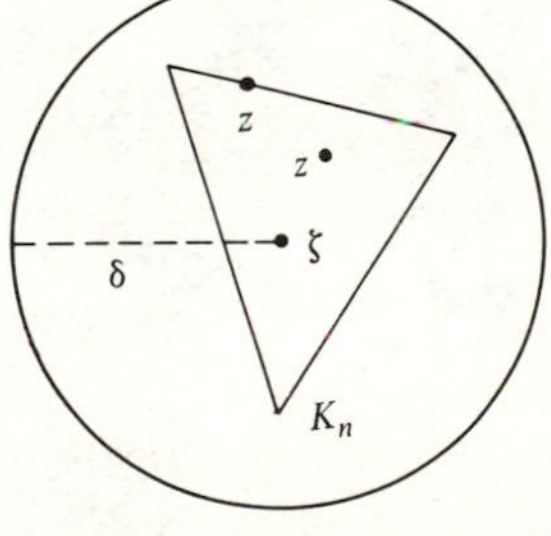

FIGURE 5.15   CAUCHY INTEGRAL THEOREM

$$^* \quad \xi = \frac{f(z) - f(\zeta)}{z - \zeta} - f'(\zeta).$$

where, in the last step, we made use of Examples 1 and 2 in Section 19. Thus, so far we have

$$\int_{K_n} f(z)\, dz = \int_{K_n} \xi(z - \zeta)\, dz. \tag{6}$$

Next, we digress briefly to mention two simple plane-geometrical facts that we need. The first one concerns the perimeters of the triangles $K_1, K_2, \ldots$. Denoting the perimeter of C by $|C|$ and that of each K_i by $|K_i|$, one may easily derive the following inequalities:

$$|K_1| = \frac{|C|}{2}, \; |K_2| = \frac{|K_1|}{2} = \frac{|C|}{2^2}, \ldots,$$

and, by induction,

$$|K_n| = \frac{|C|}{2^n}.$$

The second item concerns the following fact from plane geometry:

> The distance between any two points on or inside any triangle is less than or equal to one-half its perimeter.

Thus, in the notation of our proof,

$$|z - \zeta| \leq \frac{|K_n|}{2}.$$

Continuing with the proof, we apply Theorem 4.5(5) and the preceding two facts on Equation (6) to obtain the following:

$$\left| \int_{K_n} f(z)\, dz \right| = \left| \int_{K_n} \xi(z - \zeta)\, dz \right|$$

$$\leq \varepsilon \frac{|K_n|}{2} |K_n|$$

$$= \frac{\varepsilon |K_n|^2}{2}$$

$$= \frac{\varepsilon [|C|/2^n]^2}{2}$$

$$= \frac{\varepsilon}{2} \frac{|C|^2}{4^n}.$$

We have thus established that

$$\left| \int_{K_n} f(z)\, dz \right| < \frac{\varepsilon}{2} \frac{|C|^2}{4^n}.$$

Finally, using Equation (4) and the above inequality, we obtain

$$\left| \int_C f(z)\, dz \right| \le 4^n \left| \int_{K_n} f(z)\, dz \right|$$

$$< 4^n \frac{\varepsilon}{2} \frac{|C|^2}{4^n}$$

$$= \frac{|C|^2}{2}\varepsilon.$$

But the last relation is true for any $\varepsilon > 0$ and, since $|C|^2/2$ is a fixed finite number, we conclude that

$$\int_C f(z)\, dz = 0,$$

and Case 1 has been proved.

CASE 2: *The path C is a closed polygon.* In this case we appeal to Theorem G, which asserts that the polygon C can be decomposed into a finite number of triangles $C_1, C_2, \ldots, C_n$, in such a way that every side of each triangle that does not coincide with some side of C (or, a part thereof) will lie inside C. Now, traversing each of these triangles in the orientation induced by that of C and using the result of Case 1, we have

$$\int_{C_1} + \int_{C_2} + \cdots + \int_{C_n} = 0.$$

But every side of these triangles that is interior to the polygon is traversed twice and in opposite directions. Hence the value of the integrals along these interior sides (see Figure 5.16) contributes nothing to the above sum of integrals, and only the sides of the original polygon count in this integration. Therefore,

$$\int_C f(z)\, dz = 0,$$

and the theorem is thus established for Case 2.

CASE 3: *C is an arbitrary closed path.* First, we discuss the following four items, which constitute the basis for the proof of this case.

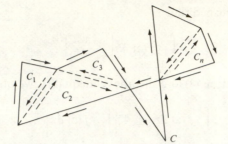

FIGURE 5.16   CAUCHY INTEGRAL THEOREM

1. We recall the definition

$$\int_C f(z)\, dz = \lim_{\mu \to 0} \sum_{k=1}^{n} f(\zeta_k)(z_k - z_{k-1}).$$

For convenience of notation we denote the sum on the right by S_n, and so the above equation takes the form

$$\int_C f = \lim S_n.$$

2. By Theorem D, there is a real number $\lambda > 0$ such that any point z on C has a distance greater than λ from any point w on the boundary of R. Consider now the set Q that consists of all points in R whose distance from the boundary of R is greater than or equal to $\lambda/2$; clearly, Q is a closed set and contains C in its interior. Since $f(z)$ is continuous on Q (f is analytic on Q), it follows, by Theorem F, that f is uniformly continuous on Q. Consequently, for any given $\varepsilon > 0$ there is a $\delta > 0$ such that, for any two points z and ζ in Q for which

$$|z - \zeta| < \delta,$$

it is true that

$$|f(z) - f(\zeta)| < \varepsilon.$$

3. Suppose that $\varepsilon > 0$ is chosen arbitrarily. We proceed to find a partition

$$P : z_0, z_1, z_2, \ldots, z_n$$

of C so that the following conditions will be satisfied:
   (a) $|\int_C f - S_n| < \varepsilon$; this is possible by item 1.
   (b) The subpaths $\widehat{z_{k-1}z_k}$ are all of length less than $\lambda/2$; this is possible, since C is a path and hence has finite length. See Remark 1, p. 152.
   (c) The length of each subpath is also less than δ, where δ is determined by uniform continuity, as in item 2.

Note that conditions (a) to (c) imposed on the partition P guarantee, among other things, that the chords which may be drawn to join z_{k-1} to z_k, for $k = 1, 2, \ldots, n$, all lie within Q and hence within R. Moreover, since the length of each subpath is less than δ, it follows that the length of each chord is less than δ; hence, if z is any point on the kth chord, then

$$|z - z_k| < \delta \qquad \text{and therefore} \qquad |f(z) - f(z_k)| < \varepsilon.$$

Consequently, for each $k = 1, 2, \ldots, n$ and any point z on the kth chord, there is a complex number $\varepsilon_k(z)$ such that

$$|\varepsilon_k(z)| < \varepsilon$$

independently of z and such that

$$f(z) = f(z_k) + \varepsilon_k(z).$$

4. Finally, we note that the points of the partition P form a polygon, call it Π, as one joins z_0 to z_1, z_1 to z_2, $\ldots$, z_{n-1} to z_n and z_n to z_0.

In items 1 to 4, we now have all the basic machinery needed to complete the Proof. By Case 2 of the proof, we have

$$\int_\Pi = 0$$

and hence the proof will be complete if we prove that

$$\left| \int_C - \int_\Pi \right| \to 0$$

as the partition P becomes arbitrarily fine, i.e., as $\mu \to 0$.

To that end we have the following (for convenience, let $z_n = z_0$):

$$\int_\Pi f(z)\, dz = \sum_{k=1}^{n} \int_{z_{k-1}}^{z_k} f(z)\, dz$$

$$= \sum_{k=1}^{n} \int_{z_{k-1}}^{z_k} [f(z_k) + \varepsilon_k(z)]\, dz$$

$$= \sum_{k=1}^{n} \int_{z_{k-1}}^{z_k} f(z_k)\, dz + \sum_{k=1}^{n} \int_{z_{k-1}}^{z_k} \varepsilon_k(z)\, dz$$

$$= \sum_{k=1}^{n} f(z_k)(z_k - z_{k-1}) + \sum_{k=1}^{n} \int_{z_{k-1}}^{z_k} \varepsilon_k(z)\, dz$$

$$= S_n + \sum_{k=1}^{n} \int_{z_{k-1}}^{z_k} \varepsilon_k(z)\, dz.$$

Thus, from the first and last expressions in the preceding development, we have

$$\left| \int_{\Pi} - S_n \right| \leq \sum_{k=1}^{n} \left| \int_{z_{k-1}}^{z_k} \varepsilon_k(z)\, dz \right|$$

$$\leq \sum_{k=1}^{n} |\varepsilon_k(z)|\,|z_k - z_{k-1}|$$

$$< \varepsilon \sum_{k=1}^{n} |z_k - z_{k-1}|$$

$$= \varepsilon |\Pi|,$$

where $|\Pi|$ denotes the length of the perimeter of the polygon Π, which is a finite number. Finally, using the last relation in conjunction with item 3(a), we have

$$\left| \int_C - \int_{\Pi} \right| \leq \left| \int_C - S_n \right| + \left| \int_{\Pi} - S_n \right|$$

$$< \varepsilon + |\Pi|\varepsilon$$

$$= (|\Pi| + 1)\varepsilon.$$

Since ε is arbitrary and since $(|\Pi| + 1)$ is a finite quantity,

$$\left| \int_C - \int_{\Pi} \right| = 0.$$

Therefore,

$$\int_C f(z)\, dz = \int_{\Pi} f(z)\, dz = 0,$$

and the proof is complete.

CHAPTER 6
Complex Power Series

Section 23   Brief introduction to sequences and series of complex numbers; limit, convergence, and divergence of sequences and series. Brief review of series of real numbers, convergence tests, and basic series from calculus.

Section 24   Fundamental concepts of power series. Radius and circle of convergence. The ratio and root tests.

Section 25   This section includes four results of theoretical nature: Within its circle of convergence, a power series (1) converges uniformly; (2) converges to a continuous function; (3) can be integrated term by term; and (4) converges to an analytic function and can be differentiated term by term.

Section 26   Taylor series expansion of an analytic function. The results of Section 25 are applied to operate with power series. Term by term differentiation and integration. Rational operations on series. Substitution in power series expansions.

Our primary objective in this chapter is to introduce, discuss, and establish a very intimate relation which exists between convergent power series, on the one hand, and analytic functions, on the other. The relation to which we are referring is twofold and can be simply described as follows : *Every power series with a nonzero radius of convergence represents an analytic function and every analytic function can be represented by a convergent power series.* The discussion of this fundamental connection begins with Section 24, while its direct or indirect implications influence most of our work for the remainder of this book.

In Section 23, the reader is introduced to sequences and series of complex numbers. The treatment of these two basic concepts is carried out only to the extent that a firm basis is formed for the subsequent discussion, and no in-depth study of these topics is given here. Directly or indirectly, the reader's knowledge of sequences and series from calculus is used to some degree ; for, as we shall see shortly, the basic notions of convergence, divergence, and limit, as they apply to sequences and series of complex numbers, are essentially the same as in the case of real numbers.

Section 23
Sequences and Series of Complex Numbers

A **sequence** of complex numbers is a function that assigns to each positive integer n a complex number. Thus, if f is such a function, then its values may be denoted

$$f(1), f(2), \ldots, f(n), \ldots .$$

However, since the domain of every such function is the set of positive integers, we simplify the notation and use the customary notation

$$\{z_n\} = \{z_1, z_2, z_3, \ldots, z_n, \ldots\}. \tag{1}$$

For example, the sequence

$$\{i, -1, -i, 1, i, -1, -i, 1, i, \ldots\}$$

is a convenient representation of the function that assigns to the positive integers

$$1, 2, 3, \ldots$$

the complex numbers

$$i, i^2, i^3, \ldots,$$

respectively; as usual, the sequence will be denoted, briefly, $\{i^n\}$.

The numbers $z_1, z_2, \ldots$, in (1) are called the **terms** of the sequence; in particular, the term z_n is called the **general term** (or, the **nth term**) of the sequence. Two sequences $\{z_n\}$ and $\{w_n\}$ are said to be **equal** if and only if their respective terms are equal:

$$z_n = w_n, \qquad \text{for all } n = 1, 2, 3, \ldots.$$

A sequence $\{z_n\}$ in which all the terms are one and the same number, i.e., $z_k = z_{k+1}$ for all $k = 1, 2, \ldots$, is called a **constant sequence**.

As in the case of sequences of real numbers, the primary question concerning a given sequence is whether the sequence converges or diverges. A sequence $\{z_n\}$ is called **convergent** provided that a number Z exists whose *every neighborhood* contains all but finitely many terms of the sequence. If such a Z exists, we write

$$\lim_{n \to \infty} z_n = Z$$

and we say that the sequence converges to the **limit** Z, or that Z is the limit of $\{z_n\}$. If a sequence does not converge, it is called **divergent.** In more precise terms, we may define convergence of a sequence as follows:

> A sequence $\{z_n\}$ is **convergent** if and only if there is a number Z with the following property: given any $N(Z, \varepsilon)$, there is an integer M (which usually depends on the size of ε) such that, whenever $n > M$, z_n is in $N(Z, \varepsilon)$.

See Figure 6.1. The reader should compare the above definition with that of the limit of a function (p. 43). He will thus note that the two definitions are

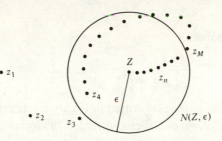

FIGURE 6.1 LIMIT OF A SEQUENCE

essentially the same, a fact that should not be surprising, since a sequence is but a function.

In view of the last remark, it is easy to see that the following two theorems are essentially restatements of Theorems 2.1 and 2.2, adapted to the case of a function whose domain is the set of positive integers, i.e., a sequence.

Theorem 6.1   (*Uniqueness of Limit of a Sequence*)
If a sequence converges, then it has a unique limit.

Proof:
See Exercise 23.14.

Theorem 6.2
Consider the sequence $\{z_n\}$ and suppose that $z_n = x_n + iy_n$, for $n = 1, 2, \ldots$.
Then, as $n \to \infty$,

$$\lim z_n = a + ib$$

if and only if

$$\lim x_n = a \qquad and \qquad \lim y_n = b.$$

Proof:
See Exercise 23.15.

The notions defined above are illustrated in the examples that follow.

EXAMPLE 1

The sequence $\{1^n\} = \{1, 1, 1, \ldots, 1, \ldots\}$ is a constant sequence and it converges trivially to $Z = 1$, since every neighborhood of 1 contains all the terms of the sequence.

EXAMPLE 2

Consider the sequence

$$\left\{\frac{i^n}{n}\right\} = \left\{i, -\frac{1}{2}, -\frac{i}{3}, \frac{1}{4}, \cdots \right\}$$

The first few of its terms are plotted in Figure 6.2. Intuitively, it is very easy to see that, as n becomes larger and larger, the terms of the sequence close in on the origin as they describe a "discrete spiral" in a counterclockwise rotation; hence the sequence converges. One may argue formally that every neighborhood of $Z = 0$ contains all but a finite number of the terms of the sequence. For, given any $N(0, \varepsilon)$, it is easy to find a positive integer M such that $1/M < \varepsilon$. Then, for all $n \geq M$, a_n is in $N(0, \varepsilon)$, which by definition, implies that $Z = 0$ is the limit of the sequence.

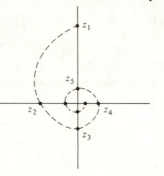

FIGURE 6.2 EXAMPLE 2

EXAMPLE 3

The sequence $\{(-1)^n + i\}$ is divergent. Its terms alternate infinitely often between the numbers $1 + i$ and $-1 + i$. Clearly, neither one of these two numbers (nor any other number) can be the limit of this sequence; we may justify this assertion as follows. Given *any point* in the plane, one can find a small enough neighborhood of that point that will not include either $1 + i$ or $-1 + i$ in its interior. But this, in turn, means that an infinite number of terms of the sequence lie outside that neighborhood (justify this!) and, therefore, the given point cannot be the limit of the sequence. Since this is true for *any point* in the plane, the sequence has no limit and hence diverges.

EXAMPLE 4

The sequence $\{(ni)^3\} = \{-i, -8i, -27i, \ldots, -n^3 i, \ldots\}$ is easily seen to diverge. However, unlike the divergent sequence of the preceding example, here the sequence diverges because its terms become arbitrarily large in magnitude and hence cannot approach any limit.

REMARK

Examples 3 and 4 illustrate the two most common types of divergent sequences: (1) the case in which a sequence diverges because its terms oscillate between two (or more) points, and (2) the case in which divergence is due to the fact that, as n becomes larger and larger, the terms increase in absolute value beyond any bound.

We close our treatment of complex sequences with a fundamental theorem, due to Cauchy, that characterizes convergent sequences. In order to appreciate one of the many useful aspects of this theorem, we recall the definition of a convergent sequence (p. 232) and we note that, according to the definition, before we can discuss convergence of a sequence we must have a point Z at our disposal that will play the role of a candidate for the limit of the sequence. Then and only then can we proceed to determine whether or not Z is indeed the limit. The following theorem, known as the **Cauchy convergence principle**, allows one to prove convergence of a sequence without knowledge of the limit to which the sequence converges. In informal terms, the theorem asserts that a sequence converges provided its terms, except possibly the first M of them, stay within an arbitrarily small distance of each other. More precisely, we have the following.

Theorem 6.3 *(Cauchy Convergence Principle)*
A sequence $\{z_n\}$ is convergent if and only if for any $\varepsilon > 0$ there is a positive integer M (which, in general, depends on ε) such that $|z_m - z_n| < \varepsilon$, for all $m > n > M$.

We omit the proof of this theorem. The reader interested in the proof is referred to Knopp, *Elements of the Theory of Functions* (New York: Dover, 1952), p. 73.

Next, we turn our attention to series of complex numbers. Let $\{z_n\}$ be a sequence and, from it, generate another sequence,

$$\{S_1, S_2, S_3, \ldots, S_n, \ldots\},$$

whose terms are defined as follows:

$$S_1 = z_1$$

$$S_2 = z_1 + z_2$$

$$S_3 = z_1 + z_2 + z_3$$

$$\vdots$$

$$S_n = z_1 + z_2 + \cdots + z_{n-1} + z_n$$

$$\vdots$$

and so forth. An examination of the process by which the terms of the sequence $\{S_n\}$ are generated shows that its limit, if it exists, must consist of the "infinite sum"

$$z_1 + z_2 + z_3 + \cdots + z_n + \cdots,$$

where all the terms of the underlying sequence $\{z_n\}$ have been added. Thus, in terms of symbols, we may write

$$\lim_{n \to \infty} S_n = \sum_{n=1}^{\infty} z_n. \tag{2}$$

The right-hand side of Equation (2) is called an **infinite series** or, just a **series** of complex numbers. The numbers $z_1, z_2, \ldots$ are called the **terms** of the series and, in particular, z_n is called the **general term**. The quantities S_1, $S_2, \ldots$ are called the **partial sums** of the series.

Again, Equation (2) says that *a series is the limit of the sequence of its partial sums*. This identification is particularly useful in defining the notion of convergence of a series. A series $\sum z_n$ will be called **convergent** if and only if the sequence $\{S_n\}$ of its partial sums converges. Indeed, if

$$\lim_{n \to \infty} S_n = S,$$

then the series converges to S, which we will then call the **sum** of the series. If a series does not converge (which will be the case if and only if the sequence of its partial sums diverges), then it is called **divergent**.

EXAMPLE 5

Consider the sequence $\left\{\dfrac{3i}{2^n}\right\} = \left\{\dfrac{3i}{2}, \dfrac{3i}{4}, \dfrac{3i}{8}, \ldots\right\}$. From its terms, let us generate another sequence $\{S_n\}$ whose terms are given by

$$S_1 = \frac{3i}{2}, \; S_2 = 3i\left(\frac{1}{2} + \frac{1}{2^2}\right), \; S_3 = 3i\left(\frac{1}{2} + \frac{1}{2^2} + \frac{1}{2^3}\right), \ldots,$$

and so forth. Clearly, the S_n are the partial sums of the series

$$\sum_{n=1}^{\infty} \frac{3i}{2^n} = 3i\left(\frac{1}{2} + \frac{1}{2^2} + \cdots + \frac{1}{2^n} + \cdots\right).$$

This series converges, since the parentheses on the right contain a geometric series whose first term is $a = \frac{1}{2}$ and whose ratio is $r = \frac{1}{2}$. Using the formula $a/(1 - r)$, which gives the sum of a geometric series, we find that the above series converges to $3i$.

In the theorems that follow, we establish a number of results that often provide convenient tools for determining convergence or divergence of a series. The first theorem finds its basis in the fact that every partial sum S_n, being nothing but a complex number, can be decomposed into its real and imaginary parts. Specifically, if each of the terms of a given series is written in the form $z_n = x_n + iy_n$, then we may write the partial sum S_n as

$$S_n = R(S_n) + I(S_n)i,$$

where

$$R(S_n) = \sum_{k=1}^{n} x_k \quad \text{and} \quad I(S_n) = \sum_{k=1}^{n} y_k.$$

In terms of this notation, Equation (2), which identifies a series with the limit of the sequence of its partial sums, may be written

$$\sum_{n=1}^{\infty} z_n = \lim_{n \to \infty} [R(S_n) + iI(S_n)].$$

Then the following criterion is an easy consequence of Theorem 6.2.

Theorem 6.4

Suppose that $\sum_{n=1}^{\infty} z_n$ is a series with partial sums

$$S_n = R(S_n) + iI(S_n).$$

Then the series converges if and only if the sequences $\{R(S_n)\}$ and $\{I(S_n)\}$ both converge.

Proof:
See Exercise 23.17.

The next theorem is a very useful tool in identifying certain types of divergent series. The customary form in which the theorem is stated and proved is as follows: *If a series $\sum z_n$ converges, then, as $n \to \infty$, $\lim z_n = 0$.* Equivalently, we have the following contrapositive form.

Theorem 6.5

Suppose that for a given series $\sum z_n$, as $n \to \infty$, $\lim z_n \neq 0$.
Then the series diverges.

Proof:
See Appendix 6(A).

The converse of the above theorem is false. A simple example illustrating this fact is the harmonic series

$$\sum_{n=1}^{\infty} \frac{1}{n}.$$

Clearly, as $n \to \infty$, $\lim (1/n) = 0$. However, it is a well-known fact that the series diverges; see the review of series of real numbers that follows. Thus, again,

$$\lim z_n = 0 \text{ does not imply that } \sum z_n \text{ converges.}$$

The last topic that we will discuss in this section concerns the concept of absolute convergence, which we define as follows: A series $\sum z_n$ is said to be **absolutely convergent** if and only if $\sum |z_n|$ converges. The property of absolutely convergent series described in the next theorem will be of use in a number of subsequent developments of theoretical interest. Its practical aspect is that, whenever applicable, it reduces the investigation of the convergence of a complex series to that of the convergence of a series of nonnegative terms.

Theorem 6.6

If a series converges absolutely, then it converges; i.e., if $\sum |z_n|$ converges, then $\sum z_n$ converges.

Proof:

See Appendix 6(A).

The converse of the above theorem is false. In other words, convergence of $\sum z_n$ does not necessarily imply convergence of $\sum |z_n|$ as the next example illustrates.

EXAMPLE 6

Consider the series $\sum_{n=1}^{\infty} \frac{(-1)^n}{n}$, which is an alternating series of real numbers. By the use of the alternating series test (see Review following this example), we find that the series converges. On the other hand, the series $\sum_{n=1}^{\infty} \left| \frac{(-1)^n}{n} \right|$ is, in fact, the harmonic series, which is known to diverge.

A series $\sum z_n$, which converges while $\sum |z_n|$ diverges, is said to be **conditionally convergent**.

BRIEF REVIEW OF SERIES OF REAL NUMBERS

Certain fundamental facts concerning series of real numbers are included at this point for review purposes, as well as for easy reference. As we pointed out earlier in this section, the notion of absolute convergence reduces, via Theorem 6.6, many problems associated with the convergence of complex series to problems involving series of real numbers. This is not difficult to see, since, for any complex series $\sum z_n$, the *absolute series* $\sum |z_n|$ is a series of nonnegative *real* numbers.

First, we list the most commonly used tests for convergence. All series involved in this review are series of real numbers.

The Ratio Test

Let $\sum u_n$ be a given series of nonnegative terms and suppose that

$$\lim_{n \to \infty} \frac{u_{n+1}}{u_n} = \rho.$$

Then
1. The series converges if $\rho < 1$.
2. The series diverges if $\rho > 1$.
3. The series may converge or may diverge if $\rho = 1$.

The Root Test

Let $\sum u_n$ be a given series of nonnegative terms and suppose that

$$\lim_{n \to \infty} \sqrt[n]{u_n} = \rho.$$

Then
1. The series converges if $\rho < 1$.
2. The series diverges if $\rho > 1$.
3. The test is inconclusive if $\rho = 1$.

The Integral Test

Given a series $\sum u_n$ of nonnegative terms, suppose that the function $y = f(x)$ is obtained by replacing n in the general term of the series by the continuous variable x. Suppose further that, for $x \geq 1$, the function $f(x)$ is decreasing. Then the series

$$\sum u_n$$

and the integral

$$\int_1^\infty f(x)\, dx$$

both converge or both diverge.

The Comparison Test

Let $\sum u_n$ be a series of nonnegative terms.
1. If another series $\sum c_n$ is known to converge and if $u_n \leq c_n$ for all n beyond a certain integer, then $\sum u_n$ also converges.
2. If another *positive* series $\sum d_n$ is known to diverge and if $u_n \geq d_n$ for all n beyond a certain integer, then $\sum u_n$ also diverges.

The Alternating Series Test

Given a series $\sum (-1)^n u_n$, where all $u_n \geq 0$, suppose that
1. $\lim u_n = 0$, as $n \to \infty$.
2. $u_{n+1} \leq u_n$, for all n larger than some integer M.
Then, the given series converges.

We conclude this brief review with a set of exercises involving some of the most common types of series of real numbers to which we shall have occasion to refer in our work with complex series.

1. Verify that the **geometric series** $\sum_{n=0}^{\infty} ar^n$ converges if $|r| < 1$ and diverges if $|r| \geq 1$.

   HINT: Multiply the partial sum $s_n = a + ar + \cdots + ar^{n-1}$ by r. Then, find the difference $s_n - rs_n$ and derive the formula

   $$s_n = \frac{a(1 - r^n)}{1 - r}.$$

   Finally, take limits as $n \to \infty$.
2. Use the integral test to verify that the **p-series** $\sum_{n=1}^{\infty} (1/n^p)$ converges for $p > 1$ and diverges for $0 < p \leq 1$.
3. Note that the **harmonic series** $\sum_{n=1}^{\infty} (1/n)$ is a special case of the p-series and determine whether it converges or diverges.
4. Use the ratio test to verify that $\sum_{n=0}^{\infty} (1/n!)$ converges.
5. Recall the Maclaurin series expansion $e^x = \sum_{n=0}^{\infty} \frac{x^n}{n!}$ and use it to verify

   that the series of the preceding exercise converges to the number e.
6. Use the method suggested in the preceding exercise to verify that

   $$\sum_{n=0}^{\infty} \frac{(-1)^n}{n!} = \frac{1}{e}.$$

7. Use the alternating series test to show that the series

   $$\sum_{n=1}^{\infty} \frac{(-1)^n}{n}$$

   converges.

EXERCISE 23

A

In Exercises 23.1–23.6, write out the first few terms of the given sequence and plot them in the complex plane. Then, determine whether the sequence converges or diverges.

23.1. $\{(2i)^n\}$.

23.2. $\{2i^n\}$.

23.3. $\left\{\dfrac{1}{n(1-i)^n}\right\}$.

23.4. $\left\{\dfrac{2n-i}{n+2i}\right\}$.

23.5. $\left\{n-\dfrac{i}{n}\right\}$.

23.6. $\{|z|^n\}$, for all z.

In Exercises 23.7–23.12, examine each of the given series for absolute convergence, conditional convergence, or divergence.

23.7. $\sum \dfrac{2i}{n^3}$.

23.8. $\sum \dfrac{i^{2n}}{n}$.

23.9. $\sum i^n$.

23.10. $\sum \dfrac{i^{4n}}{(2n)!}$.

23.11. $\sum \dfrac{(1-i)^{2n}}{n!}$.

23.12. $\sum \left(\dfrac{1}{n+i}-\dfrac{1}{n+1+i}\right)$.

B

23.13. The series

$$\sum_1^\infty \left(\frac{i}{2}\right)^n$$

converges absolutely. (Verify this.) Hence, according to Theorem 6.6, it converges. Assuming that all manipulations suggested below are justifiable, complete the various parts of this exercise to show that the sum of this series is $(-1+2i)/5$.

(a) Rearrange the terms of the series to write it in the form

$$\sum_1^\infty \frac{1}{2^{4n}}-\sum_1^\infty \frac{1}{2^{4n-2}}+i\sum_1^\infty \frac{1}{2^{4n-3}}-i\sum_1^\infty \frac{1}{2^{4n-1}}.$$

(b) Note that each of the above four series is a geometric series; find their respective sums using the formula $a/(1-r)$.

(c) Finish the problem by performing the necessary arithmetic.

23.14. Imitate the proof of Theorem 2.1 to effect a proof of Theorem 6.1.

23.15. Imitate the proof of Theorem 2.2 to effect a proof of Theorem 6.2.

C

23.16. By completing the three parts of this exercise, prove that if $\{z_n\}$ converges, then it is bounded; i.e., for some positive number B, $|z_n| \le B$ for all n.

    (a) Argue that, by definition, there is a point Z in the plane such that $|z_n - Z| < 1$ for all but finitely many z_n's.

    (b) From (a) conclude that $|z_n| \le 1 + |Z|$ for all but finitely many z_n's.

    (c) Accommodate the finitely many z_n's that are left outside the circle $|z| = |Z| + 1$ by increasing its radius to a sufficiently large B, so that $|z_n| \le B$ for all z_n.

23.17. Review the paragraph preceding Theorem 6.4, and then prove that theorem as a consequence of Theorem 6.2.

23.18. Show that if $\sum_{n=1}^{\infty} z_n$ converges, then, for any integer N, $\sum_{n=N}^{\infty} z_n$ also converges.

23.19. Show that if $\sum_{n=1}^{\infty} z_n$ converges, then, for any constant c, $\sum_{n=1}^{\infty} (cz_n)$ also converges.

23.20. Prove or disprove that the converse of Exercise 23.18 is true.

Section 24
Power Series

A **power series** is an infinite series of the form

$$\sum_{n=0}^{\infty} a_n(z-c)^n = a_0 + a_1(z-c) + a_2(z-c)^2 + \cdots + a_n(z-c)^n + \cdots,$$

where z is a complex variable, $a_0, a_1, \ldots$ are complex constants called the **coefficients** of the series, and c is an arbitrary but fixed complex number called the **center** of the series.

Most of our work in this and the next two sections will involve functions that can be expanded in a Taylor series; in connection with this, the basic notions associated with such series, as we know them from the calculus of real functions, will be recalled and adapted to complex functions. For example, in connection with the notion of convergence of a power series, once again we shall concern ourselves with the "radius of convergence," although instead of the "interval of convergence" we shall now talk about the "circle of convergence."

The concept of convergence of a power series is defined as follows. The power series

$$\sum_{n=0}^{\infty} a_n(z-c)^n \tag{1}$$

is said to be **convergent at a point** $z = z_0$ if and only if the series

$$\sum_{n=0}^{\infty} a_n(z_0 - c)^n,$$

which is a series of complex numbers, converges. Otherwise, the series in expression (1) will be said to be **divergent** at z_0. If the power series in (1) is convergent at every point of a set S, we will say that the series is **convergent on** S and if it diverges at every point of S, we will say that it is **divergent on** S.

EXAMPLE 1

Consider the power series $\sum_{n=1}^{\infty} \dfrac{z^n}{n^2}$.

1. At the point $z = i$ we obtain the series $\sum_{n=1}^{\infty} \dfrac{i^n}{n^2}$, which can be shown to be absolutely convergent (it is a p-series) and hence convergent. It follows that the given power series converges at $z = i$.

2. At $z = 3$ we obtain the series $\sum_{n=1}^{\infty} \dfrac{3^n}{n^2}$, which, by use of the ratio test, is found to diverge. Hence the given power series diverges at $z = 3$. We will see shortly that, in fact, the given series converges on the closed unit disk $|z| \leq 1$ and diverges for all other z.

We begin now the systematic study of power series with the investigation of their behavior with respect to convergence or divergence. In so doing, we choose to follow two distinct developments which are complementary in the sense that whenever either one demonstrates a shortcoming, the other one alleviates the problem.

The Theoretical Approach

This development supplies the foundation on which we shall base the discussion of the "practical approach." The main result here is the Cauchy–Hadamard theorem, which answers the question of convergence or divergence of a power series in all cases. However, from a practical point of view, this approach has a definite shortcoming in that it does not lend itself readily to problem solving unless its user is quite familiar with the concept of "limit superior" and some of its basic properties. We discuss this approach along with the necessary preliminaries in Appendix 6(B).

The Practical Approach

In this development, we discuss and use two relatively simple tools which lend themselves, quite readily, to the solution of most of the problems with which we shall concern ourselves. We are referring to the familiar *ratio test* and *root test* as they are employed in finding the radius of convergence of a power series.

The Cauchy–Hadamard theorem (p. 277) establishes the following facts, on which we base the present development.

Associated with each power series

$$\sum_{n=0}^{\infty} a_n(z - c)^n$$

there is a unique number ρ, $0 \le \rho \le \infty$, called the **radius of convergence** of the series, with respect to which the series has the following properties:

1. If $\rho = 0$, then the power series converges only at the point $z = c$, i.e., its center, and diverges for all other z.
2. If $0 < \rho < \infty$, then the power series converges absolutely, and hence converges, for all z with $|z - c| < \rho$ and diverges for all z with $|z - c| > \rho$.
3. If $\rho = \infty$, then the series converges absolutely, and hence converges, for every z such that $|z - c| < \infty$; i.e., for all finite z.

The circle $|z - c| = \rho$ is called the **circle of convergence** of the power series. See Figure 6.3.

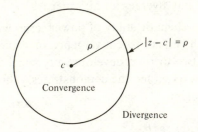

FIGURE 6.3 CIRCLE OF CONVERGENCE

The reader has probably noted already that, in the above discussion, nothing is said about convergence or divergence of the series at points *on* the circle of convergence. As we shall see in the examples that follow, the reason for this is that there are series which converge at *every point* of their circle of convergence, there are others which converge at *no such point*, and still others which converge at *some but not all* of the points of their circle of convergence.

With these facts at our disposal, we are now in a position to utilize the formulas of the next two theorems in order to determine where a given power series converges and where it diverges. Note, however, that the truth of these two theorems—and, hence, their usefulness—depends on the hypothesis that a certain limit exists; the fact that the limit is not guaranteed to exist for every series constitutes the weakness of this approach to which we alluded earlier.

Theorem 6.7

Suppose that, for a given series $\sum_{n=0}^{\infty} a_n(z - c)^n$, the

$$\lim_{n \to \infty} \frac{|a_n|}{|a_{n+1}|}$$

exists and is equal to ρ, where $0 \le \rho \le \infty$.
Then ρ is the radius of convergence of the given series.

Proof:
See Exercise 24.17.

Theorem 6.8

Suppose that, for a given series $\sum_{n=0}^{\infty} a_n(z - c)^n$, the

$$\lim_{n \to \infty} \frac{1}{|a_n|^{1/n}}$$

exists and is equal to ρ, where $0 \le \rho \le \infty$.
Then ρ is the radius of convergence of the given series.

Proof:
See Exercise 24.18.

The use of the above two theorems is illustrated in the examples that follow.

EXAMPLE 2

Consider the series $\sum_{n=0}^{\infty} z^n$.

Here $a_n = 1$ for all n. Using the formula from either Theorem 6.7 or Theorem 6.8, one easily finds $\rho = 1$. Hence the series converges absolutely for $|z| < 1$ and diverges for $|z| > 1$. We now show that the series diverges at every point of its circle of convergence $|z| = 1$. Clearly, any such point can be written in the form $z = e^{it}$, from which $z^n = e^{nit}$ and hence, for any point on the circle, the given series becomes

$$\sum_{n=0}^{\infty} e^{nit}.$$

Now, since, as $n \to \infty$, $\lim e^{nit} \neq 0$, it follows from Theorem 6.5 that the last series diverges. Therefore, the given series $\sum z^n$ diverges at every point of its circle of convergence.

EXAMPLE 3

Consider the power series $\sum_{n=1}^{\infty} z^n/n^3$.

By use of the formula of Theorem 6.7, we find that

$$\rho = \lim_{n \to \infty} \frac{|a_n|}{|a_{n+1}|} = \lim_{n \to \infty} \frac{1/n^3}{1/(n+1)^3} = 1.$$

Therefore, since the center of the series is at $c = 0$, the series converges on $|z| < 1$ and diverges on $|z| > 1$. Concerning the points on the circle of convergence, $|z| = 1$, we note that for any such point

$$\sum \left| \frac{z^n}{n^3} \right| = \sum \frac{|z|^n}{n^3} = \sum \frac{1}{n^3}.$$

But this is a p-series with $p > 1$; hence it is convergent. We conclude then that the given series converges at every point on its circle of convergence.

Combining the above results, we see that the given power series converges for $|z| \leq 1$ and diverges for $|z| > 1$.

EXAMPLE 4

Working as in Example 3, we find that the power series $\sum \frac{z^n}{n}$ has a radius of convergence $\rho = 1$.

Concerning the points on the circle of convergence, $|z| = 1$, we note that if we let $z = 1$, the series becomes $\sum 1/n$, which is the harmonic series, known to diverge. On the other hand, if we let $z = -1$, we obtain the series $\sum (-1)^n/n$, which can be shown to converge by use of the alternating series test. It follows, then, that no general assertion can be made concerning the points on the circle of convergence. Therefore, one may only assert that the given power series converges on $|z| < 1$ and diverges on $|z| > 1$.

EXAMPLE 5

Consider the series

$$\sum_{n=1}^{\infty} \left[\frac{n+1}{n} \right]^{n^2} (z-1)^n.$$

By use of the formula of Theorem 6.8, we find that

$$\rho = \lim_{n \to \infty} \frac{1}{\left[\left(\frac{n+1}{n}\right)^{n^2}\right]^{1/n}} = \lim_{n \to \infty} \frac{1}{\left(\frac{n+1}{n}\right)^n} = \frac{1}{\lim_{n \to \infty} \left(\frac{n+1}{n}\right)^n} = \frac{1}{e}.$$

It follows that the circle of convergence of the given power series is

$$C : |z - 1| = e^{-1},$$

and hence the series converges in Int (C) and diverges in Ext (C).

EXAMPLE 6

Using Theorem 6.7, we find that the radius of convergence of the power series $\sum n!(z + i)^n$ is

$$\rho = \lim_{n \to \infty} \frac{n!}{(n+1)!} = 0.$$

Therefore, the above power series converges only at its center $c = -i$ and diverges everywhere else. See Exercise 24.12.

EXAMPLE 7

Again, by Theorem 6.7, we find that the power series $\sum \dfrac{z^n}{(2n)!}$ has a radius of convergence $\rho = \infty$. Therefore, the series converges for all z. See Exercise 24.12.

EXERCISE 24

A

The general term of a power series is given in each of Exercises 24.1–24.10. In each case, find the radius of convergence and specify the circle of convergence.

24.1. $z^n/2^n$.

24.2. $e^n(z + 2)^n$.

24.3. $n^2 z^n$.

24.4. $(z - i)^n/3^n$.

24.5. $e^n z^n/n!$.

24.6. $e^n(z + i)^n/n$.

24.7. $2n(z + 1)^n/(2n - 1)$.

24.8. $n!(z + \pi i)^n/2^n$.

24.9. $(2n)! z^n/(n!)^2$.

24.10. $n(n + 1)(z + e)^n/(n^2 - 2)$.

24.11. Examine the series of Exercise 24.1 for convergence at each of the points 2, -2, $2i$, and $-2i$ by direct substitution.

24.12. Carry out the details of the limiting process in Example 6 and in Example 7 of this section.

24.13. Consider the power series $\sum\limits_{n=0}^{\infty} \dfrac{z^n}{2^n}$. Differentiate the series term by term to obtain the series $\sum\limits_{n=0}^{\infty} \dfrac{nz^{n-1}}{2^n}$. Integrate the given series term by term from $z = 0$ to $z = \zeta$ to obtain the series $\sum\limits_{n=0}^{\infty} \dfrac{\zeta^{n+1}}{2^n(n+1)}$. Verify that each of these three power series has a radius of convergence equal to 2. (See Exercises 24.14 and 24.16.)

B

24.14. Consider the three power series

$$\sum a_n z^n, \qquad \sum \frac{a_n}{n+1} z^{n+1}, \qquad \sum a_n n z^{n-1}.$$

Use the formula of Theorem 6.7 to verify that all three series have the same radius of convergence, assuming that the limit in the hypothesis of the theorem exists for the first of the above series.

24.15. Verify that if the radius of convergence of the series $\sum a_n z^n$ is ρ, then the radius of convergence of $\sum a_n z^{2n}$ is $\sqrt{\rho}$.

C

24.16. The fact described in Exercise 24.14 will be discussed in detail in Sections 25 and 26. As a prelude to that discussion, verify that each of the three series in that exercise can be obtained from one of the other two either by differentiation or by integration. Attempt a formulation of an appropriate theorem.

24.17. Prove Theorem 6.7 using the ratio test (p. 239).

24.18. Prove Theorem 6.8 using the root test (p. 239).

Section 25
Power Series as Analytic Functions

This section consists of four theorems that are devoted to the proof of the first part of a twofold relation between power series and analytic functions that was mentioned in the opening remarks of the present chapter. Specifically, we prove in this section that

> if a power series converges at every point of a circular region R, then it converges to a function that is analytic at least in R.

Put in different terms,

> a power series with a positive radius of convergence can be represented by an analytic function.

The proof of this fact is accomplished via a sequence of four theorems whose assertions we may outline, briefly and informally, as follows: Any power series having a circle of convergence C:

1. Converges uniformly on and inside any circle in Int (C).
2. Converges to a function continuous at every z in Int (C).
3. Can be integrated term by term along any path in Int (C).
4. Converges to a function analytic at every z in Int (C) and can be differentiated term by term at any such z.

These results certainly demonstrate the extremely desirable behavior of power series within their circle of convergence.

For simplicity of notation and without loss of generality, in this section, we shall restrict our entire discussion to the cases of power series with center at $c = 0$. The entire development is readily extendable to the general case by the substitution of $(z - c)^n$ in place of z^n. Also, in order that we view the entire development without interruption, we will postpone discussion of examples until the next section. The reader who wishes to postpone the study of the proofs may proceed to the next section after a careful study of the statements of Theorems 6.11 and 6.12.

The first result of this section makes use of the concept of "uniform convergence," which we have not encountered heretofore; we define it as follows. The power series $\sum a_n z^n$ is said to **converge uniformly on a set** E if and only if, given any $\varepsilon > 0$, there is an integer M such that

$$\left| \sum_{k=0}^{\infty} a_k z^k - \sum_{k=0}^{n-1} a_k z^k \right| < \varepsilon, \qquad \text{for all } n > M \text{ and every } z \text{ in } E.$$

It is not difficult to see that uniform convergence of a series is stronger than ordinary convergence, at least in the sense that uniform convergence implies convergence of a series, but not vice versa. Note that, in the above definition, the fundamental inequality can also be written

$$\left| \sum_{k=n}^{\infty} a_k z^k \right| < \varepsilon.$$

Theorem 6.9

Suppose that the power series $\sum_{n=0}^{\infty} a_n z^n$ has a radius of convergence $\rho > 0$. Then the series converges uniformly on and inside any circle $C: |z| = r < \rho$.

Proof:

According to the definition of uniform convergence, the theorem will have been proved if, for any $\varepsilon > 0$, we can produce an integer M such that

$$\left| \sum_{k=n}^{\infty} a_k z^k \right| < \varepsilon, \qquad \text{for all } n > M \text{ and all } z \text{ such that } |z| \leq r.$$

Let C be any circle as prescribed by the theorem and draw a circle K concentric with C and having radius R such that $0 < r < R < \rho$; see Figure 6.4. Let $\alpha = r/R$ and note that $0 < \alpha < 1$.

Now, by hypothesis, the series converges for all z such that $|z| = R$. Hence, by Theorem 6.5, its terms become arbitrarily small in absolute value. In particular, for sufficiently large n and for any z on K,

$$|a_n z^n| = |a_n| R^n < 1 - \alpha. \tag{1}$$

On the other hand, since $0 < \alpha < 1$, one may find n large enough so that for any given $\varepsilon > 0$,

$$\alpha^n < \varepsilon. \tag{2}$$

Take M sufficiently large so that (1) and (2) will be satisfied simultaneously for all $n > M$, and consider any z on or inside C. Then, for $m \geq 1$ and $n > M$, we have

$$\left| \sum_{k=n}^{n+m} a_k z^k \right| \leq \sum_{k=n}^{n+m} |a_k|\, |z|^k \leq \sum_{k=n}^{n+m} |a_k| r^k$$

$$= \sum_{k=n}^{n+m} |a_k| \alpha^k R^k < (1 - \alpha) \sum_{k=n}^{n+m} \alpha^k$$

$$= \alpha^n - \alpha^{n+m+1} = \alpha^n (1 - \alpha^{m+1})$$

$$< \varepsilon(1 - \alpha^{m+1}).$$

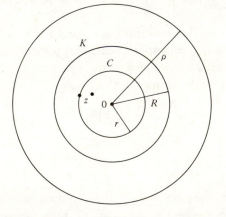

FIGURE 6.4 THEOREM 6.9

Finally, letting $m \to \infty$ in the first and last expressions above, we note that $\alpha^{m+1} \to 0$; hence

$$\left| \sum_{k=n}^{\infty} a_k z^k \right| < \varepsilon,$$

and the theorem is proved.

Theorem 6.10

Suppose that the power series $\sum_{n=0}^{\infty} a_n z^n$ has a circle of convergence C of radius $\rho > 0$.

Then the series converges to a function $f(z)$, which is continuous at every z in Int (C).

Proof:

According to Theorem 6.9, the given series converges uniformly at every point z in Int (C). We may thus define the function to be

$$f(z) = \sum_{n=0}^{\infty} a_n z^n, \qquad \text{for every } z \text{ in Int } (C).$$

That this is indeed a single-valued function follows from the fact that given any such point z, the series, and hence $f(z)$, yields a unique and finite number due to convergence. Hence it remains to be shown that $f(z)$ is continuous on Int (C).

So, let ζ be any point in Int (C). Then (see Figure 6.5) one may find a circle K, concentric with C and having radius r such that $0 < |\zeta| < r < \rho$. Now, let $\varepsilon > 0$ be chosen at random. Then, since the series converges uniformly on and inside K, an integer M exists such that

$$\left| f(z) - \sum_{k=0}^{n} a_k z^k \right| < \frac{\varepsilon}{3}, \qquad \text{for all } n > M \text{ and all } z \text{ on or in } K.$$

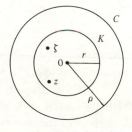

Figure 6.5 Theorem 6.10

In particular, since ζ is one such z,

$$\left| f(\zeta) - \sum_{k=0}^{n} a_k \zeta^k \right| < \frac{\varepsilon}{3}.$$

But $\sum_{k=0}^{n} a_k z^k$ is a polynomial, hence continuous at ζ. Then, by definition,

$$\left| \sum_{k=0}^{n} a_k z^k - \sum_{k=0}^{n} a_k \zeta^k \right| < \frac{\varepsilon}{3},$$

for all z sufficiently close to ζ.

Finally,

$$|f(z) - f(\zeta)| \leq \left| f(z) - \sum_{k=0}^{n} a_k z^k \right|$$

$$+ \left| \sum_{k=0}^{n} a_k z^k - \sum_{k=0}^{n} a_k \zeta^k \right|$$

$$+ \left| \sum_{k=0}^{n} a_k \zeta^k - f(\zeta) \right|$$

$$< \frac{\varepsilon}{3} + \frac{\varepsilon}{3} + \frac{\varepsilon}{3} = \varepsilon.$$

Thus the definition of continuity of f at ζ is satisfied and the proof is complete, since ζ was arbitrarily chosen from Int (C).

Theorem 6.11

Suppose that the series $\sum_{n=0}^{\infty} a_n z^n$ has a circle of convergence C of radius $\rho > 0$.

Then the series can be integrated term by term along any path K, lying in Int (C); *i.e.,*

$$\int_K \left(\sum_{n=0}^{\infty} a_n z^n \right) dz = \sum_{n=0}^{\infty} \int_K a_n z^n \, dz.$$

Proof:

First, we note that every integral in the right-hand side of the conclusion of the theorem exists, since the integrand $a_n z^n$ is continuous everywhere and K is a path.

Second, by Theorem 6.10, the given series represents a function $f(z)$ continuous in Int (C) and, therefore, the integral in the left-hand side of the conclusion exists.

Third, by Theorem 6.9, the series converges uniformly to $f(z)$. Hence, given $\varepsilon > 0$, there is an integer M such that for all $n > M$,

$$\left| \sum_{k=0}^{n-1} a_n z^n - f(z) \right| < \varepsilon, \qquad \text{for all } z \text{ in Int } (C).$$

Now, for every z in Int (C), every $n > M$, and any path K in Int (C),

$$\int_K \left(\sum_{n=0}^{\infty} a_n z^n \right) dz = \int_K f(z) \, dz$$

$$= \int_K \left(\sum_{k=0}^{n-1} a_k z^k \right) dz + \int_K \left(f(z) - \sum_{k=0}^{n-1} a_k z^k \right) dz. \qquad (1)$$

Concerning the last two integrals above, we have

$$\int_K \left(\sum_{k=0}^{n-1} a_k z^k \right) dz = \int_K (a_0 + a_1 z + \cdots + a_n z^{n-1}) \, dz$$

$$= \sum_{k=0}^{n-1} \int_K a_k z^k \, dz, \qquad (2)$$

and by Theorem 4.5(5),

$$\left| \int_K \left(f(z) - \sum_{k=0}^{n-1} a_k z^k \right) dz \right| \le \varepsilon L, \qquad \text{where } L = \text{length of } K. \qquad (3)$$

Clearly, as $\varepsilon \to 0$, the integral in (3) tends to 0. At the same time, as $\varepsilon \to 0$, M and hence n tend to ∞ and, therefore, the finite summation of integrals in (2) becomes

$$\sum_{k=0}^{\infty} \left(\int_K a_k z^k \, dz \right).$$

Finally, substituting in (1), we have

$$\int_K \left(\sum_{n=0}^{\infty} a_n z^n \right) dz = \sum_{n=0}^{\infty} \int_K a_n z^n \, dz,$$

and the theorem follows.

Theorem 6.12

Suppose that the power series $\sum_{n=0}^{\infty} a_n z^n$ has a circle of convergence C of radius $\rho > 0$.

Then

1. The series converges to a function $f(z)$ that is analytic throughout Int (C).

2. *The derivative of* $f(z)$ *is given by* $f'(z) = \sum\limits_{n=0}^{\infty} \dfrac{d}{dz}(a_n z^n)$; *i.e., the series can be differentiated term by term inside its circle of convergence.*

3. *The derived series in part 2 converges uniformly to* $f'(z)$ *at every point on and inside any circle* T *concentric with* C *and of radius* $r < \rho$.

Proof:

1. By Theorem 6.9, the series converges uniformly to a continuous function $f(z)$. We prove that $f(z)$ is analytic at every z in Int (C).

According to Theorem 6.11,

$$\int_K f(z)\,dz = \sum_{n=0}^{\infty} \int_K a_n z^n\,dz$$

for any path K in Int (C) and, in particular, for any such K which is simple and closed. But if K is simple and closed, then every integral in the above summation is zero, since the integrand in each case is analytic everywhere. Therefore,

$$\int_K f(z)\,dz = 0,$$

and this is true for every simple closed path K in Int (C). Hence, by Morera's theorem, $f(z)$ is analytic throughout Int (C).

2. We prove now that, for any z in Int (C),

$$f'(z) = \sum_{n=0}^{\infty} \frac{d}{dz}(a_n z^n).$$

Let ζ be any point in Int (C). Then, since ζ is an interior point, a simple closed path K can be found which lies entirely in Int (C) and such that ζ is in Int (K). Then, using Theorem 5.8, we have

$$f'(\zeta) = \frac{1}{2\pi i} \int_K \frac{f(z)}{(z-\zeta)^2}\,dz$$

$$= \frac{1}{2\pi i} \int_K \left[\sum_{n=0}^{\infty} \frac{a_n z^n}{(z-\zeta)^2}\,dz \right]$$

$$= \sum_{n=0}^{\infty} \frac{1}{2\pi i} \int_K \frac{a_n z^n}{(z-\zeta)^2}\,dz*$$

$$= \sum_{n=0}^{\infty} \frac{d}{dz}(a_n \zeta^n).$$

---

* See the Remark at the end of this section.

Since ζ is arbitrary in Int (C), assertion (2) has been proved.

3. Finally, we show that the derived series just obtained converges uniformly to $f'(z)$ in Int (C). To this end, given $\varepsilon > 0$, we produce an integer M such that, for all $n > M$ and for any ξ on or inside any circle $T : |z| = r < \rho$,

$$\left| f'(\xi) - \sum_{k=0}^{n-1} \frac{d}{d\xi}(a_k \xi^k) \right| < \varepsilon.$$

So, let K be a circle $|z| = \lambda$, with $r < \lambda < \rho$. By Theorem 6.9, there is an integer M such that for $n > M$ and for any z on or inside K

$$\left| f(z) - \sum_{k=0}^{n-1} a_k z^k \right| < \varepsilon,$$

or, which is the same,

$$\left| \sum_{k=n}^{\infty} a_k z^k \right| < \varepsilon.$$

Then

$$\left| f'(\xi) - \sum_{k=0}^{n-1} \frac{d}{d\xi}(a_k \xi^k) \right| = \left| \sum_{k=n}^{\infty} \frac{d}{d\xi}(a_k \xi^k) \right|$$

$$= \left| \sum_{k=n}^{\infty} \frac{1}{2\pi i} \int_K \frac{a_k z^k}{(z-\xi)^2} \, dz \right|$$

$$= \frac{1}{2\pi} \left| \int_K \frac{\sum\limits_{k=n}^{\infty} a_k z^k}{(z-\xi)^2} \, dz \right|$$

$$\leq \frac{1}{2\pi} \frac{\left| \sum\limits_{k=n}^{\infty} a_k z^k \right|}{|z-\xi|^2} L, \qquad L = \text{length of } K$$

$$< \frac{1}{2\pi} \frac{\varepsilon}{(\lambda-r)^2} L$$

$$= (\text{constant}) \cdot \varepsilon,$$

where, in obtaining the last inequality, we used the fact that, for z on K,

$$\lambda = |z| > r \geq |\xi| \quad \text{and hence} \quad |z-\xi| \geq |z| - |\xi| \geq \lambda - r > 0.$$

Since ε is arbitrary, the uniform convergence of the derived series has been established. This proves the last assertion of the theorem and the proof is complete.

REMARK

The justification of the step marked by an asterisk in the proof of Theorem 6.12 is not supplied in this book, because the development and the proof of the necessary result along with its preliminaries is beyond the scope of this book. We state, however, the said result, which will allow the interchange of summation and integration at that point of the proof. The discussion and the proof of this result can be found in Knopp *Theory of Functions*, Part I (New York: Dover, 1945), Chap. 6.

Theorem
Consider an infinite series of continuous functions

$$\sum_{n=0}^{\infty} f_n(z)$$

and suppose that the series converges uniformly along a path K. Then

$$\int_K \left(\sum_{n=0}^{\infty} f_n(z) \right) dz = \sum_{n=0}^{\infty} \int_K f_n(z)\, dz.$$

As we remarked early in this section, we will postpone discussion of examples on the material of this section until we develop at least part of the next section. Similarly, exercises on this section will be included in Exercise 26 of the next section.

Section 26
Analytic Functions as Power Series

In the preceding section we saw that a power series with a nonzero radius of convergence can be represented by an analytic function $f(z)$ at every point within its circle of convergence. Moreover, the series can be differentiated term by term and the derived series converges to $f'(z)$. Finally, we saw that the power series can be integrated term by term along any path lying entirely within its circle of convergence.

In the present section we discuss the fact that every analytic function can be represented by a power series. Specifically, we have the following.

Theorem 6.13.   (*Taylor's Theorem*)
Suppose that $f(z)$ is analytic at a point c in the plane.
Then there is a power series

$$\sum_{n=0}^{\infty} a_n(z - c)^n$$

whose coefficients are given by the formula

$$a_n = \frac{f^{(n)}(c)}{n!}, \qquad n = 0, 1, 2, \dots,$$

and which converges to $f(\zeta)$ for every ζ in every neighborhood of c throughout which f is analytic:

$$f(\zeta) = \sum_{n=0}^{\infty} \frac{f^{(n)}(c)}{n!}(\zeta - c)^n.$$

Proof:
See Appendix 6(A).

The series of the above theorem is called the **Taylor series** of f at c; if $c = 0$, then the series is referred to as the **Maclaurin series** of f. The formula by which the coefficients are found will be called the **Taylor formula.**

The radius of convergence of the Taylor series may be found by the methods of Section 24. However, this is hardly necessary, since it turns out that *the radius is actually the distance between the center c of the series and the singularity of f (if such exists) which lies closest to c.* For example, we will see shortly that if we expand the function

$$f(z) = \frac{1}{1 + z}$$

in a Taylor series with center at $c = i$, then its radius of convergence will be $\rho = \sqrt{2}$, which is precisely the distance between the center and the point $z = -1$, which is the only singularity of f. By the same token, if a function is entire, then its Taylor series will have an infinite radius of convergence.

It is proved in Chapter 9 that the power series expansion of an analytic function is unique in the sense that if two series

$$\sum_{n=0}^{\infty} a_n(z - c)^n \qquad \text{and} \qquad \sum_{n=0}^{\infty} b_n(z - c)^n$$

are obtained for one and the same function, then $a_n = b_n$, for all $n \geq 0$.

EXAMPLE 1

In this example we develop the power series expansions of three functions, giving only an outline of the process. The reader should supply the necessary details and justifications.

1. Find the Maclaurin series for $f(z) = e^z$. In this case, it is easy to see that

$$f^{(n)}(z) = e^z, \qquad \text{hence} \qquad f^{(n)}(0) = 1$$

and, therefore,

$$a_n = \frac{1}{n!}, \qquad \text{for all } n = 0, 1, 2, \ldots .$$

It follows that

$$e^z = \sum_{n=0}^{\infty} \frac{z^n}{n!}, \qquad \rho = \infty.$$

2. Consider the function

$$f(z) = \frac{1}{1 - z}.$$

We find

$$f^{(n)}(z) = \frac{n!}{(1 - z)^{n+1}}.$$

Now, if the Maclaurin series of f is sought, we calculate

$$f^{(n)}(0) = n!$$

from which we obtain $a_n = 1$ for all $n \geq 0$, and hence

$$\frac{1}{1 - z} = \sum_{n=0}^{\infty} z^n, \qquad \rho = 1.$$

If, on the other hand, the Taylor series of f with center at $c = -i$ is sought, we calculate

$$f^{(n)}(-i) = \frac{n!}{(1 + i)^{n+1}}$$

and using the Taylor formula once again we obtain

$$\frac{1}{1 - z} = \sum_{n=0}^{\infty} \frac{(z + i)^n}{(1 + i)^{n+1}}, \qquad \rho = \sqrt{2}.$$

See Figure 6.6 and the Remark following Example 2.

3. We find that the Taylor series of $f(z) = 1/z$ in powers of $(z - 1)$, i.e., with center at $c = 1$. We have the following:

$$f^{(n)}(z) = \frac{(-1)^n n!}{z^{n+1}}$$

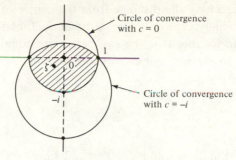

FIGURE 6.6 EXAMPLE 1(2)

from which

$$f^{(n)}(1) = (-1)^n n!$$

and, therefore,

$$\frac{1}{z} = \sum_{n=0}^{\infty} (-1)^n (z-1)^n, \qquad \rho = 1.$$

EXAMPLE 2

In this example we list the power series expansions of a number of functions and in each case we indicate the radius of convergence. The reader should verify the expansions by use of the Taylor formula and justify the assertion concerning the radius of convergence.

1. $\sin z = \sum_{n=0}^{\infty} (-1)^n \dfrac{z^{2n+1}}{(2n+1)!}, \ \rho = \infty.$

2. $\dfrac{1}{1+z} = \sum_{n=0}^{\infty} (-1)^n z^n, \ \rho = 1.$

3. $\dfrac{1}{1+z} = \sum_{n=0}^{\infty} (-1)^n \dfrac{(z+i)^n}{(1-i)^{n+1}}, \ \rho = \sqrt{2}.$

4. $e^z = e \sum_{n=0}^{\infty} \dfrac{1}{n!}(z-1)^n, \ \rho = \infty.$

5. $\dfrac{1}{z^2} = \sum_{n=0}^{\infty} (-1)^n (n+1) \dfrac{(z-i)^n}{i^{n+2}}, \ \rho = 1.$

REMARK

In Example 1(2), the function $f(z) = 1/(1-z)$ was expanded in series, first with $c = 0$ and then with $c = -i$. The circles of convergence of the two expansions are drawn in Figure 6.6. Note that the radius of convergence in each case is precisely the distance from the center of the expansion to the point $z = 1$, which is the only singularity of the function.

It is interesting to note that the two series are equal to each other at every point ζ in the region common to the interiors of the two circles of convergence; i.e.,

$$\sum_{n=0}^{\infty} \zeta^n = \sum_{n=0}^{\infty} \frac{(\zeta + i)^n}{(1 + i)^{n+1}}$$

for every ζ satisfying $|\zeta| < 1$ and $|\zeta + i| < \sqrt{2}$ simultaneously. At all other points in the plane the two series cannot be equal, since one or both of them must diverge.

In addition to the Taylor formula, a number of other methods are used for expanding functions in power series. In the remainder of this section we introduce and illustrate some of these methods. Two of them are consequences of the theorems in the preceding section while others essentially consist of performing basic algebraic operations on convergent power series. The theoretical justification of these operations on series is beyond the scope of our work in this book; we shall merely illustrate them and use them whenever they are applicable. For a complete treatment of this entire subject see Knopp, *Infinite Sequences and Series* (New York: Dover, 1956).

OPERATIONS WITH SERIES

Term by Term Differentiation and Integration

The following two facts are consequences of our work in Section 25.

1. A power series can be differentiated term by term within its circle of convergence:

$$\frac{d}{dz}\left[\sum_{n=0}^{\infty} a_n(z - c)^n\right] = \sum_{n=0}^{\infty} \frac{d}{dz}[a_n(z - c)^n].$$

2. A power series can be integrated term by term along any path K lying entirely within its circle of convergence:

$$\int_K \left[\sum_{n=0}^{\infty} a_n(z - c)^n\right] dz = \sum_{n=0}^{\infty} \int_K a_n(z - c)^n \, dz.$$

We illustrate the use of these two very effective tools in the next three examples.

EXAMPLE 3

Use the series expansion for sin z given in Example 2 to find the Maclaurin series for cos z using term by term differentiation.

We find

$$\cos z = \frac{d}{dz}(\sin z)$$

$$= \frac{d}{dz}\left[\sum_{n=0}^{\infty} \frac{(-1)^n}{(2n+1)!} z^{2n+1}\right]$$

$$= \sum_{n=0}^{\infty} \frac{d}{dz}\left[\frac{(-1)^n}{(2n+1)!} z^{2n+1}\right]$$

$$= \sum_{n=0}^{\infty} \frac{(-1)^n}{(2n)!} z^{2n}.$$

According to Theorem 6.12, the radius of convergence of the derived series is the same as that of the sine series. Hence

$$\cos z = \sum_{n=0}^{\infty} \frac{(-1)^n}{(2n)!} z^{2n}, \qquad \rho = \infty.$$

EXAMPLE 4

Apply term by term differentiation on the series of Example 1(2) to find the Maclaurin series for the function

$$f(z) = \frac{1}{(1-z)^2}.$$

We have

$$\frac{1}{(1-z)^2} = \frac{d}{dz}\left(\frac{1}{1-z}\right)$$

$$= \frac{d}{dz}\left(\sum_{n=0}^{\infty} z^n\right)$$

$$= \sum_{n=0}^{\infty} \frac{d}{dz}(z^n)$$

$$= \sum_{n=0}^{\infty} nz^{n-1}$$

$$= \sum_{n=0}^{\infty} (n+1)z^n, \qquad \rho = 1.$$

EXAMPLE 5

From Example 2(2), we know that

$$\frac{1}{1+z} = \sum_{n=0}^{\infty} (-1)^n z^n, \qquad \rho = 1.$$

We also know that the series can be integrated term by term along any path within the circle $|z| = 1$. In particular, taking any such path from 0 to any point w with $|w| < 1$ and integrating both sides of the above equation, we find that

$$\text{Log}\,(1+z)\Big|_0^w = \sum_{n=0}^{\infty} \frac{(-1)^n}{n+1} z^{n+1}\Big|_0^w$$

from which we obtain

$$\text{Log}\,(1+w) = \sum_{n=0}^{\infty} \frac{(-1)^n}{n+1} w^{n+1}.$$

This series is the Maclaurin expansion of the function $f(z) = \text{Log}\,(1+z)$ and its radius of convergence is $\rho = 1$.

Rational Operations on Power Series

Suppose that

$$f(z) = \sum_{n=0}^{\infty} a_n(z-c)^n \qquad \text{with } \rho = r_1,$$

$$g(z) = \sum_{n=0}^{\infty} b_n(z-c)^n \qquad \text{with } \rho = r_2,$$

and, for definiteness, suppose that $r_1 \le r_2$. Then the two power series may be added, subtracted, multiplied, or divided in the ordinary sense of operating with polynomials, to yield the power series expansion, respectively, of the sum, difference, product, and quotient of the functions f and g. In each case the resulting power series has a radius of convergence r_1, the smaller of the original two radii. However, in the case of the quotient, the radius must be appropriately restricted, if necessary, so that the circle of convergence will not contain, in its interior, points at which the denominator function vanishes; see Example 6.

The use of these operations is illustrated in the next two examples.

EXAMPLE 6

The series expansion

$$\frac{1}{1+z} = \sum_{n=0}^{\infty} (-1)^n z^n, \qquad \rho = 1,$$

which was developed in an earlier example, can be derived by "long division" as follows. The numerator of the above function can be thought of as a function f whose Maclaurin series is

$$f(z) = 1 + 0z + 0z^2 + \cdots + 0z^n + \cdots, \qquad \rho = \infty;$$

similarly, the denominator can be thought of as a function g whose Maclaurin series is

$$g(z) = 1 + z + 0z^2 + \cdots + 0z^n + \cdots, \qquad \rho = \infty.$$

Dividing these two series, one easily obtains

$$\frac{1}{1+z} = 1 - z + z^2 - z^3 + \cdots + (-1)^n z^n + \cdots = \sum_{n=0}^{\infty} (-1)^n z^n,$$

which is precisely the series obtained earlier.

Clearly, the circle of convergence must be restricted to $|z| = 1$ in order for it to avoid enclosing in its interior the point $z = -1$ at which the denominator vanishes. Therefore, again, $\rho = 1$.

EXAMPLE 7

Expand the function

$$h(z) = \frac{(z-1)^2}{z}$$

in a Taylor series with center at $c = 1$, i.e., in powers of $(z - 1)$.

We may consider $h(z)$ as the product of two functions

$$f(z) = (z - 1)^2 \qquad \text{and} \qquad g(z) = \frac{1}{z}$$

of which $f(z)$ is already in powers of $(z - 1)$.

Now, from Example 1(3), we have

$$g(z) = \frac{1}{z} = \sum_{n=0}^{\infty} (-1)^n (z - 1)^n, \qquad \rho = 1.$$

Therefore, multiplying the series representing f and g, we obtain

$$\frac{(z-1)^2}{z} = (z-1)^2 \sum_{n=0}^{\infty} (-1)^n (z - 1)^n$$

$$= \sum_{n=0}^{\infty} (-1)^n (z - 1)^{n+2}, \qquad \rho = 1.$$

It should be emphasized that even if we did not have the series for $g(z)$ from a previous example, it would still be to our advantage to develop it for this problem instead of resorting to the Taylor formula for expanding the given function $h(z)$.

The Principle of Substitution

Suppose that $f(z) = \sum_{n=0}^{\infty} a_n z^n$ and that the series converges for all z such that $|z| < \rho$. Suppose further that $g(z)$ is a function analytic throughout the disk $|z| < \rho$. Then $f(g(z)) = \sum_{n=0}^{\infty} a_n [g(z)]^n$ and the series converges at all z such that $|g(z)| < \rho$.

This is a relatively deep result in the theory of infinite series; we shall make use of it for fairly simple cases in which $g(z)$ is of the form az^n.

It will soon be evident from the illustrations in the next three examples that the principle of substitution will allow us to develop a multitude of series expansions from a relatively small number of known series, by means of appropriate substitutions.

EXAMPLE 8

We find the Maclaurin series for

$$f(z) = \frac{1}{2 + 4z}.$$

In using the substitution principle, one must first associate the given function with another function "of the same general form" whose series expansion is available or easy to develop. Here, of course, $f(z)$ reminds us of

$$\frac{1}{1+z} = \sum_{n=0}^{\infty} (-1)^n z^n, \qquad |z| < 1.$$

We then proceed as follows:

$$\frac{1}{2+4z} = \frac{1}{2}\frac{1}{1+2z}$$

$$= \frac{1}{2}\sum_{n=0}^{\infty} (-1)^n (2z)^n, \qquad |2z| < 1,$$

$$= \frac{1}{2}\sum_{n=0}^{\infty} (-1)^n 2^n z^n, \qquad |z| < \frac{1}{2},$$

$$= \sum_{n=0}^{\infty} (-1)^n 2^{n-1} z^n, \qquad \rho = \frac{1}{2},$$

and the given function has been expanded in a Maclaurin series.

EXAMPLE 9

Occasionally, preliminary algebraic manipulations may be needed before one can associate the given function with a familiar form. Suppose, for instance, that the function

$$f(z) = \frac{1}{3 - z}$$

is to be expanded in powers of $(z - 2i)$. We proceed as follows:

$$\frac{1}{3 - z} = \frac{1}{(3 - 2i) - (z - 2i)}$$

$$= \frac{1}{3 - 2i} \frac{1}{1 - [(z - 2i)/(3 - 2i)]}$$

$$= \frac{1}{3 - 2i} \sum_{n=0}^{\infty} \left(\frac{z - 2i}{3 - 2i}\right)^n, \qquad \left|\frac{z - 2i}{3 - 2i}\right| < 1$$

$$= \sum_{n=0}^{\infty} \frac{(z - 2i)^n}{(3 - 2i)^{n+1}}, \qquad |z - 2i| < \sqrt{13}.$$

In the first step of the above process, our objective was to create the quantity $(z - 2i)$ in the denominator since we wish our expansion to be in powers of $(z - 2i)$. In the second step, by use of algebraic manipulations, we created an expression of the form $\dfrac{1}{1 - g(z)}$. The third step consists of the substitution in the familiar expansion for $\dfrac{1}{1 - z}$ and, finally, the last step is just algebra. Note that the radius of convergence $\rho = \sqrt{13}$ is precisely the distance between the center of the expansion $c = 2i$ and the only singularity of the given function, $z = 3$.

EXAMPLE 10

1. Since $e^z = \sum \dfrac{z^n}{n!}$ with $\rho = \infty$, it follows readily that

$$e^{z^2} = \sum \frac{z^{2n}}{n!}, \qquad \rho = \infty.$$

2. Similarly, since $\cos z = \sum \dfrac{(-1)^n}{(2n)!} z^{2n}$, it is immediate that

$$\cos (3z) = \sum \frac{(-1)^n 9^n}{(2n)!} z^{2n}, \qquad \rho = \infty.$$

3. Expand e^z with center at $c = 2i$. We have

$$e^z = e^z e^{2i-2i}$$

$$= e^{2i} e^{z-2i}$$

$$= e^{2i} \sum \frac{(z - 2i)^n}{n!}, \qquad \rho = \infty.$$

EXERCISE 26

A

26.1. Supply the details and justify the various steps in all three parts of Example 1.

26.2. Verify the five series expansions in Example 2 by use of the Taylor formula.

In Exercises 26.3–26.12, expand the given function in a power series about the respective center and determine the radius and the circle of convergence in each case.

26.3. $\dfrac{z}{1 + z}$, $c = 0$.

26.4. $\dfrac{z^2}{2 + z}$, $c = 0$.

26.5. e^{z+1}, $c = 1$.

26.6. $\dfrac{1}{z}$, $c = i$.

26.7. $\dfrac{1}{z}$, $c = 4$.

26.8. e^{-z}, $c = 0$.

26.9. $\sinh z$, $c = 0$.

26.10. $\operatorname{Log} z$, $c = i$.

26.11. $\dfrac{1 - z}{1 + 2z}$, $c = 0$.

26.12. $\dfrac{1 - z}{1 + 2z}$, $c = 1$.

26.13. Find the Maclaurin series of $\cosh z$ first by use of the Taylor formula and then by term by term differentiation of the series for $\sinh z$; see Exercise 26.9.

26.14. Use the Taylor formula to find the first three nonzero terms of the Maclaurin series for $\tan z$. What is its radius of convergence?

B

26.15. Multiply the Maclaurin series for e^z and $\cos z$ to find the first few terms for the function $e^z \cos z$.

26.16. Find the first few terms of the Maclaurin series for $\dfrac{\sin z}{1 - z}$ (a) by division of series; (b) by multiplication of series. Determine the circle of convergence of your answer.

26.17. Find the Taylor series for $f(z) = (z - 1)^2 e^z$ with center at $c = 1$.

26.18. Verify that $\dfrac{1 - z}{z - 2} = \displaystyle\sum_{n=0}^{\infty} (z - 1)^{n+1}$. What is ρ?

26.19. Verify that $\dfrac{1}{1 + z} = -\displaystyle\sum_{n=0}^{\infty} \dfrac{(z + 3)^n}{2^{n+1}}$. What is ρ?

26.20. Expand $\dfrac{\sin z}{z - \pi}$ with center at $c = \pi$.

26.21. Find the Maclaurin series for $f(z) = \dfrac{1}{(z - i)(z + 3)}$.

HINT: Use partial fraction decomposition to write

$$f(z) = \frac{A}{z - i} + \frac{B}{z + 3}$$

and then expand each term separately.

c

26.22. The **summing of a power series** is the process by which one finds the analytic function, called the **sum function,** which represents a convergent power series; see Section 25. The method consists of a succession of operations performed on a power series with a known sum function $f(z)$ until the series becomes identical with a given series which we wish to sum. Then, the same succession of operations applied on $f(z)$ will yield the sum function to which the given series converges.

In each part of this exercise a power series is given which is to be summed.

(a) $\displaystyle\sum_{n=0}^{\infty} (-1)^{n+1} n z^n$.

HINT: The alternation of signs and the factor z^n suggests use of

$$\frac{1}{1 + z} = \sum_{n=0}^{\infty} (-1)^n z^n, \qquad \rho = 1.$$

Differentiate and multiply by $-z$ to obtain $\dfrac{z}{(1 + z)^2}$, which is the answer.

(b) $\displaystyle\sum_{n=0}^{\infty} (n + 1)z^n$.

NT: Think $\dfrac{1}{1 - z}$ and differentiate. See Example 4.

(c) $\displaystyle\sum_{n=0}^{\infty} \dfrac{z^{n+3}}{n!}$.

HINT: Powers of z and $n!$ suggest e^z.

(d) $\displaystyle\sum_{n=0}^{\infty} \dfrac{(-1)^n}{(2n + 1)!} z^{2n}$.

(e) $\displaystyle\sum_{n=1}^{\infty} \dfrac{(-1)^n 2n}{(2n + 1)!} z^{2n-1}$.

26.23. Verify that, for $z \neq 0$ and $\zeta \neq z$, $\dfrac{1}{z - \zeta} = \dfrac{1}{z} \cdot \dfrac{1}{1 - \dfrac{\zeta}{z}}$.

Then show that, for $w \neq 1$,

$$\frac{1}{1 - w} = 1 + w + w^2 + \cdots + w^{n-1} + \frac{w^n}{1 - w}.$$

Finally, combine the above two identities and, letting $w = \zeta/z$, establish relation (1) in the proof of Theorem 6.13.

REVIEW EXERCISES — CHAPTER 6

1. Mark the following statements *True* or *False*.
   (a) A series is the limit of the sequence of its partial sums.
   (b) A power series converges at every point inside and on its circle of convergence.
   (c) Every power series converges to an analytic function.
   (d) Every function continuous at a point c possesses a Taylor series expansion with center at c.
   (e) If a series converges, then it converges absolutely.
   (f) If a power series has an infinite radius of convergence, then the series converges to a constant function.

2. Find the radius of convergence for the power series whose general term is

   (a) $\dfrac{n^2}{2^n}(z - i)^n$.

   (b) $\dfrac{(2n)!}{n^3}(z + 1)^n$.

   (c) $\dfrac{n(2n + 1)}{n^2}z^n$.

   (d) $\dfrac{n!}{n^n}(z - 2 + i)^n$.

3. Find the Taylor series expansion of each of the following functions with center at the respective point c:

   (a) e^z, $c = i$.

   (b) $\cos z$, $c = \pi/2$.

   (c) $\dfrac{1}{z + 2i}$, $c = 0$.

   (d) $\dfrac{1}{(z + 1)(z - 1)}$, $c = 0$.

   (e) $\dfrac{1}{(1 - z)^3}$, $c = 0$.

   (f) e^{z-1}, $c = 2$.

4. Sum the series $\displaystyle\sum_{n=0}^{\infty} \dfrac{(-1)^n(2n + 2)}{(2n + 1)!} z^{2n+1}$; see Exercise 26.22.

5. Find the first few terms of the Maclaurin series of $f(z) = e^{1/(1-z)}$.

6. Find the Maclaurin series of $f(z) = \dfrac{1}{(2 - z^3)^2}$.

7. Prove that the series $\displaystyle\sum_{n=1}^{\infty} \left(\dfrac{n - 1}{n} + \dfrac{1}{n^2} \right)$ diverges.

8. Find the series expansion of the function

$$f(z) = e^{z^2 + 2z}$$

   with center at $c = -1$.

9. Find the Maclaurin series for $\dfrac{1}{z^2 + 4}$.

10. Find the Maclaurin series for each of the following functions:

   (a) $\displaystyle\int_0^z e^{t^2}\, dt$.

   (b) $\displaystyle\int_0^z \cos(t^2)\, dt$.

   (c) $\displaystyle\int_0^z \dfrac{\sin t}{t}\, dt$.

   (d) $\displaystyle\int_0^z \dfrac{e^t - 1}{t}\, dt$.

11. Find the first few nonzero terms of the Maclaurin series for $\tan z$ by a method other than that suggested in Exercise 26.14.

12. Find the first few nonzero terms of the Maclaurin series for $\sec z$.

A P P E N D I X   6

Part A
Proofs of Theorems

Theorem 6.5

If a series $\sum_{n=0}^{\infty} z^n$ converges, then, as $n \to \infty$, $\lim z_n = 0$.

Proof:

It must be shown that, for any $\varepsilon > 0$, $|z_n| < \varepsilon$ for all n beyond a certain one.

Since the series converges it has a sum, S, and, by definition, $S = \lim S_n$, where S_n are the partial sums of the given series. Hence, given any $\varepsilon > 0$, we have

$$|S_n - S| < \frac{\varepsilon}{2}$$

for all n beyond a certain one. Then, also,

$$|S_{n+1} - S| < \frac{\varepsilon}{2}.$$

Therefore, by use of the triangle inequality,

$$|S_{n+1} - S_n| \leq |S_{n+1} - S| + |S - S_n| < \varepsilon.$$

But $S_{n+1} - S_n = z_{n+1}$. Thus, for all n beyond a certain one,

$$|z_{n+1}| < \varepsilon,$$

and the theorem is proved.

Theorem 6.6

If a series converges absolutely, then it converges; i.e., if $\sum |z_n|$ converges, then $\sum z_n$ converges.

Proof:

Let $\{T_n\}$ and $\{S_n\}$ denote the sequences of partial sums of $\sum |z_n|$ and $\sum z_n$, respectively.

By hypothesis and Theorem 6.3, given any $\varepsilon > 0$, there is an integer M such that

$$|T_m - T_n| < \varepsilon, \qquad \text{for all } m > n > M,$$

a fact that can also be written in the form

$$|z_{n+1}| + |z_{n+2}| + \cdots + |z_m| < \varepsilon. \tag{1}$$

Then, by use of the triangle inequality, we have the following:

$$|S_m - S_n| = |z_{n+1} + z_{n+2} + \cdots + z_m|$$

$$\leq |z_{n+1}| + |z_{n+2}| + \cdots + |z_m|$$

$$< \varepsilon.$$

Thus, again, for all $m > n > M$, $|S_m - S_n| < \varepsilon$.

Therefore, according to Theorem 6.3, the sequence $\{S_n\}$ converges. Hence the series $\sum z_n$ converges and the theorem follows.

Theorem 6.13 *(Taylor's Theorem)*
Suppose that $f(z)$ is analytic at a point c in the plane. Then there is a power series

$$\sum_{n=0}^{\infty} a_n(z - c)^n$$

whose coefficients are given by the formula

$$a_n = \frac{f^{(n)}(c)}{n!}, \qquad n = 0, 1, 2, \ldots,$$

and which converges to $f(\zeta)$ for every ζ in every neighborhood of c throughout which f is analytic:

$$f(\zeta) = \sum_{n=0}^{\infty} \frac{f^{(n)}(c)}{n!}(\zeta - c)^n.$$

Proof:
For simplicity of notation and without loss of generality, the theorem is proved for the case $c = 0$. The extension to the case of an arbitrary center c is immediate.

By hypothesis, f is analytic at $z = 0$ and hence it is analytic in some neighborhood $N(0, \rho)$; see Figure 6.7. Note that the size of ρ can be taken to be at most equal to the distance between the center $c = 0$ and the singularity of f nearest to 0.

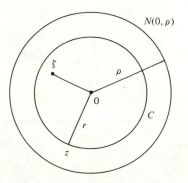

FIGURE 6.7 TAYLOR'S THEOREM

Let ζ be an arbitrary but fixed point in $N(0, \rho)$. Then take a circle C centered at 0 and having radius r such that $|\zeta| < r < \rho$. Of course, ζ is in Int (C) and f is analytic on and inside C. Elementary algebraic manipulations (see Exercise 26.23) yield the identity

$$\frac{f(z)}{z - \zeta} = \frac{f(z)}{z} + \zeta \frac{f(z)}{z^2} + \cdots + \zeta^{n-1} \frac{f(z)}{z^n} + \zeta^n \frac{f(z)}{(z - \zeta)z^n}. \tag{1}$$

Upon dividing (1) by $2\pi i$ and integrating each term along C, positively oriented, with respect to z, we obtain

$$\frac{1}{2\pi i} \int_C \frac{f(z)}{z - \zeta} dz = \frac{1}{2\pi i} \int_C \frac{f(z)}{z} dz + \frac{\zeta}{2\pi i} \int_C \frac{f(z)}{z^2} dz + \cdots$$

$$+ \frac{\zeta^{n-1}}{2\pi i} \int_C \frac{f(z)}{z^n} dz + \frac{\zeta^n}{2\pi i} \int_C \frac{f(z)}{(z - \zeta)z^n} dz. \tag{2}$$

Denoting the last term in (2) by R_n and using the Cauchy integral formulas (Theorems 5.7 and 5.8), we may write (2) as follows:

$$f(\zeta) = f(0) + \zeta f'(0) + \cdots + \zeta^{n-1} \frac{f^{(n-1)}(0)}{(n-1)!} + R_n. \tag{3}$$

C L A I M: As $n \to \infty$, $|R_n| \to 0$.

From Theorem E, p. 219, and its corollary we know that there is a positive number M such that

$$|f(z)| \le M, \qquad \text{for all } z \text{ on } C.$$

Also, we note that the length of C is $2\pi r$. Then

$$\left| \frac{\zeta^n f(z)}{(z - \zeta)z^n} \right| = \frac{|\zeta|^n |f(z)|}{|z - \zeta| |z|^n} \le \frac{M}{|z - \zeta|} \left| \frac{\zeta}{z} \right|^n.$$

Therefore, by use of Theorem 4.5(5), we find that

$$|R_n| = \left| \frac{1}{2\pi i} \int_C \frac{\zeta^n f(z)}{(z - \zeta)z^n} dz \right|$$

$$\le \frac{1}{2\pi} \frac{M}{|z - \zeta|} \left| \frac{\zeta}{z} \right|^n 2\pi r$$

$$= \frac{Mr}{|z - \zeta|} \left| \frac{\zeta}{z} \right|^n. \tag{4}$$

Now, since

$$|z - \zeta| \ge |z| - |\zeta| = r - |\zeta|,$$

we have

$$\frac{1}{|z - \zeta|} \leq \frac{1}{r - |\zeta|}.$$

Therefore, substituting in (4), we obtain

$$|R_n| \leq \frac{Mr}{r - |\zeta|} \left|\frac{\zeta}{z}\right|^n = (\text{constant}) \cdot \left[\frac{|\zeta|}{r}\right]^n.$$

But $|\zeta|/r < 1$; hence, as $n \to \infty$, $[|\zeta|/r]^n \to 0$, and the claim is justified.

Finally, from (3), as $n \to \infty$, we obtain

$$f(\zeta) = \sum_{n=0}^{\infty} \frac{f^{(n)}(0)}{n!}\zeta^n. \tag{5}$$

Since ζ was arbitrarily chosen in $N(0, \rho)$, it follows that (5) holds for every such ζ and the proof is complete.

Part B
More on Sequences and Series

LIMIT SUPERIOR AND LIMIT INFERIOR

Let $\{x_n\}$ be a sequence of *real numbers* and suppose that a point P (which may or may not be a term of the sequence) has the following property: Every neighborhood of P contains infinitely many terms of $\{x_n\}$. Then P is said to be a **limit point** of the sequence.

The following elementary properties of limit points are easy to establish; see Example 1 below and the exercises at the end of this appendix.

Properties of Limit Points

1. A sequence may have no limit point, it may have exactly one, or it may have several limit points.
2. If a sequence converges, then it has exactly one limit point that coincides with its limit; this, of course, implies that if a sequence has fewer or more than one limit point, then it diverges.
3. From the preceding property it follows that the limit of a sequence, if such exists, is a limit point of that sequence. However, a limit point of a sequence is not necessarily the limit of the sequence.

EXAMPLE 1

1. The sequence

$$\left\{\frac{n+1}{n}\right\} = \left\{2, \frac{3}{2}, \frac{4}{3}, \dots, \frac{n+1}{n}, \dots\right\}$$

has exactly one limit point: $P = 1$. Note that the sequence converges to that point; i.e., the limit point is also the limit of the sequence. See property 2.

2. The sequence

$$\left\{(-1)^n \frac{n}{n+1}\right\} = \left\{0, -\frac{1}{2}, \frac{2}{3}, -\frac{3}{4}, \dots\right\}$$

has two limit points: $P = 1, -1$. Note that the sequence diverges.

3. The sequence

$$\left\{\frac{1}{2}, \frac{3}{2}, \frac{5}{2}, \frac{1}{3}, \frac{4}{3}, \frac{7}{3}, \dots, \frac{1}{n}, \frac{n+1}{n}, \frac{2n+1}{n}, \dots\right\}$$

has three limit points: $P = 0, 1, 2$.

4. The sequence

$$\{n\} = \{1, 2, 3, \dots, n, \dots\}$$

has no limit point. Note that the sequence is unbounded and hence it diverges.

Suppose now that we have a sequence $\{x_n\}$ of real numbers. Let S be the set of all the limit points of $\{x_n\}$. We define the **limit superior** of the sequence $\{x_n\}$, denoted

$$\overline{\lim} \, x_n \qquad \text{or} \qquad \lim \sup x_n$$

to be the supremum of S (p. 214), provided that it exists. Thus, by definition,

$$\overline{\lim} \, x_n = \sup S.$$

If the sequence is bounded above, then the set of its limit points is also bounded above. Therefore, by the completeness axiom (p. 214), sup S exists and, hence,

$$\overline{\lim} \, x_n \text{ exists.}$$

If, on the other hand, $\{x_n\}$ is unbounded above, then, *by convention*, we denote the fact that, as $n \to \infty$, $\lim x_n = +\infty$, by writing $\sup S = +\infty$. This is a widely used convention that facilitates our study of a number of topics;

if it is taken only as a convenient agreement and not as implying that the *ideal number* $+\infty$ has been admitted as a member of the system of real numbers, no harm is done. Thus, again,

$$\overline{\lim}\ x_n \text{ exists.}$$

Since the two cases just considered ($\{x_n\}$ bounded and $\{x_n\}$ unbounded) exhaust all possibilities, we have the following:

> Given a sequence $\{x_n\}$ of real numbers, there exists a unique number λ (possibly $+\infty$) such that $\overline{\lim}\ x_n = \lambda$.

N O T E: The uniqueness of λ follows from the fact that the supremum of a set, if it exists, is unique. The proof of this fact is based exclusively on the definition of a supremum and is left to the reader as an exercise.

In what follows, we will be interested in two fundamental properties of the limit superior and we discuss them briefly at this point. They are both true regardless of whether the limit superior is a finite or an infinite number. See the exercises at the end of this appendix.

Properties of the Limit Superior

Given a sequence $\{x_n\}$ of real numbers, suppose that $\overline{\lim}\ x_n = L$. Then, with respect to the terms of the sequence, L possesses the following two properties; see Figure 6.8.
  1. If α is any real number such that $L < \alpha$, then, for all but a finite number of n's, $x_n < \alpha$.
  2. If β is any real number such that $\beta < L$, then $x_n > \beta$ for an infinite number of n's.

A completely symmetric development will lead to the concept of **limit inferior** of a sequence $\{x_n\}$, which, in informal terms, is the smallest of all the limit points of the sequence and is denoted

$$\underline{\lim}\ x_n \qquad \text{or} \qquad \lim \inf x_n.$$

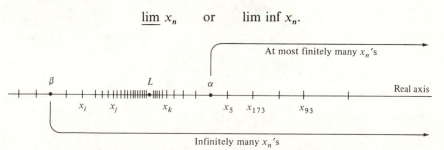

FIGURE 6.8 TWO PROPERTIES OF LIM SUP

The reader will find it very helpful in understanding the above development to attempt an analogous development for the concept of limit inferior; see the exercises at the end of this appendix.

THE CAUCHY–HADAMARD THEOREM

The proof of the Cauchy–Hadamard theorem is preceded by a lemma which, in fact, is the root test and bears the brunt of the proof.

Lemma   (*The Root Test*)
Given a series $\sum_{n=1}^{\infty} z_n$ of complex numbers, suppose that

$$\overline{\lim} \, |z_n|^{1/n} = L.$$

Then
1. If $L < 1$, the series converges absolutely.
2. If $L > 1$, the series diverges.

Proof:
1. Suppose that $L < 1$.
Choose an arbitrary real number α such that $L < \alpha < 1$. Then, by property 1, p. 275,

$$|z_n|^{1/n} < \alpha$$

for all except at most finitely many n's. Therefore, for all n greater than some N,

$$|z_n| < \alpha^n;$$

i.e.,

$$|z_N| < \alpha^N, |z_{N+1}| < \alpha^{N+1}, |z_{N+2}| < \alpha^{N+2}, \ldots.$$

Now, the series

$$\alpha^N + \alpha^{N+1} + \alpha^{N+2} + \cdots$$

is a geometric series with $|\alpha| < 1$. Therefore, it converges and, hence, by the comparison test,

$$|z_N| + |z_{N+1}| + \cdots = \sum_{n=N}^{\infty} |z_n|$$

converges. Finally, by Exercise 23.20 the series

$$\sum_{n=1}^{\infty} |z_n|$$

converges, which, in turn, implies that

$$\sum_{n=1}^{\infty} z_n$$

converges absolutely.

2. Suppose that $L > 1$.

Then, by property 2, the relation

$$|z_n|^{1/n} > 1$$

and, hence,

$$|z_n| > 1$$

holds for infinitely many n's. It follows then that as $n \to \infty$, $\lim z_n \neq 0$ and, therefore, by Theorem 6.5,

$$\sum_{n=1}^{\infty} z_n$$

diverges.

This completes the proof of the lemma.

Theorem (*Cauchy–Hadamard Theorem*)

Given a power series $\sum_{n=0}^{\infty} a_n(z - c)^n$, suppose that

$$\overline{\lim} \, |a_n|^{1/n} = \lambda, \qquad for \; 0 \leq \lambda \leq \infty.$$

Define $\rho = 1/\lambda$ with the convention that if $\lambda = 0$, then $\rho = \infty$ and if $\lambda = \infty$, then $\rho = 0$.

Then the given series has the following properties:
1. *If $\rho = 0$, then the series diverges for all $z \neq c$.*
2. *If $0 < \rho < \infty$, then the series converges for all z such that $|z - c| < \rho$ and diverges for all z such that $|z - c| > \rho$.*
3. *If $\rho = \infty$, then the series converges for all z in the finite plane.*
In all cases, the series converges absolutely at its center c.

Proof:
The unique existence of λ was established earlier in this section.
According to the above lemma, the given series converges if

$$1 > \overline{\lim} \, |a_n(z - c)^n|^{1/n}$$

$$= \overline{\lim} \, |a_n|^{1/n}|z - c|$$

$$= \lambda|z - c|;$$

in other words, it converges for all z such that

$$|z - c| < \frac{1}{\lambda} = \rho.$$

Again, by the above lemma, the given series diverges if

$$1 < \overline{\lim} \, |a_n(z - c)^n|^{1/n}.$$

Similarly, then, we find that the series diverges for all z such that

$$|z - c| > \frac{1}{\lambda} = \rho.$$

Clearly, all three conclusions of the theorem are satisfied, and the proof is complete.

EXERCISES—SET 1

1. Give two examples (other than those discussed in the text) for each of the following cases:
   (a) A sequence with no limit point in the finite plane.
   (b) A sequence with two finite limit points.
   (c) A sequence with two infinite limit points.
   (d) A sequence with two limit points: one finite and one infinite.
   (e) A sequence with four limit points.
2. Compare the concept of a limit point with that of a cluster point (p. 215). Then answer the following two questions. If you answer *yes*, prove your answer; if you answer *no*, give an example.
   (a) Is a cluster point necessarily a limit point?
   (b) Is a limit point necessarily a cluster point?
3. Prove both parts of property 2, p. 273.
4. Prove both parts of property 3, p. 273.
5. Review the note on p. 275 and prove that if the supremum of a set S exists, then it is unique. (Assume, to the contrary, that there are two suprema α and β such that $\alpha < \beta$. Then, using the two properties in the definition of sup S, arrive at a contradiction.)
6. Imitate the discussion beginning with the definition of limit superior and extending through its properties, to develop the symmetric concept of the limit inferior of a sequence; see p. 275.
7. Prove property 1, p. 275. (If the sequence is unbounded, the assertion is trivially true; justify this! If it is bounded by, say, the number B, then assume that the interval $\alpha \leq x \leq B$ contains infinitely many x_n's and use the Bolzano–Weierstrass theorem to reach a contradiction.)
8. Prove property 2, p. 275. (Assume that only finitely many x_n's are greater than β and reach a contradiction.)

CHAPTER 7
Laurent Series. Residues

Section 27
Laurent Series

The expansion of a function $f(z)$ in a Taylor series represents the function within its circle of convergence, which is, however, almost always only part of the region of analyticity of f. For instance, the series $\sum z^n$ converges to $f(z) = 1/(1 - z)$ only on the disk $|z| < 1$, even though f is analytic everywhere except at $z = 1$. The natural question then is : Is there a series expansion that represents f in a more complicated region or, perhaps, at all points at which f is analytic? It is the primary objective of this section to give some answers to questions of this general nature by developing the *Laurent series* of analytic functions. As we will see, Laurent series are generalizations of Taylor series in that they have a (finite or infinite) number of negative integral powers of $(z - c)$ in addition to the positive integral powers of $(z - c)$ which they *may* contain. We will also see that a Laurent series of a function $f(z)$ converges, in general, in a circular annulus $r < |z - c| < \rho$ (see Figure 7.1), so that we will now be concerned with the *annulus of convergence* instead of with the circle of convergence.

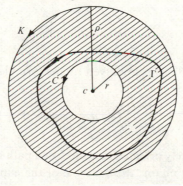

FIGURE 7.1 ANNULUS OF CONVERGENCE

The following theorem formalizes the above introductory remarks in more precise terms.

Theorem 7.1   (*Laurent's Theorem*)
Suppose that $f(z)$ is analytic at every point of the closed annulus

$$A : r \leq |z - c| \leq \rho.$$

Then there is a series of positive and negative powers of $(z - c)$ that represents f at every point ζ in the (open) annulus $r < |z - c| < \rho$:

$$f(\zeta) = \sum_{n=0}^{\infty} a_n(\zeta - c)^n + \sum_{n=1}^{\infty} \frac{b_n}{(\zeta - c)^n}.$$

The coefficients of the series are given by the formulas

$$a_n = \frac{1}{2\pi i} \int_K \frac{f(z)}{(z - c)^{n+1}} \, dz, \qquad n = 0, 1, 2, \ldots,$$

and

$$b_n = \frac{1}{2\pi i} \int_C \frac{f(z)}{(z - c)^{-n+1}} \, dz, \qquad n = 1, 2, 3, \ldots,$$

where $K : |z - c| = \rho$ and $C : |z - c| = r$, both positively oriented; see Figure 7.1.

Proof:
See Appendix 7.

The series expansion in the above theorem is called the **Laurent series** of f at c and the open annulus $r < |z - c| < \rho$ is called the **annulus of convergence** of the series. Note that one may also write the series in the form

$$\sum_{n=-\infty}^{\infty} c_n(z - c)^n,$$

in which case the coefficients are given by the formula

$$c_n = \frac{1}{2\pi i} \int_\Gamma \frac{f(z)}{(z - c)^{n+1}} \, dz, \qquad n = 0, \pm 1, \pm 2, \ldots,$$

where Γ is any positively oriented simple closed path that lies in the annulus of convergence and contains the center c of the expansion in its interior; see Figure 7.1.

REMARK 1

By a method similar to one used in the proof of Theorem 6.9, it can be shown that the Laurent series of a function $f(z)$ converges uniformly to f at any point, indeed, on any closed set in the annulus of convergence. A consequence of this is that, as in the case of Taylor series, a Laurent series can be differentiated or integrated term by term within its annulus of convergence.

REMARK 2

It is shown in Chapter 9 that if a Laurent series expansion of a function on a given annulus exists, then it is unique. This fact guarantees that once a Laurent series has been developed for a given function $f(z)$, then the expansion is *the* Laurent series of f.

REMARK 3

Note that if all $b_n = 0$ in the formulas of Theorem 7.1, then the Laurent series reduces to a Taylor series. In that sense, a Taylor series is a special case of a Laurent series.

It is important to note that, given a function $f(z)$ and a point c in the plane, it is possible that f may have more than one Laurent series with center at c *depending on the annulus of convergence* on which the Laurent series is to represent f. It should be emphasized that this fact is not in contradiction to Remark 2. (Explain why.) In general, the number of distinct Laurent series that a function f will admit will depend on the center c and the number of singularities of f. For example, the function

$$f(z) = \frac{3}{z(z - i)}$$

has three distinct series expansions with center at $c = -i$:
1. A Taylor series converging in the open disk $|z + i| < 1$.
2. A Laurent series having annulus of convergence $1 < |z + i| < 2$.
3. A Laurent series converging in the annulus $2 < |z + i| < \infty$.

The three regions of convergence appear shaded in the three parts of Figure 7.2, respectively.

In the examples that follow, we illustrate various techniques for developing the Laurent series of an analytic function. The determination of the coefficients from the formulas of Theorem 7.1 is, at best, cumbersome and so one usually resorts to more direct methods. In particular, the method of substitution and operations on series, as they were used in Section 26, will be employed almost exclusively.

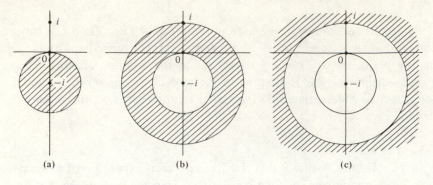

(a)   (b)   (c)

FIGURE 7.2 THREE DISTINCT SERIES FOR $f(z) = 3/[z(z - i)]$

EXAMPLE 1

1. Find the Laurent series expansion of the function $f(z) = 1/z$ with center at $c = 1$ and annulus of convergence $A : 1 < |z - 1| < \infty$. Clearly, f is analytic in A, so that, in the notation of Theorem 7.1, r and ρ can be any real numbers greater than 1. Now, we seek a series in powers of $(z - 1)$ that will converge to $f(z)$ for all z such that $|z - 1| > 1$; see Figure 7.3(a). From the last inequality we obtain

$$\left| \frac{1}{z - 1} \right| < 1.$$

Therefore, the quantity $1/(z - 1)$ can be substituted for z in any series that converges for $|z| < 1$. We then have

$$\frac{1}{z} = \frac{1}{(z - 1) + 1}$$

$$= \frac{1}{z - 1} \frac{1}{1 + \dfrac{1}{z - 1}}$$

$$= \frac{1}{z - 1} \sum_{n=0}^{\infty} (-1)^n \left(\frac{1}{z - 1} \right)^n, \qquad \left| \frac{1}{z - 1} \right| < 1,$$

$$= \sum_{n=0}^{\infty} (-1)^n \frac{1}{(z - 1)^{n+1}}, \qquad 1 < |z - 1|.$$

Thus the Laurent series of $f(z)$ is

$$\frac{1}{z} = \frac{1}{z - 1} - \frac{1}{(z - 1)^2} + \frac{1}{(z - 1)^3} - \cdots$$

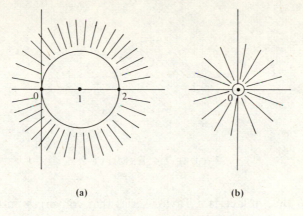

(a)   (b)

FIGURE 7.3 EXAMPLE 1

and its annulus of convergence is $1 < |z - 1| < \infty$ as prescribed by the problem.

2. Note that the function $f(z) = 1/z$ is already a Laurent series with center at $c = 0$, whose annulus of convergence is $0 < |z| < \infty$; see Figure 7.3(b).

EXAMPLE 2

Find a series expansion of the function

$$f(z) = \frac{1}{(z - 1)(z + 1)}$$

in the annulus $0 < |z - 1| < 2$ "between" the two singular points $z = 1$ and $z = -1$; see Figure 7.4.

First, by use of the substitution principle we find

$$\frac{1}{z + 1} = \frac{1}{2 + (z - 1)}$$

$$= \frac{1}{2} \frac{1}{1 + \dfrac{z - 1}{2}}$$

$$= \frac{1}{2} \sum_{n=0}^{\infty} (-1)^n \left(\frac{z - 1}{2}\right)^n, \qquad \left|\frac{z - 1}{2}\right| < 1,$$

$$= \sum_{n=0}^{\infty} (-1)^n \frac{(z - 1)^n}{2^{n+1}}, \qquad |z - 1| < 2.$$

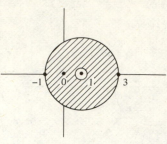

FIGURE 7.4 EXAMPLE 2

Note that this, in fact, is a Taylor series that converges on the interior of the circle $|z - 1| = 2$.

Next, multiplying the above series by $\dfrac{1}{z - 1}$, which is already a Laurent series in powers of $(z - 1)$, we obtain

$$\frac{1}{(z - 1)(z + 1)} = \sum_{n=0}^{\infty} (-1)^n \frac{(z - 1)^{n-1}}{2^{n+1}}$$

$$= \frac{1}{2}\frac{1}{z - 1} - \frac{1}{4} + \frac{1}{8}(z - 1) - \cdots.$$

Of course, the point $z = 1$ must now be removed from the region of convergence, which then becomes $0 < |z - 1| < 2$, as prescribed by the problem.

EXAMPLE 3

We develop the series expansion of $f(z) = e^z/z^2$ with center at $c = 0$ by the method used in Example 2.

Thus we begin with the familiar

$$e^z = \sum_{n=0}^{\infty} \frac{z^n}{n!}, \qquad |z| < \infty,$$

and then we multiply both sides of it by $1/z^2$ to obtain

$$\frac{e^z}{z^2} = \sum_{n=0}^{\infty} \frac{z^{n-2}}{n!}$$

$$= \frac{1}{z^2} + \frac{1}{z} + \frac{1}{2!} + \frac{z}{3!} + \cdots$$

which converges for all $z \neq 0$, i.e., on the annulus $0 < |z| < \infty$.

EXAMPLE 4

The principle of substitution may be employed once again to give us the Laurent series

$$\cos\left(\frac{1}{z}\right) = \sum_{n=0}^{\infty} \frac{1}{(2n)!z^{2n}}.$$

Its annulus of convergence is $0 < |z| < \infty$ and is obtained from the region of convergence $|z| < \infty$ of the cosine series by excluding $z = 0$.

EXAMPLE 5

We expand the function $f(z) = \dfrac{5z + 2i}{z(z + i)}$ in the annulus $1 < |z - i| < 2$; see Figure 7.5(a).

We outline the process; the reader should supply the details.

First, by partial fraction decomposition we find that

$$f(z) = \frac{2}{z} + \frac{3}{z + i}.$$

The Laurent series for $2/z$ in the annulus $1 < |z - i| < \infty$ [see Figure 7.5(b)] is

$$\frac{2}{z} = 2 \sum_{n=0}^{\infty} \frac{(-1)^n i^n}{(z - i)^{n+1}}.$$

On the other hand, the Taylor series of $\dfrac{3}{z + i}$ in the disk $|z - i| < 2$ [see Figure 7.5(c)] is

$$\frac{3}{z + i} = 3 \sum_{n=0}^{\infty} \frac{(-1)^n (z - i)^n}{(2i)^{n+1}}.$$

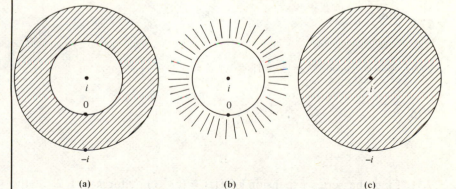

(a)   (b)   (c)

FIGURE 7.5 EXAMPLE 5

Finally, the series expansion of $f(z)$ is the sum of these two series and the region of convergence is the region common to both, i.e., $1 < |z - i| < 2$, which is the annulus throughout which both series converge simultaneously. Note that the annulus in question is precisely the one prescribed by the problem.

EXERCISE 27

A

In Exercises 27.1–27.10, find the series expansion of the given function in the specified region.

27.1. $\dfrac{1}{z + 3}$, $|z| > 3$.

27.2. $\dfrac{1}{(z + 2)(z - 1)}$, $1 < |z - 2| < 4$.

27.3. $\sin\left(\dfrac{1}{z}\right)$, $0 < |z| < \infty$.

27.4. $\dfrac{1}{(z + 2)(z - 1)}$, $4 < |z - 2| < \infty$.

27.5. $\dfrac{e^z - (z + 1)}{z^3}$, $0 < |z| < \infty$.

27.6. $\dfrac{\cos(z - 1)}{z - 1}$, $0 < |z - 1| < \infty$.

27.7. e^{1/z^2}, $0 < |z| < \infty$.

27.8. $\dfrac{\sinh z}{z^2}$, $0 < |z| < \infty$.

27.9. $\dfrac{1}{z^3 - 2z^2 + z}$, $0 < |z - 1| < 1$.

27.10. $\dfrac{1}{z} + \dfrac{1}{z - 1} + \dfrac{1}{z - i}$, $0 < |z| < 1$.

B

27.11. Find the series expansion of $f(z) = (z - c)^{-k}$, for $k = 1, 2$, in the annulus $|c| < |z| < \infty$.

27.12. Find all possible series expansions, with center at $c = 0$, for

$$f(z) = \frac{1}{z^2(z - 1)(z - 2)}.$$

27.13. Find the Taylor series for $f(z) = 1/z$ with center at $c = 1$. Then, use your answer to find the Laurent series for the function

$$f(z) = z^{-1}(z - 1)^{-2}$$

in the annulus $0 < |z - 1| < 1$.

27.14. Find the same Laurent series for the function $f(z)$ of the preceding exercise by using the following procedure. Set $z = w + 1$, expand the resulting function in powers of w, and then resubstitute $w = z - 1$ to find your answer.

c

27.15. Supply the missing details in the proof of Theorem 7.1.

27.16. Expand each of the following functions about the origin and find the region of convergence in each case.

(a) $\dfrac{e^z - 1}{z}$.

(b) $\dfrac{\sin z}{z}$.

(c) $\dfrac{\cos (z^2) - 1}{z^2}$.

(d) $\dfrac{e^z - (z + 1)}{z^2}$.

27.17. Note that each of the series expansions in the preceding exercise is a Taylor series which, therefore, converges at its center $c = 0$. But this, in turn, means that each of the functions is defined at $z = 0$. Find the value of each function at $z = 0$ so that its Taylor series will represent it at the origin.

Section 28
Singularities and Zeros of an Analytic Function

From Section 8 we recall the following definition: A point z_0 is a **singularity** of a function $f(z)$, provided that f fails to be analytic at z_0, while *every neighborhood* of z_0 contains at least one point at which f is analytic. Basically, there are two types of singularities:

1. Nonisolated singularity.
2. Isolated singularity.

A point z_0 is a **nonisolated singularity** of a function f if and only if z_0 is a singularity of f and *every neighborhood* of z_0 contains at least one singularity of f other than z_0. For example, the function

$$f(z) = \text{Log } z$$

has a nonisolated singularity at every point of the nonpositive real axis. In general, every function with which there is associated a branchcut possesses nonisolated singularities. Since, by definition, every deleted neighborhood of a nonisolated singularity of a function f contains at least one other singularity of f, it follows that if a function has one nonisolated singularity, then it has infinitely many singularities, *although not necessarily nonisolated;* see Exercise 28.17.

Suppose now that z_0 is a singularity of a function $f(z)$. Then z_0 will be called an **isolated singularity** of f, provided that a deleted neighborhood of z_0 exists, throughout which f is analytic. For example, the function

$$f(z) = \frac{4i}{z^2 + 1}$$

has two isolated singularities, one at $+i$ and one at $-i$. This is not difficult to see, since a deleted neighborhood of radius 1 (or less) can be drawn about either of these two points through which f is analytic.

Isolated singularities are further classified as follows. Suppose that z_0 is an isolated singularity of a function $f(z)$ Then $f(z)$ is analytic throughout a deleted neighborhood $N^*(z_0, \rho)$, i.e., throughout the annulus

$$0 < |z - z_0| < \rho.$$

Therefore, f possesses a Laurent series expansion

$$f(z) = \sum_{n=-\infty}^{\infty} c_n(z - z_0)^n. \tag{1}$$

The following three possibilities present themselves.

C A S E 1: *No negative powers of $(z - z_0)$ appear in (1).* In this case, z_0 is called a **removable singularity.** For instance, the function

$$f(z) = \frac{\sin z}{z}$$

has a removable singularity at $z_0 = 0$ since its series expansion about that point is

$$\frac{\sin z}{z} = 1 - \frac{z^2}{3!} + \frac{z^4}{5!} - \cdots, \tag{2}$$

which contains no negative powers of z; i.e., the series is actually a Taylor series. Note, then, that the series in Equation (2) is well defined at $z = 0$, where it attains the value 1. This fact suggests that the *superficial singularity* at $z = 0$ may be *removed* by properly defining the function at that point. The process by which this is done is quite natural. Thus, by taking limits in (2), as $z \to 0$, we obtain

$$\lim_{z \to 0} \frac{\sin z}{z} = 1,$$

which, in turn, suggests that we define

$$f(0) = 1,$$

and thus the apparent singularity has been removed.

In general, if $f(z)$ has a removable singularity at z_0, then, upon defining the function to have the value c_0 at z_0, $f(z_0) = c_0$, one may argue that f is analytic at z_0; see Exercise 28.21.

C A S E 2: *Only a finite number of negative powers of $(z - z_0)$ with nonzero coefficients appear in (1).* In this case, (1) takes the form

$$f(z) = \sum_{n=-N}^{\infty} c_n(z - z_0)^n$$

$$= \frac{c_{-N}}{(z - z_0)^N} + \cdots + c_0 + c_1(z - z_0) + \cdots, \qquad (3)$$

where N is a positive integer and $c_{-N} \neq 0$. Then z_0 is called a **pole of order** N. For example, we will soon be able to verify that

$$f(z) = \frac{1}{z^3} \qquad \text{has a pole of order 3 at } z = 0,$$

$$g(z) = \frac{1}{(z + 3i)^{15}} \qquad \text{has a pole of order 15 at } z = -3i,$$

$$h(z) = \frac{1}{(z - i)(z + 2)} \qquad \text{has a pole of order 1 at each of } z = i \text{ and } z = -2,$$

$$k(z) = \frac{e^z - 1}{z^2} \qquad \text{has a pole of order 1 at } z = 0.$$

In Chapter 9 it is proved that if a function has a pole at z_0, then, as $z \to z_0$, $\lim f(z) = \infty$.

The part of the Laurent series of a function $f(z)$ that contains the negative powers of $(z - z_0)$ is called the **principal part** of f at z_0.

Again, suppose that $f(z)$ has a pole of order N at z_0. Then f has a Laurent expansion given in (3) with $c_{-N} \neq 0$. Multiplying (3) by $(z - z_0)^N$, one obtains

$$(z - z_0)^N f(z) = c_{-N} + c_{-N+1}(z - z_0) + c_{-N+2}(z - z_0)^2 + \cdots. \qquad (4)$$

The right-hand side of Equation (4) is, obviously, a Taylor series having a positive radius of convergence. Therefore, the function

$$g(z) = (z - z_0)^N f(z)$$

is analytic and the pole of f has become a removable singularity of g. Moreover, in view of (4), we see that

$$\lim_{z \to z_0} (z - z_0)^N f(z) = c_{-N} \neq 0.$$

Thus, if $f(z)$ has a pole of order N at z_0, then $g(z)$, as defined above, has a removable singularity at z_0 and its limit as z approaches z_0 is nonzero. It turns out that the converse of the preceding statement is also true; see Exercise 28.15. We then have the following theorem, which completely characterizes a pole.

Theorem 7.2

Suppose that $f(z)$ is analytic throughout a deleted neighborhood

$$0 < |z - z_0| < \rho$$

of a point z_0.
Then

$$z_0 \text{ is a pole of } f \text{ of order } N$$

if and only if $(z - z_0)^N f(z)$ has a removable singularity at the point z_0 and

$$\lim_{z \to z_0} (z - z_0)^N f(z) \neq 0.$$

Proof:
See Exercise 28.15.

In the next section, where we shall be interested in identifying poles of analytic functions, Theorem 7.2 will prove to be a very convenient tool.

C A S E 3: *The principal part of (1) contains an infinite number of negative powers of $(z - z_0)$, with nonzero coefficients.* In this case, z_0 is called an

essential singularity of the function. For example, the function

$$f(z) = \sin\left(\frac{1}{z}\right)$$

has an essential singularity at $z = 0$, since

$$\sin\left(\frac{1}{z}\right) = \frac{1}{z} - \frac{1}{3!z^3} + \frac{1}{5!z^5} - \cdots.$$

The Casorati–Weierstrass theorem, which is proved in Chapter 9, shows that the behavior of a function near one of its essential singularities is extremely complicated. More precisely, the theorem shows that if z_0 is an essential singularity of $f(z)$, then the value of $f(z)$, as $z \to z_0$, can be made to approach any conceivable limit by an appropriate choice of the values attained by z as it approaches z_0. This extraordinary fact demonstrates the high degree of instability of a function near one of its essential singularities. Closely associated with this behavior is the following theorem, which we state without proof.

Theorem   (*Picard's Theorem*)
In every neighborhood of each of its essential singularities, a function takes on every conceivable finite value, with one possible exception, an infinite number of times.

For an illustration of this theorem, see Exercise 28.16.

EXAMPLE

1. The function $f(z) = \dfrac{z + 1}{z^2 + 1}$ has two poles: $z = i$ and $z = -i$, each or order 1. Let us verify this for $z = i$ by use of Theorem 7.2. First, we form the function

$$(z - i)f(z) = \frac{z + 1}{z + i}$$

as prescribed by the theorem. It is then easy to see that this function has a removable singularity at $z = i$ (it is analytic there) and, furthermore, as $z \to i$, $\lim (z - i)f(z) = (1 + i)/2i \neq 0$. Since these two conditions are satisfied, then the theorem asserts that f has a pole of order 1 at $z = i$.

2. The function $f(z) = \dfrac{e^z - 1}{z^2}$ appears to have a pole of order 2 at $z = 0$.

However, a careful examination reveals that the pole is, in fact, of order 1 since

$$\frac{e^z - 1}{z^2} = \frac{1}{z^2}\left(\sum_{n=0}^{\infty} \frac{z^n}{n!} - 1\right)$$

$$= \frac{1}{z} + \frac{1}{2!} + \frac{z}{3!} + \cdots.$$

3. Again, although the function $f(z) = \sinh z/z^3$ appears to have a pole of order 3 at $z = 0$, its pole is actually of order 2. For, as the reader may verify by supplying the missing details,

$$f(z) = \frac{1}{z^2} + \frac{1}{3!} + \frac{z^2}{5!} + \cdots.$$

We turn now briefly to the concept of the zeros of an analytic function. If $\Phi(z)$ is an arbitrary function, then a number ζ is called a *zero* of Φ if and only if $\Phi(\zeta) = 0$. A special form of this definition is used to define a zero of an analytic function.

Suppose that $f(z)$ is analytic at a point z_0. Then f has a Taylor series expansion

$$f(z) = \sum_{n=0}^{\infty} a_n(z - z_0)^n.$$

If $a_0 = a_1 = a_2 = \cdots = a_{N-1} = 0$ and $a_N \neq 0$, then, of course,

$$f(z) = a_N(z - z_0)^N + a_{N+1}(z - z_0)^{N+1} + \cdots$$

and, if this is the case, f is said to have a **zero of order** N or of **multiplicity** N at z_0.

REMARK 1

If $g(z)$ is a function analytic at z_0, then, of course, it possesses a series expansion

$$g(z) = \sum_{n=0}^{\infty} a_n(z - z_0)^n.$$

If, in addition, $g(z_0) \neq 0$, i.e., if $a_0 \neq 0$, then the function

$$f(z) = (z - z_0)^N g(z)$$

has a zero of order N at z_0. We may thus formulate the following criterion:

A function $f(z)$ has a zero of order N at z_0, provided that f can be put in the form $f(z) = (z - z_0)^N g(z)$, where $g(z)$ is analytic at z_0 and $g(z_0) \neq 0$.

REMARK 2

As a consequence of Remark 1, it can be shown that if z_0 is a zero of a function $f(z)$, analytic in some neighborhood of z_0, and if $f(z)$ is not the zero function, then there is a deleted neighborhood of z_0 throughout which $f(z) \neq 0$. In short,

the zeros of an analytic function are isolated.

EXERCISE 28

A

In Exercises 28.1–28.10, determine the type of each singularity of the given function. If the singularity is removable, define the function at that point so that it will be analytic there.

28.1. $\dfrac{z^2 + 1}{z}$.

28.2. $\cos\left(\dfrac{1}{z}\right)$.

28.3. $\dfrac{e^z - \cos(z - 1)}{z - 1}$.

28.4. $\dfrac{1}{(z - 1)(z - 2)^2}$.

28.5. $\dfrac{\cos z - 1}{z^2}$.

28.6. $\dfrac{\cos(z + i) - 1}{(z + i)^4}$.

28.7. $\dfrac{2}{z^2} + \dfrac{3}{z - \pi}$.

28.8. $\dfrac{z^2 - 3z + 2}{z - 2}$.

28.9. $\dfrac{z^3 + 2z^2 - 1}{z + 1}$.

28.10. $\dfrac{e^{2z} - 1}{z^4}$.

In Exercises 28.11–28.14, use either the definition of a zero or the criterion of Remark 1, this section, to verify that, in each case, the given point is a zero of the respective function. Also, in each case, find the order of the zero.

28.11. $z^2 - 1$, $z_0 = -1$.

28.12. $z^4 - 2z^3 + 2z - 1$, $z_0 = 1$.

28.13. $\sin z$, $z_0 = 0$.

28.14. $\cos z$, $z_0 = \pi/2$.

B

28.15. In connection with Theorem 7.2, study the discussion preceding the statement of the theorem. Then, prove the converse:

> If $(z - z_0)^N f(z)$ has a removable singularity at z_0 and if $\lim (z - z_0)^N f(z) \neq 0$, as $z \to z_0$, then $f(z)$ has a pole of order N at z_0.

28.16. Give an illustration of Picard's theorem by verifying that the equation

$$e^{1/z} = i$$

is satisfied by an infinite number of values of z in every neighborhood of $z = 0$.

HINT: Write $i = e^{(\pi/2 + 2k\pi)i}$.

28.17. Show that the function

$$f(z) = \frac{1}{\sin\left(\dfrac{\pi}{z}\right)}$$

has infinitely many singularities, only one of which is nonisolated.

C

28.18. Determine the type of singularity of the function

$$f(z) = \frac{1}{z^2(e^z - 1)}$$

at $z = 0$.

28.19. Investigate the function $f(z) = z \csc z$ for singularities at $z = 0$. If any exist, identify their types.

28.20. Prove that if z_0 is a pole of order N of a function $f(z)$, then z_0 is a zero of order N of the function $1/f(z)$.

28.21. Review Section 28 through Case 1 and prove the assertion made there to the effect that if z_0 is a removable singularity of $f(z)$, then, by an appropriate definition, f can be shown to be analytic at z_0.

Section 29
Theory of Residues

The topic of our discussion in this section is one of the most important and most often used tools that the "applied scientist" extracts from the theory of complex variables. Residue theory is employed in a wide variety of applications ranging from the evaluation of real integrals to stability of linear systems to image evaluation in photographic science.

In this section we develop and illustrate some of the fundamental techniques employed in complex integration by use of residue theory. The basic element of the theory is to be found in Theorem 7.1 and the discussion immediately following the statement of that theorem.

Let $f(z)$ be an analytic function and suppose that z_0 is an isolated singularity of f. Then, by definition, f is analytic in a deleted neighborhood of z_0

$$A : 0 < |z - z_0| < r,$$

which is a circular annulus centered at z_0. By virtue of Theorem 7.1, f has a Laurent expansion that converges to f for all z in A; see Figure 7.6. The coefficients of this expansion are given by the formula

$$c_n = \frac{1}{2\pi i} \int_C \frac{f(z)}{(z - z_0)^{n+1}} \, dz,$$

where C is any positively oriented, simple closed path lying entirely in A and enclosing z_0 in its interior. In particular, for $n = -1$ we have

$$c_{-1} = \frac{1}{2\pi i} \int_C f(z) \, dz.$$

The number c_{-1}, which is the coefficient of the term $\dfrac{1}{z - z_0}$ in the Laurent expansion of f over A, is called the **residue** of f at z_0; we will denote it

$$\text{Res} \, [f, z_0]$$

so that, by definition,

$$\text{Res} \, [f, z_0] = c_{-1}.$$

In the context of the preceding discussion we may then write

$$\int_C f(z) \, dz = 2\pi i \, \text{Res} \, [f, z_0],$$

and this formula is the fundamental element in the applications of the theory of residues to which we alluded earlier in this section. By use of the multiple annulus theorem (p. 194), a generalization of Equation (1) is immediate, as the next result shows.

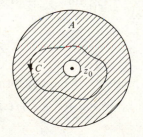

FIGURE 7.6 RESIDUE AT z_0

Theorem 7.3 *(Residue Theorem)*

Suppose that $f(z)$ is analytic on and inside a positively oriented, simple closed path C, except at a finite number of points $z_1, z_2, \ldots, z_n$, each of which is an isolated singularity of f.

Then

$$\int_C f(z) \, dz = 2\pi i \, (\text{Res} \, [f, z_1] + \cdots + \text{Res} \, [f, z_n]).$$

Proof:

See Appendix 7.

The effectiveness of the above theorem depends, of course, on how efficiently one can evaluate the residues of f at the various singularities. The first step is to recognize the type of singularity—a task that, in most cases, is not difficult. However, caution must be exercised to avoid reaching hasty conclusions based on appearances; e.g., see parts (2) and (3) in the example in Section 28. Having recognized the type of singularity, then the following course of action is effective in most cases.

1. If z_i is a removable singularity of f, then, of course, Res $[f, z_i] = 0$, since the series expansion of f about z_i is in fact a Taylor series, and hence $c_{-1} = 0$.

2. If z_i is an essential singularity of f, then a direct expansion of f in series about z_i will be necessary in most cases in order to find the residue.

3. If z_i is a pole, then, first, determine the order of the pole by the methods of Section 28; in many instances, this process will also yield the value of the residue. If not, then use the formula of Theorem 7.4 or the formula of its corollary.

Theorem 7.4

Suppose that $f(z)$ has a pole of order n at a point z_0.

Then

$$\text{Res} \, [f, z_0] = \frac{1}{(n-1)!} \lim_{z \to z_0} \frac{d^{(n-1)}}{dz^{(n-1)}}[(z - z_0)^n f(z)].$$

Proof:

See Appendix 7.

For $n = 1, 2,$ and 3, respectively, the above formula becomes

$$\text{Res} \, [f, z_0] = \lim_{z \to z_0} [(z - z_0) f(z)],$$

$$\text{Res} \, [f, z_0] = \lim_{z \to z_0} \frac{d}{dz}[(z - z_0)^2 f(z)],$$

and

$$\text{Res}\,[f, z_0] = \frac{1}{2} \lim_{z \to z_0} \frac{d^2}{dz^2}[(z - z_0)^3 f(z)].$$

We illustrate the above two theorems and the formulas therein by means of the following examples.

EXAMPLE 1

Evaluate $\displaystyle\int_C \frac{e^z}{z^2}\, dz$, where C is the unit circle $|z| = 1$, positively oriented.

From

$$\frac{1}{z^2} e^z = \frac{1}{z^2}\left(1 + z + \frac{z^2}{2!} + \cdots\right) = \frac{1}{z^2} + \frac{1}{z} + \frac{1}{2!} + \frac{z}{3!} + \cdots$$

we find that $\text{Res}\left[\dfrac{e^z}{z^2}, 0\right] = 1$. Therefore,

$$\int_C \frac{e^z}{z^2}\, dz = 2\pi i \cdot \text{Res}\left[\frac{e^z}{z^2}, 0\right] = 2\pi i.$$

EXAMPLE 2

Use residues to evaluate the integral of $f(z) = \dfrac{1}{(z - 1)(z + 1)}$ along $C : |z| = 3$, positively oriented.

Clearly, f has poles at $z = 1$ and $z = -1$, each of order 1. Using Theorem 7.4 with $n = 1$, we find that

$$\text{Res}\,[f, 1] = \lim_{z \to 1} \frac{1}{z + 1} = \frac{1}{2}$$

and

$$\text{Res}\,[f, -1] = \lim_{z \to -1} \frac{1}{z - 1} = -\frac{1}{2}.$$

Therefore, by Theorem 7.3,

$$\int_C \frac{dz}{(z - 1)(z + 1)} = 2\pi i\,(\text{Res}\,[f, 1] + \text{Res}\,[f, -1]) = 0.$$

EXAMPLE 3

Evaluate the integral of $f(z) = \dfrac{e^{iz} - \sin z}{(z - \pi)^3}$ along $C : |z - 3| = 1$, positively oriented.

First, by use of Theorem 7.2, we determine the order of the pole at $z = \pi$, which is $n = 3$ since the two conditions of the theorem are met:

$$(z - \pi)^3 f(z) = e^{iz} - \sin z$$

is analytic at $z = \pi$, and

$$\text{as } z \to \pi, \lim (z - \pi)^3 f(z) = e^{i\pi} - \sin \pi \neq 0.$$

We then proceed to find that

$$\text{Res } [f, \pi] = \frac{1}{2} \lim_{z \to \pi} \frac{d^2}{dz^2} (e^{iz} - \sin z) = \frac{1}{2}.$$

Therefore,

$$\int_C \frac{e^{iz} - \sin z}{(z - \pi)^3} \, dz = 2\pi i \text{ Res } [f, \pi] = \pi i,$$

and the given integral has been evaluated.

EXAMPLE 4

Evaluate the integral of $f(z) = \dfrac{e^{2z} - z^3}{(z + 1)(z + 2)(z - 1)}$ along $C : |z| = 3$, positively oriented.

The reader should verify, by use of Theorem 7.2, that f has poles of order 1 at each of the three points $z = -1, -2,$ and 1, all of which lie in Int (C). Then, by use of Theorem 7.4, one finds that

$$\text{Res } [f, -1] = \lim_{z \to -1} \frac{e^{2z} - z^3}{(z + 2)(z - 1)} = -\frac{1}{2}(e^{-2} + 1),$$

$$\text{Res } [f, -2] = \lim_{z \to -2} \frac{e^{2z} - z^3}{(z + 1)(z - 1)} = \frac{1}{3}(e^{-4} + 8),$$

and

$$\text{Res } [f, 1] = \lim_{z \to 1} \frac{e^{2z} - z^3}{(z + 1)(z + 2)} = \frac{1}{6}(e^2 - 1).$$

Hence

$$\int_C f(z) \, dz = \frac{\pi i}{3}(e^2 + 2e^{-4} - 3e^{-2} + 12).$$

The next result establishes a rule for finding the residue at a **simple pole** (i.e., a pole of order 1), which is an easy consequence of Theorem 7.4. Quite often, this rule is much easier to apply than the formula of Theorem 7.4.

Corollary

Suppose that

1. $f(z)$ and $g(z)$ are both analytic at a point z_0.

2. $f(z_0) \neq 0$.

3. $g(z)$ has a zero of order 1 at z_0.

Then

$$\operatorname{Res}\left[\frac{f(z)}{g(z)}, z_0\right] = \frac{f(z_0)}{g'(z_0)}.$$

Proof:

See Exercise 29.21.

EXAMPLE 5

1. Consider the rational function $h(z) = \dfrac{z^2 + 2z + 5}{z - i}$.

The functions composing the numerator and denominator of $h(z)$ are both analytic at $z = i$ and the denominator has a zero of order 1 at that point. Also, the numerator does not vanish at $z = i$. Therefore, using the above corollary, we find that

$$\operatorname{Res}\left[h(z), i\right] = \frac{-1 + 2i + 5}{1} = 4 + 2i.$$

2. From Exercise 28.13, we know that $\sin z$ has a zero of order 1 at the point $z = 0$. Also, the functions e^{iz} and $\sin z$ are analytic at $z = 0$, and e^{iz} does not vanish there. Therefore, again by use of the above corollary, we find that

$$\operatorname{Res}\left[\frac{e^{iz}}{\sin z}, 0\right] = \frac{e^{0i}}{\cos 0} = 1.$$

3. The reader will find it instructive to verify that all the necessary conditions are satisfied for use of the above corollary in the following calculation:

$$\operatorname{Res}\left[\frac{e^{iz}\cos z}{z(z - \pi)}, \pi\right] = \frac{e^{i\pi}\cos \pi}{2\pi - \pi} = \frac{1}{\pi}.$$

EXERCISE 29

A

In Exercises 29.1–29.10, find the residue of the given function at each of its singularities.

29.1. $\dfrac{e^{2z} - 1}{z}$.

29.2. $\dfrac{z^2 + 1}{z - 1}$.

29.3. $\dfrac{(1 - z^2)e^{2z}}{z^4}$.

29.4. $\dfrac{\sinh z}{z^2}$.

29.5. $\dfrac{z^2 - 1}{(z - 2)(z + 1)(z - \pi)}$.

29.6. $\dfrac{\sin (z - 1)}{(z - 1)^3}$.

29.7. $\dfrac{ze^z}{z^4 - z^2}$.

29.8. $\dfrac{\tan z}{z^3}$, for $|z| < 1$.

29.9. $\dfrac{e^z}{z^4 - z^2}$.

29.10. $\dfrac{\cos (z - \pi) + 1}{z^5}$.

In Exercises 29.11–29.16, evaluate the integral of the given function along the respective path, oriented positively.

29.11. $\dfrac{\tan z}{z^3}$, $|z| = 1$.

29.12. $\dfrac{(z^2 + 1)e^z}{(z + i)(z - 1)^3}$, $|z| = \dfrac{1}{2}$.

29.13. $\dfrac{(z^2 + 1)e^z}{(z + i)(z - 1)^3}$, $|z - 2| = 3$.

29.14. $\dfrac{1}{z^2(z - 1)(z + \pi i)}$, $|z| = 2$.

29.15. $e^{1/z}$, $|z| = 6$.

29.16. $\dfrac{z}{z^4 - 1}$, $|z| = 4$.

B

In Exercises 29.17–29.20, proceed as in Exercises 29.11–29.16.

29.17. $\dfrac{e^z}{z^3 - 2z^2 + z}$, $|z| = 2$.

29.18. $\dfrac{e^{3z} + \cos 2z}{z(z - 1)}$, $|z - 1| = 2$.

29.19. $\tanh z$, $|z| = 1$.

29.20. $\dfrac{z^3}{(z^4 - 1)^2}$, $|z| = 2$.

C

29.21. Prove the corollary on p. 299.

HINT: $g(z) = (z - z_0)h(z)$, where $h(z_0) \neq 0$. Then use the formula of Theorem 7.4 for $n = 1$, with f replaced by f/g.

29.22. Evaluate the integral of $f(z) = \sec z/z^{\cdot}$ around the circle $|z| = 2$, positively oriented. (See the Corollary on p. 299.)

REVIEW EXERCISES — CHAPTER 7

1. Find a series expansion for $f(z) = \dfrac{1}{z - z^2}$

(a) on $0 < |z| < 1$.   (b) on $|z| > 1$.
(c) on $0 < |z - 1| < 1$.   (d) on $|z - 1| > 1$.

2. Find a series expansion for $f(z) = \dfrac{1}{z^2 - 3z + 2}$

(a) on $|z| < 1$.   (b) on $1 < |z| < 2$.
(c) on $|z| > 2$.   (d) on $0 < |z - 1| < 1$.
(e) on $0 < |z - 2| < 1$.

3. If $C : |z| = 10$, traversed positively, use residues to evaluate

$$\int_C \frac{z^2 - z + 1}{(z - 1)(z + 3)(z - 4)}\, dz.$$

4. If $C : |z - z_0| = 1$, positively oriented, use residues to verify that

$$\int_C \frac{dz}{z - z_0} = 2\pi i.$$

5. If $C : |z| = 2$, positively oriented, verify that

$$\int_C \frac{\sinh z}{z^6}\, dz = \frac{\pi i}{60}.$$

6. (a) Verify that $g(z) = \cosh z$ has a *simple zero* (i.e., a zero of order 1) at $\pi i/2$.

(b) Use part (a) and the corollary to Theorem 7.4 to find

$$\text{Res}\left[\tanh z, \frac{\pi i}{2}\right].$$

7. Find Res $[\sinh z, 0]$.

8. Use residues to evaluate $\int_C \tanh z \, dz$, where $C : |z - 2i| = 1$, positively oriented.

9. Evaluate $\displaystyle\int_C \frac{e^{iz}}{z^4 + 2z^2 + 1} \, dz$, where $C : |z| = 2$, traversed positively.

10. Determine the type of singularity of $f(z) = \dfrac{e^{2z} + \cos 2z - 2}{z^2}$ at $z = 0$.

11. Find Res $[(z - 1) \csc z, 0]$. See Exercise 28.13.

12. Find all singularities of $\dfrac{1}{\cos z - 1}$ and classify them as isolated or non-isolated.

13. Find the Laurent series of z^{-3} on $|z - 1| > 1$.

14. Find Res $[\tan z, \pi/2]$.

15. Find Res $[\cot z, \pi/2]$.

16. The various parts of this exercise, if properly completed, will constitute a proof of the **argument principle**: *Suppose that (1) $f(z)$ is analytic on and inside a positively oriented simple closed path C, except at a finite number of poles in* Int (C); *(2) $f(z)$ has no zeros on C; (3) N_z is the number of zeros of f in* Int (C) *and N_p is the number of poles of f in* Int (C), *where in determining N_z and N_p the orders of the poles and zeros are counted. Then*

$$\frac{1}{2\pi i} \int_C \frac{f'(z)}{f(z)} \, dz = N_z - N_p.$$

(a) If ζ is a zero of f of order m in Int (C), then using Remark 1, p. 292, show that

$$\frac{f'(z)}{f(z)} = \frac{m}{z - \zeta} + \frac{g'(z)}{g(z)},$$

where $g'(z)/g(z)$ is analytic at ζ.

(b) From (a), conclude that $f'(z)/f(z)$ has a pole of order 1 at ζ and, hence, Res $[f'(z)/f(z), \zeta] = m$.

(c) If ξ is a pole of f of order n in Int (C), then show, by use of Theorem 7.2, that

$$\frac{f'(z)}{f(z)} = \frac{-n}{z - \xi} + \frac{h'(z)}{h(z)},$$

where $h'(z)/h(z)$ is analytic at ξ.

(d) As in (b), conclude that Res $[f'(z)/f(z), \xi] = -n$.

(e) Repeat the above processes for each pole and each zero of f in Int (C) to determine N_p and N_z and thus complete the proof.

17. Use the result of the preceding exercise to evaluate the integral of each of the following functions around $|z| = 3$, traversed in the positive sense.

(a) $f(z) = \dfrac{1}{z}$.

(b) $f(z) = \dfrac{2z}{z^2 + 1}$.

(c) $f(z) = \dfrac{z}{z^2 - 1}$.

(d) $f(z) = \dfrac{1/(z + 1)^2}{z/(z + 1)}$.

APPENDIX 7

Proofs of Theorems

Theorem 7.1 *(Laurent's Theorem)*
Suppose that $f(z)$ is analytic at every point of the closed annulus

$$A : r \le |z - c| \le \rho.$$

Then there is a series of positive and negative powers of $(z - c)$ that represents f at every point ζ in the (open) annulus $r < |z - c| < \rho$:

$$f(\zeta) = \sum_{n=0}^{\infty} a_n(\zeta - c)^n + \sum_{n=1}^{\infty} \frac{b_n}{(\zeta - c)^n}.$$

The coefficients of the series are given by the formulas

$$a_n = \frac{1}{2\pi i} \int_K \frac{f(z)}{(z - c)^{n+1}}\, dz, \qquad n = 0, 1, 2, \ldots$$

and

$$b_n = \frac{1}{2\pi i} \int_C \frac{f(z)}{(z - c)^{-n+1}}\, dz, \qquad n = 1, 2, 3, \ldots,$$

where $K : |z - c| = \rho$ and $C : |z - c| = r$, both positively oriented; see Figure 7.7.

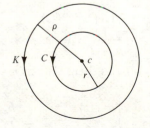

FIGURE 7.7 LAURENT'S THEOREM

Proof:

For simplicity of notation and without loss of generality, the theorem is proved for the case $c = 0$. The extension to the case of an arbitrary center c is immediate.

The basic arguments in this proof are identical with those used in the proof of Taylor's theorem; see p. 271. For this reason, the proof is given in outline; the reader will find it instructive to supply the missing details.

Let ζ be an arbitrary but fixed point such that $r < |\zeta| < \rho$. Then, by the Cauchy integral formula (p. 198),

$$f(\zeta) = \frac{1}{2\pi i} \int_K \frac{f(z)}{z - \zeta}\, dz - \frac{1}{2\pi i} \int_C \frac{f(z)}{z - \zeta}\, dz. \tag{1}$$

The first integral on the right-hand side of (1) may now be treated in exactly the same manner as in the proof of Taylor's theorem [see Equations (1), (2) and (3) on p. 272] to yield

$$\frac{1}{2\pi i} \int_K \frac{f(z)}{z - \zeta}\, dz = \sum_{n=0}^{\infty} a_n \zeta^n, \tag{2}$$

where

$$a_n = \frac{1}{2\pi i} \int_K \frac{f(z)}{z^{n+1}}\, dz.$$

Now, in order to obtain the formulas for b_n we first note that

$$-\frac{1}{z - \zeta} = \frac{1}{\zeta} + \frac{z}{\zeta^2} + \cdots + \frac{z^{n-1}}{\zeta^n} + \frac{z^n}{\zeta^n(\zeta - z)} \tag{3}$$

is an identity whose truth can be established by a method similar to that suggested in Exercise 26.23. Again, as in the proof on p. 272, divide both sides of (3) by $2\pi i$, multiply by $f(z)$, and integrate along the path C in the positive sense. Then, denoting the last term of the resulting expression by

$$R_n = \frac{1}{2\pi i} \int_C \frac{z^n f(z)}{\zeta^n(\zeta - z)}\, dz,$$

show that, as $n \to \infty$, $|R_n| \to 0$. But then, also, as $n \to \infty$,

$$-\frac{1}{2\pi i} \int_C \frac{f(z)}{z - \zeta}\, dz = \sum_{n=1}^{\infty} \frac{b_n}{\zeta^n}, \tag{4}$$

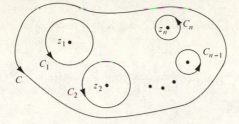

FIGURE 7.8 RESIDUE THEOREM

where

$$b_n = \frac{1}{2\pi i} \int_C \frac{f(z)}{z^{-n+1}} \, dz.$$

Finally, combining Equations (1), (2), and (4), complete the proof of the theorem for the case $c = 0$.

Theorem 7.3   (*Residue Theorem*)
Suppose that $f(z)$ is analytic on and inside a positively oriented simple closed path C, except at a finite number of points $z_1, z_2, \ldots, z_n$, each of which is an isolated singularity of f.
Then

$$\int_C f(z) \, dz = 2\pi i (\operatorname{Res}[f, z_1] + \cdots + \operatorname{Res}[f, z_n]).$$

Proof:
Since each z_k is an isolated singularity of f in Int (C), it is possible to find circles C_k, $k = 1, 2, \ldots, n$, such that each is contained entirely in Int (C), it is centered at the respective z_k, it contains no other singularity in its interior, except the z_k which is its center, and it does not pass through any of the other singularities of f; see Figure 7.8. Then, for each positively oriented C_k, we have

$$\int_{C_k} f(z) \, dz = 2\pi i \operatorname{Res}[f, z_k].$$

Finally, by the multiple annulus theorem (p. 194), we have

$$\int_C f(z) \, dz = \int_{C_1} f(z) \, dz + \int_{C_2} f(z) \, dz + \cdots + \int_{C_n} f(z) \, dz$$

$$= 2\pi i (\operatorname{Res}[f, z_1] + \operatorname{Res}[f, z_2] + \cdots + \operatorname{Res}[f, z_n])$$

and the theorem follows.

Theorem 7.4

Suppose that $f(z)$ has a pole of order n at a point z_0.
Then

$$\text{Res}\,[f, z_0] = \frac{1}{(n-1)!} \lim_{z \to z_0} \frac{d^{(n-1)}}{dz^{(n-1)}}[(z - z_0)^n f(z)].$$

Proof:

By hypothesis, f has a pole of order n at z_0. Therefore,

$$f(z) = \sum_{k=-n}^{\infty} c_k(z - z_0)^k, \qquad c_{-n} \neq 0.$$

Multiplying both sides of this equality by $(z - z_0)^n$, we obtain

$$(z - z_0)^n f(z) = \sum_{k=-n}^{\infty} c_k(z - z_0)^{k+n}$$

$$= c_{-n} + c_{-n+1}(z - z_0) + \cdots + c_{-1}(z - z_0)^{n-1} + \cdots$$

$$+ \sum_{k=0}^{\infty} c_k(z - z_0)^{k+n},$$

which is a Taylor series and, hence, can be differentiated term by term any number of times. Thus, after taking $n - 1$ derivatives, we have

$$\frac{d^{(n-1)}}{dz^{(n-1)}}[(z - z_0)^n f(z)] = (n - 1)!c_{-1} + \sum_{k=0}^{\infty} \frac{d^{(n-1)}}{dz^{(n-1)}}[c_k(z - z_0)^{k+n}].$$

Now, each term of the last series has a factor $(z - z_0)$. Therefore, if we let $z \to z_0$, the whole series vanishes; i.e.,

$$\lim_{z \to z_0} \frac{d^{(n-1)}}{dz^{(n-1)}}[(z - z_0)^n f(z)] = (n - 1)!c_{-1},$$

from which the assertion of the theorem follows.

PART III
Topics for Further Study

CHAPTER 8
Applications

Section 30 Evaluation of integrals by use of residues: integrals of rational functions of trigonometric functions. Improper integrals of certain types of rational functions.

Section 31 The concept of integration around a branch point is introduced by means of an example.

Section 32 Review of the concepts of the point at infinity and the extended complex plane. Neighborhood at infinity. Singularities at infinity.

Section 33 Discussion of certain special mappings as they are used in applications. Flow around a cylinder; the Joukowski airfoil; flow at the end of a channel; the Schwarz–Christoffel map.

Section 30
Evaluation of Real Integrals

It is often necessary, especially in the various areas of applied mathematics, to employ methods of complex integration in order to evaluate certain types of real integrals that do not yield to the calculus of real variables, and range from the innocent-looking

$$\int_0^{2\pi} \frac{dt}{2 + \sin t}$$

to the more formidable

$$\int_0^{\infty} \cos(x^2)\, dx$$

and to much more complicated ones. The theory of residues provides a very efficient, elegant, and relatively simple method for evaluating many types of nonelementary real integrals. In this section we propose to discuss primarily through examples, some of the simpler types of evaluations of such integrals. Some generalizations will be touched upon, but no proofs will be given here establishing the more general formulas.*

* For a extensive discussion of this subject, the reader is referred to Whittaker and Watson, *Modern Analysis* (New York: Cambridge University Press, 1965), Chap. 6.

INTEGRALS OF RATIONAL FUNCTIONS OF cos *t* AND sin *t*

Integrals of this type have the general form

$$J = \int_0^{2\pi} f(\cos t, \sin t)\, dt,$$

where *f* is a rational function of cos *t* and sin *t*; i.e., it involves only integral powers of trigonometric functions and is finite for $0 \le t \le 2\pi$.

The basic substitution that is employed in the evaluation of such integrals is

$$z = \cos t + i \sin t, \qquad 0 \le t \le 2\pi. \tag{1}$$

Note that, as *t* varies from 0 to 2π, *z* describes the unit circle $C : |z| = 1$ in the positive sense. From Equation (1) we also obtain

$$\frac{1}{z} = \cos t - i \sin t,$$

which combined with (1) yields

$$\cos t = \frac{1}{2}\left(z + \frac{1}{z}\right) \qquad \text{and} \qquad \sin t = \frac{1}{2i}\left(z - \frac{1}{z}\right). \tag{2}$$

Finally, from (1) we have

$$dz = (-\sin t + i \cos t)\, dt = (iz)\, dt$$

and, therefore,

$$dt = \frac{dz}{iz} = -\frac{i}{z}\, dz. \tag{3}$$

Now, if the integrand of *J* is a rational function of sin *t* and cos *t*, then substitution from (2) and (3) will yield an integral

$$\int_C g(z)\, dz,$$

where *g* is a rational function (p. 81) of *z* and hence can be evaluated by the methods developed in Part II; see Exercise 30.16. Some special cases of this procedure are illustrated in the next two examples.

EXAMPLE 1

We evaluate the integral $\int_0^{2\pi} \dfrac{dt}{2 + \cos t}.$

Substituting from (2) and (3), we have

$$\int_0^{2\pi} \frac{dt}{2 + \cos t} = -i \int_C \frac{dz}{z \left[2 + \frac{1}{2} \left(z + \frac{1}{z} \right) \right]}$$

$$= -2i \int_C \frac{dz}{z^2 + 4z + 1}$$

$$= -2i \int_C \frac{dz}{(z + 2 + \sqrt{3})(z + 2 - \sqrt{3})}, \qquad (4)$$

where C is the positively oriented circle $|z| = 1$. The singularities

$$z_1 = -2 - \sqrt{3} \qquad \text{and} \qquad z_2 = -2 + \sqrt{3}$$

of the last integrand, $f(z)$, are poles each of order 1, and are such that z_1 is in Ext (C) and z_2 is in Int (C). Now, using Theorem 7.4 with $n = 1$, we find that

$$\text{Res} \, [f(z), -2 + \sqrt{3}] = \frac{1}{2\sqrt{3}}.$$

Therefore, continuing from (4) and using the residue theorem, we obtain

$$-2i(2\pi i \, \text{Res} \, [f(z), -2 + \sqrt{3}]) = \frac{2\pi}{\sqrt{3}},$$

which is the value of the trigonometric integral that we set out to evaluate.

EXAMPLE 2

Let us evaluate the integral $\int_0^{2\pi} \frac{\sin^2 t}{5 - 4 \cos t} dt$.

Substituting from (2) and (3), we find that

$$\int_0^{2\pi} \frac{\sin^2 t}{5 - 4 \cos t} dt = \int_C \frac{\left[\left(z - \frac{1}{z} \right) \Big/ 2i \right]^2}{5 - 4 \left[\frac{1}{2} \left(z + \frac{1}{z} \right) \right]} \left(-\frac{i}{z} \right) dz$$

$$= -\frac{i}{4} \int_C \frac{z^4 - 2z^2 + 1}{z^2 (2z^2 - 5z + 2)} dz$$

$$= -\frac{i}{4} \int_C \frac{z^4 - 2z^2 + 1}{z^2 (z - 2)(2z - 1)} dz,$$

where again $C : |z| = 1$, positively oriented. Clearly, the integrand $f(z)$ has a pole of order 2 at $z = 0$ [in Int (C)], a simple pole at $z = \frac{1}{2}$ [in Int (C)] and a simple pole at $z = 2$ [in Ext (C)]. We then find that

$$\text{Res}\,[f, 0] = \tfrac{5}{4} \quad \text{and} \quad \text{Res}\,[f, \tfrac{1}{2}] = -\tfrac{3}{4}.$$

Finally, by use of the residue theorem,

$$\int_0^{2\pi} \frac{\sin^2 t}{5 - 4\cos t}\, dt = -\frac{i}{4} \int_C \frac{z^4 - 2z^2 + 1}{z^2(z - 2)(2z - 1)}\, dz$$

$$= -\frac{i}{4}\left[2\pi i\left(\frac{5}{4} - \frac{3}{4}\right)\right]$$

$$= \frac{\pi}{4}.$$

The method illustrated in Example 1 can be generalized to the case of any integral of the form

$$\int_0^{2\pi} \frac{dt}{a + b\cos t} \quad \text{or} \quad \int_0^{2\pi} \frac{dt}{a + b\sin t}$$

under the condition that $|a| > |b|$; see Exercise 30.11. Then, substitutions from (2) and (3) will result in an integral of the form

$$(\text{constant}) \cdot \int_C \frac{dz}{(z - z_1)(z - z_2)},$$

where C is the unit circle, as in the examples, and z_1 and z_2 are the roots of a quadratic equation which will always result in the denominator of the integrand. It is always the case that, of the above two roots, one will be in Int (C) and one will be in Ext (C). The evaluation of the last integral may then be carried out as usual.

CERTAIN TYPES OF IMPROPER INTEGRALS

We first consider integrals of the type

$$\int_{-\infty}^{\infty} f(x)\, dx,$$

where
1. $f(x)$ is a rational function whose denominator does not vanish at any (real) x.
2. As $x \to \pm\infty$, $\lim xf(x) = 0$.

Note that, since $f(x)$ is a rational function, condition 2 is equivalent to the condition that the denominator of f is of power at least two degrees higher than that of the numerator.

Given an integral of this form, consider the complex function $f(z)$. Since $f(x)$ is a rational function, then so is $f(z)$. Therefore, $f(z)$ has a finite number of poles (the zeros of its denominator) $z_1, z_2, \ldots, z_n$ in the upper half-plane. [It will soon become apparent that we are not concerned with the poles of $f(z)$, if any, in the lower half-plane.] We consider now the complex integral

$$\int_C f(z)\, dz,$$

where the path $C = C_1 + C_2$ consists of the upper semicircle C_1 (see Figure 8.1) and the segment $C_2 : -R \le x \le R$ of the real axis, and it is so chosen that $z_1, z_2, \ldots, z_n$ are Int (C). In this context, one may then show that

$$\int_{-\infty}^{\infty} f(x)\, dx = 2\pi i(\text{Res}\,[f, z_1] + \cdots + \text{Res}\,[f, z_n]).$$

(See the footnote on p. 309.)

We illustrate the process by which this evaluation may be carried out in the following example.

EXAMPLE 3

Let us show that $\displaystyle\int_{-\infty}^{\infty} \frac{dx}{(x^2 + 4)^2} = \frac{\pi}{16}.$

The function

$$f(z) = \frac{1}{(z^2 + 4)^2}$$

has two poles, $z_1 = 2i$ and $z_2 = -2i$, each of order 2. We then take a path $C = C_1 + C_2$, as in Figure 8.1, with $R > 2$, so that $z_1 = 2i$, which

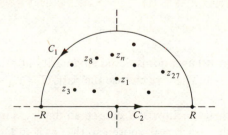

FIGURE 8.1 EXAMPLE 3

is the only singularity of f in the upper half-plane, will be in Int (C). Next, by use of integration by residues, we find that

$$\int_C \frac{dz}{(z^2 + 4)^2} = \frac{\pi}{16},$$

which can also be written

$$\int_{C_1} \frac{dz}{(z^2 + 4)^2} + \int_{-R}^{R} \frac{dx}{(x^2 + 4)^2} = \frac{\pi}{16}. \tag{5}$$

We now complete the problem by taking the limit in (5) as $R \to \infty$. To that end, we note that, since C_1 has the representation

$$z = Re^{it}, \qquad 0 \leq t \leq \pi,$$

$$\lim_{R \to \infty} \left| \int_{C_1} \frac{dz}{(z^2 + 4)^2} \right| = \lim_{R \to \infty} \left| \int_0^{\pi} \frac{iRe^{it}\, dt}{(R^2 e^{2it} + 4)^2} \right| = 0.$$

On the other hand,

$$\lim_{R \to \infty} \int_{-R}^{R} \frac{dx}{(x^2 + 4)^2} = \int_{-\infty}^{\infty} \frac{dx}{(x^2 + 4)^2}.$$

In view of (5), the assertion of the problem follows.

The next example illustrates a method for the evaluation of an improper integral whose integrand is not a rational function. The method that will be used presupposes knowledge of some basic properties of multivalued functions; see pp. 125–32.

EXAMPLE 4

Evaluate the integral $\displaystyle\int_0^{\infty} \frac{dx}{\sqrt{x}(x + 1)}$.

First, we consider the complex function

$$f(z) = \frac{1}{z^{1/2}(z + 1)},$$

and we note that it has a nonisolated singularity at $z_1 = 0$ and a pole of order 1 at $z_2 = -1$. Then we choose the path $C = C_1 + C_2 + C_3 + C_4$, where

1. C_1 is the circle $|z| = R$, with R chosen so that z_2 is in Int (C_1).
2. C_3 is the circle $|z| = r > 0$, chosen so that z_2 is in Ext (C_3).
3. Each of C_2 and C_4 is the segment $r \leq x \leq R$ of the real axis.

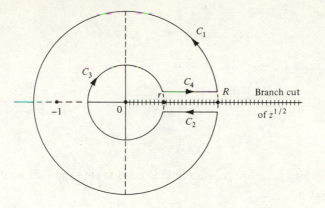

FIGURE 8.2 EXAMPLE 4

The above four smooth curves are oriented as shown in Figure 8.2. (In spite of their appearance in the drawing, C_2 and C_4 are actually *on* the real axis.) Clearly, since the only singularity of $f(z)$ in Int (C) is the simple pole at -1,

$$\int_C f(z)\, dz = 2\pi i \operatorname{Res} [f, -1] = 2\pi;$$

equivalently, we have

$$\int_{C_1} f(z)\, dz + \int_{C_2} f(z)\, dz + \int_{C_3} f(z)\, dz + \int_{C_4} f(z)\, dz = 2\pi. \qquad (6)$$

Now, letting $R \to \infty$ and $r \to 0$, it is possible to show (see Exercise 30.12) that

$$\lim_{R \to \infty} \int_{C_1} f(z)\, dz = 0 \qquad \text{and} \qquad \lim_{r \to 0} \int_{C_3} f(z)\, dz = 0.$$

Therefore, (6) now becomes

$$\lim_{\substack{R \to \infty \\ r \to 0}} \left[\int_{C_2} f(z)\, dz + \int_{C_4} f(z)\, dz \right] = 2\pi. \qquad (7)$$

Since $f(z)$ is a multivalued function with the real axis as its branch cut, we must choose the argument of z as it traverses C_4 and C_2. Taking the argument of z on C_4 to be 0, the argument of z on C_2 must then be 2π. Hence

$$\text{on } C_4 : z^{1/2} = e^{(1/2)[\log z]} = e^{(1/2)[\ln x + 0i]} = \sqrt{x},$$

whereas

$$\text{on } C_2 : z^{1/2} = e^{(1/2)[\log z]} = e^{(1/2)[\ln x + 2\pi i]} = -\sqrt{x}.$$

Then, since $C_4 = -C_2$, we have

$$\int_{C_2} f(z)\, dz = \int_{C_2} \frac{dx}{-\sqrt{x}(x+1)} = -\int_{C_2} \frac{dx}{\sqrt{x}(x+1)}$$

$$= \int_{C_4} \frac{dx}{\sqrt{x}(x+1)} = \int_{C_4} f(z)\, dz.$$

Thus, substituting appropriately in (7) and taking the indicated limits, we find that

$$2 \int_0^\infty \frac{dx}{\sqrt{x}(x+1)} = 2\pi \quad \text{and, hence,} \quad \int_0^\infty \frac{dx}{\sqrt{x}(x+1)} = \pi,$$

thereby completing the evaluation of the given integral.

A generalization of the preceding example resulting in the formula

$$\int_0^\infty \frac{dx}{x^{1-\alpha}(1+x)} = \frac{\pi}{\sin(\alpha\pi)}, \quad 0 < \alpha < 1,$$

is discussed in the footnote given on p. 309 and in Goursat, *Functions of a Complex Variable*, Vol. 2, Part 1 (New York: Dover, 1959).

We continue by considering improper integrals of the general form

$$\int_{-\infty}^\infty g(x)e^{i\alpha x}\, dx,$$

where $\alpha > 0$ and $g(x)$ is a rational function whose numerator and denominator have no common factors, the degree of the denominator exceeds that of the numerator, and whose denominator does not vanish for any (real) x. Under these conditions, the complex function

$$f(z) = g(z)e^{i\alpha z}$$

possesses only a finite number of singularities $z_1, z_2, \ldots, z_n$ in the upper half-plane, each of which is a pole; see Figure 8.1. It can then be shown (see the footnote on p. 309) that

$$\int_{-\infty}^\infty g(x)e^{i\alpha x}\, dx = 2\pi i \,(\text{Res}\,[f, z_1] + \cdots + \text{Res}\,[f, z_n]). \tag{8}$$

The above result lends itself, quite effectively, to the evaluation of certain types of improper integrals whose integrand is the product of a rational function of x and sine or cosine terms. An illustration follows.

EXAMPLE 5

Let us show that $\displaystyle\int_0^\infty \frac{x \sin x}{x^2 + 4}\,dx = \frac{\pi e^{-2}}{2}$.

We begin by considering the integral

$$\int_{-\infty}^\infty \frac{xe^{ix}}{x^2 + 4}\,dx.$$

In the notation of the above general discussion, we let

$$g(z) = \frac{z}{z^2 + 4} \qquad \text{and} \qquad f(z) = \frac{z}{z^2 + 4}e^{iz}.$$

Clearly, g is a rational function whose denominator has no zeros on the real axis; also, g has a pole of order 1 in the upper half-plane, namely, $z_1 = 2i$. Now, since Res $[f, z_1] = e^{-2}/2$, we have from (8),

$$\int_{-\infty}^\infty \frac{xe^{ix}}{x^2 + 4}\,dx = \pi i e^{-2},$$

which is the same as

$$\int_{-\infty}^\infty \left[\frac{x \cos x}{x^2 + 4} + i\frac{x \sin x}{x^2 + 4} \right] dx = \pi i e^{-2}.$$

Therefore, equating the imaginary parts in the last equation, we obtain

$$\int_{-\infty}^\infty \frac{x \sin x}{x^2 + 4}\,dx = \pi e^{-2}.$$

But the integrand is an even function.* Hence

$$2 \int_0^\infty \frac{x \sin x}{x^2 + 4}\,dx = \pi e^{-2},$$

from which the equality we set out to verify follows.

The last item in this section concerns the evaluation of the improper integrals

$$\int_0^\infty \cos (x^2)\,dx \qquad \text{and} \qquad \int_0^\infty \sin (x^2)\,dx. \qquad (9)$$

In the process of evaluating these two integrals, we shall make use of the fact that

$$\int_0^\infty e^{-x^2}\,dx = \frac{\sqrt{\pi}}{2},$$

---

* A real function $h(x)$ is said to be an **even function**, provided that $h(-x) = h(x)$ for all x in its domain. For some fundamental properties of such functions, see Exercise 30.13.

a result that can be established by use of methods of the elementary calculus
of real functions; see Exercise 30.14.

EXAMPLE 6

Let us evaluate the two integrals in expression (9).

Consider the simple closed path $C = C_1 + C_2 + C_3$, as shown in
Figure 8.3. Since

$$f(z) = e^{-z^2}$$

is an entire function, it is clear that

$$\int_C e^{-z^2} \, dz = 0.$$

Equivalently, we have

$$\int_{C_1} e^{-x^2} \, dx + \int_{C_2} e^{-z^2} \, dz + \int_{C_3} e^{-z^2} \, dz = 0. \tag{10}$$

Let us show that, as $R \to \infty$, the integral along C_2 approaches zero.
Using Exercises 19.15 and 30.15, we argue as follows. Along C_2,

$$z = Re^{it}, \qquad 0 \le t \le \pi/4.$$

Therefore,

$$\left| \int_{C_2} e^{-z^2} \, dz \right| = \left| \int_0^{\pi/4} e^{-R^2(\cos 2t + i \sin 2t)} i Re^{it} \, dt \right|$$

$$\le \int_0^{\pi/4} \left| e^{-R^2(\cos 2t + i \sin 2t)} \right| \left| i Re^{it} \, dt \right|$$

$$= R \int_0^{\pi/4} e^{-R^2 \cos 2t} \, dt$$

$$\le R \int_0^{\pi/4} e^{-R^2(1 - 4t/\pi)} \, dt *$$

$$= \frac{\pi}{4R} (1 - e^{-R^2}).$$

Clearly, as $R \to \infty$, the last expression approaches zero and, therefore, so
does the integral along C_2.

* See Exercise 30.15.

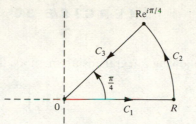

FIGURE 8.3 EXAMPLE 6

We continue now, by evaluating the integral along C_3. For z on C_3,

$$z = te^{i\pi/4}, \qquad 0 \le t \le R.$$

Hence

$$dz = e^{i\pi/4}\, dt \qquad \text{and} \qquad z^2 = t^2 e^{i\pi/2} = it^2.$$

Therefore,

$$\int_{C_3} e^{-z^2}\, dz = \int_R^0 e^{-it^2} e^{i\pi/4}\, dt = e^{i\pi/4} \int_R^0 e^{-it^2}\, dt.$$

Using the above results and taking limits in (10), as $R \to \infty$, we have

$$\int_0^\infty e^{-x^2}\, dx + \lim_{R\to\infty} e^{i\pi/4} \int_R^0 e^{-it^2}\, dt = 0,$$

which, in view of the result of Exercise 30.14, yields

$$\lim_{R\to\infty} \int_0^R e^{-it^2}\, dt = \frac{\sqrt{\pi}}{2} e^{-i\pi/4}$$

and, therefore,

$$\int_0^\infty e^{-it^2}\, dt = \frac{\sqrt{\pi}}{2\sqrt{2}}\, (1 - i). \tag{11}$$

Finally, writing

$$e^{-it^2} = \cos(t^2) - i\sin(t^2),$$

and equating real and imaginary parts in (11), we find the values of the integrals in (9):

$$\int_0^\infty \cos(t^2)\, dt = \frac{\sqrt{\pi}}{2\sqrt{2}} = \int_0^\infty \sin(t^2)\, dt.$$

EXERCISE 30

A

Evaluate the following integrals using methods illustrated in this section.

30.1. $\displaystyle\int_0^{2\pi} \frac{3\,dt}{5 + 2\cos t}.$

30.2. $\displaystyle\int_0^{2\pi} \frac{dt}{3 - \sin t}.$

30.3. $\displaystyle\int_0^{2\pi} \frac{6\,dt}{2 + \sin t}.$

30.4. $\displaystyle\int_{-\infty}^{\infty} \frac{dx}{1 + x^2}.$

30.5. $\displaystyle\int_{-\infty}^{\infty} \frac{x\,dx}{(1 + x^2)^2}.$

30.6. $\displaystyle\int_0^{\infty} \frac{dx}{x^{1/3}(x + 1)}.$

30.7. $\displaystyle\int_0^{\infty} \frac{dx}{1 + x^4}.$

30.8. $\displaystyle\int_0^{2\pi} \frac{\cos t}{3 + \sin t}\,dt.$

30.9. $\displaystyle\int_0^{\infty} \frac{x \sin x}{x^2 + 1}\,dx.$

30.10. $\displaystyle\int_0^{\infty} \frac{x \sin x}{(x^2 + 1)(x^2 + 4)}\,dx.$

B

30.11. Justify the need for the condition $|a| > |b|$ deemed necessary for the evaluation of the trigonometric integrals on p. 312.

30.12. Refer to Example 4 and show that each of the limits preceding relation (7) is indeed zero.

C

30.13. A function $h(x)$ is said to be **even**, provided that $h(-x) = h(x)$ for all x in its domain. A function $g(x)$ is said to be **odd** if and only if $g(-x) = -g(x)$ for all such x.

(a) Verify that the following functions are even:
$$f(x) = x^2; \qquad g(x) = \cos x; \qquad h(x) = 2.$$

(b) Verify that the following functions are odd:
$$f(x) = x; \qquad g(x) = \sin x; \qquad h(x) = x^3.$$

(c) Verify that the following functions are neither odd nor even:
$$f(x) = x + 3; \qquad g(x) = e^x; \qquad h(x) = x^2 + \sin x.$$

(d) Show that the product of either two odd functions or two even functions is always even.

(e) Prove that the product of an odd by an even function is an odd function.

(f) Prove that if $f(x)$ is an even function, then

$$\int_{-\alpha}^{\alpha} f(x)\, dx = 2 \int_{0}^{\alpha} f(x)\, dx$$

and that if $g(x)$ is an odd function, then

$$\int_{-\alpha}^{\alpha} g(x)\, dx = 0$$

for all α such that $0 \le \alpha \le \infty$.

30.14. Write

$$\left[\int_{0}^{\infty} e^{-t^2}\, dt \right]^2 = \left[\int_{0}^{\infty} e^{-x^2}\, dx \right]\left[\int_{0}^{\infty} e^{-y^2}\, dy \right]$$

$$= \int_{0}^{\infty} \int_{0}^{\infty} e^{-(x^2+y^2)}\, dy\, dx.$$

Evaluate the last integral using polar coordinates and then use the result to show that

$$\int_{0}^{\infty} e^{-x^2}\, dx = \frac{\sqrt{\pi}}{2},$$

as claimed on p. 317.

30.15. Use a geometrical or algebraic argument to show that if $0 \le t \le \pi/4$, then $\cos 2t \ge 1 - 4t/\pi$. Conclude that

$$e^{-R^2 \cos 2t} \le e^{-R^2(1-4t/\pi)},$$

thereby providing a justification for the step marked with an asterisk in Example 6.

30.16. Review the discussion preceding Example 1 of this section. Then
  (a) Argue that g has at most a finite number of poles $z_1,\ldots,z_n$ and no other singularities in Int (C).
  (b) Show that

$$J = 2\pi i \sum_{k=1}^{n} \text{Res}\, [g, z_k].$$

30.17. Show that

$$\int_{-\infty}^{\infty} \frac{x \cos x}{x^2 + 4}\, dx = 0$$

in the following two ways:
  (a) As a consequence of our work in Example 5.
  (b) By using parts (d), (e), and (f) of Exercise 30.13.

Section 31
Integration Around a Branch Point

An introduction to the concept of multivalued functions and the related notions of branch, branch point, and branch cut was presented in Appendix 3. Also in that appendix, we acquainted ourselves with the general idea of a Riemann surface, which enables one, via a geometrical configuration, to regard multivalued functions as being single-valued. As we saw there, the basic method consists of associating with each branch of a multivalued function a copy of the complex plane.

At first, this process may leave one with the impression that the whole development is entirely artificial; after all, one may argue, the fact that the various sheets of a Riemann surface are "copies" of the plane does not alter the fact that each one of them is *the plane*. The main objective of this short section is to demonstrate that the various sheets of the Riemann surface associated with a multivalued function are *distinct* in at least one essential way. To that end, we consider the functions

$$w_0 = r^{1/3}e^{it/3}, \qquad r > 0, 0 \le t < 2\pi,$$

$$w_1 = r^{1/3}e^{it/3}, \qquad r > 0, 2\pi \le t < 4\pi,$$

$$w_2 = r^{1/3}e^{it/3}, \qquad r > 0, 4\pi \le t < 6\pi,$$

(see p. 127) and we integrate each of them around a positively oriented circle $C : |z| = R > 0$ to find the following.

$$\int_C w_0 \, dz = \int_0^{2\pi} R^{1/3}e^{it/3}iRe^{it} \, dt = \tfrac{3}{4}R^{4/3}(e^{8\pi i/3} - 1);$$

$$\int_C w_1 \, dz = \int_{2\pi}^{4\pi} R^{1/3}e^{it/3}iRe^{it} \, dt = \tfrac{3}{4}R^{4/3}(e^{16\pi i/3} - e^{8\pi i/3});$$

$$\int_C w_2 \, dz = \int_{4\pi}^{6\pi} R^{1/3}e^{it/3}iRe^{it} \, dt = \tfrac{3}{4}R^{4/3}(1 - e^{16\pi i/3}).$$

The results obtained above lead us to some interesting conclusions, which we may outline as follows.

First, we note that, although the integrand was the same in each case and all three paths of integration were identical but were taken on different copies of the z-plane, the values of the three integrals are different. This result demonstrates the fact that the three sheets of the Riemann surface for $w = z^{1/3}$ are not equivalent in at least one essential way.

Second, comparing these results with the case of an integration around an isolated singularity, where the same value is obtained in each encirclement of such a singularity, we conclude that the nature of an isolated singularity is essentially different from that of a nonisolated singularity, at least in the way in which it affects the value of an integral.

Third, we note that the value of an integral in each case depends on the radius of the circular path which one may choose around the branch point $z = 0$. By way of comparison, we recall that this is not the case when one integrates around isolated singularities, where, within certain limitations, the "size" and the "shape" of the path are immaterial. Again, this contrast demonstrates the different nature of the two types of singularities.

Finally, we note that the values of the three integrals add up to zero. This result is not unrelated to the fact that addition of the three partial results amounts to integrating the function $w = z^{1/3}$ along a path that encircles the branch point $z = 0$ three times, each time on a different sheet of the Riemann surface of $w = z^{1/3}$.

EXERCISE 31

31.1. Evaluate the integral of each of the two branches of $w = z^{1/2}$:

$$w_0 = r^{1/2}e^{it/2}, \qquad r > 0, \quad 0 < t < 2\pi,$$

$$w_1 = r^{1/2}e^{it/2}, \qquad r > 0, \quad 2\pi < t < 4\pi.$$

Use the path $C : |z| = R > 0$.

31.2. Evaluate, as in the preceding exercise, the function $w = z^{1/4}$.

Section 32
Behavior of Functions at Infinity

A brief introduction to the main idea of this section has been discussed in Section 10 in conjunction with the reciprocal function $w = 1/z$. We saw there that the **point at infinity** is an ideal point which we denote by ∞ and which is characterized by the property that

$$|z| < \infty \text{ for every complex number } z.$$

The z-plane augmented with this ideal point is called the **extended complex plane.** The reader who is familiar with the notion of stereographic projection (p. 28) will recall that, under this projection, the point at infinity of the extended complex plane corresponds to the north pole of the Riemann sphere. Later, in Chapter 3, using the reciprocal function

$$w = \frac{1}{z}$$

and motivated by the limiting process

$$z \to 0 \qquad \text{if and only if} \qquad w \to \infty,$$

we identified the behavior of a function $f(z)$ at $z = \infty$ with the behavior of
the function $f(1/z)$ at $z = 0$. In this section we shall explore and exploit
some consequences of the developments that we reviewed above. Before we
begin, however, we should emphasize once again that, although we shall
use expressions such as "the point $z = \infty$," the point at infinity is not to
be treated as an ordinary number, especially when it comes to using algebraic
operations on it.

We begin with the following definition. A **neighborhood** of the point
$z = \infty$ is defined to be the set of all points z such that $|z| > M$, for some
real number $M > 0$, together with the point $z = \infty$ itself; it will be denoted

$$N(\infty, M).$$

Put in different terms, $N(\infty, M)$ is the region exterior to the circle $|z| = M$
together with $z = \infty$. Note that if we set $\varepsilon = 1/M$ we may, equivalently,
define $N(\infty, M)$ as consisting of all z such that $1/|z| < \varepsilon$ together with $z = \infty$.
The **deleted neighborhood** of infinity, denoted $N^*(\infty, M)$, is the set of all z
such that $|z| > M$.

We continue by discussing the notions of analyticity and singular points
at infinity. Let M be a real number such that $0 < M < \infty$ and suppose that
a function $f(z)$ is analytic for all z with $|z| > M$; i.e., suppose that $f(z)$ is
analytic at every z in $N^*(\infty, M)$. Then the function

$$f\left(\frac{1}{w}\right)$$

is analytic through a deleted neighborhood of zero, namely, $N^*(0, 1/M)$.
Therefore, $f(1/w)$ has a Laurent expansion at $w = 0$ given by

$$f\left(\frac{1}{w}\right) = \sum_{n=-\infty}^{\infty} c_n w^n, \qquad \text{for } 0 < |w| < \frac{1}{M}. \tag{1}$$

But then $f(z)$ has a Laurent expansion at $z = \infty$ given by

$$f(z) = \sum_{n=-\infty}^{\infty} c_n \frac{1}{z^n}. \tag{2}$$

As in Section 28 we distinguish three cases.

C A S E 1: *No negative powers of w appear in* (1). In this case, $f(1/w)$ has a
removable singularity at $w = 0$ and, as we saw in Section 28, the function
$f(1/w)$ is analytic at $w = 0$ and its value at that point is c_0. But then, also,

the function $f(z)$ is analytic at $z = \infty$. Now, comparing Equations (1) and (2), we see that when $f(1/w)$ has no negative powers of w, $f(z)$ has no positive powers of z. Therefore, (2) may now be written in the form

$$f(z) = \sum_{0}^{\infty} c_n z^{-n},$$

and if we define $f(\infty) = c_0$, then $f(z)$ has a **removable singularity at $z = \infty$** and, in this case, we say that $f(z)$ is **analytic at infinity.**

C A S E 2: *Only a finite number of negative powers of w with nonzero coefficients appear in* (1). In this case it is known that $f(1/w)$ has a pole of order N (for some positive integer N) at $w = 0$ and, hence, $f(z)$ has a pole of order N at $z = \infty$. Thus (1) takes on the form

$$f\left(\frac{1}{w}\right) = \frac{c_{-N}}{w^N} + \frac{c_{-N+1}}{w^{N-1}} + \cdots + \frac{c_{-1}}{w} + \sum_{n=0}^{\infty} c_n w^n, \qquad \text{with } c_{-N} \neq 0,$$

and, therefore, (2) takes on the form

$$f(z) = \sum_{n=0}^{\infty} c_n z^{-n} + c_{-1} z + c_{-2} z^2 + \cdots + c_{-N} z^N, \qquad \text{with } c_{-N} \neq 0.$$

From this we conclude that a function $f(z)$ has a **pole of order N at $z = \infty$**, provided that its Laurent expansion at infinity has a principal part of the form

$$a_1 z + a_2 z^2 + \cdots + a_N z^N, \qquad \text{with } a_N \neq 0.$$

Theorem 7.2 may now be adapted to give the following characterization of a pole at infinity: *Suppose that $f(z)$ is analytic throughout a deleted neighborhood of infinity. Then*

f has a pole of order N at $z = \infty$
if and only if $z^{-N} f(z)$ has a removable singularity at $z = \infty$ and

$$\lim_{z \to \infty} z^{-N} f(z) \neq 0.$$

C A S E 3: *An infinite number of negative powers of w appear in* (1). In this case we know that $f(1/w)$ has an essential singularity at $w = 0$ and this, in turn, implies that $f(z)$ has an essential singularity at $z = \infty$. Since the series in (1) has an infinite number of negative powers of w, the series in (2) has an infinite number of positive powers of z. Thus a function $f(z)$ has an **essential singularity at infinity**, provided that its Laurent series at $z = \infty$ contains an infinite number of positive powers of z with nonzero coefficients.

Based on the above discussion, the reader should verify the assertions made in the following.

EXAMPLE

1. Any polynomial $f(z) = a_0 + a_1 z + \cdots + a_n z^n, a_n \neq 0$, has a pole of order n at infinity. This fact becomes obvious upon considering the function $f(1/w)$ at $w = 0$.
2. The function $g(z) = 1/(z + z^2)$ is analytic at $z = \infty$, since, as one may easily verify, the function $g(1/w)$ is analytic at the origin.
3. The exponential function $h(z) = e^z$ has an essential singularity at infinity.

EXERCISE 32

A

32.1. Verify the assertions in all three parts of the above example.

32.2. Examine each of the following functions for analyticity, or lack thereof, at the point at infinity.

(a) $f(z) = \dfrac{1}{z + 1}$.   (b) $f(z) = \dfrac{z}{z + 1}$.

(c) $f(z) = e^{1/z^2}$.   (d) $f(z) = \sin z$.

32.3. Classify each of the points listed with each of the following functions as "point of analyticity" or "pole of order ..." or "essential singularity" of the respective function.

(a) $f(z) = 1/z$; $1, 0, \infty$.

(b) $f(z) = \dfrac{\cos z}{z}$; $\pi, 0, \infty$.

(c) $f(z) = \dfrac{e^z - \cos z}{z^2}$; $0, \infty$.

B

32.4. Prove the adaptation of Theorem 7.2 to the case of a pole at infinity; see p. 325.

32.5. Let $f(z)$ be analytic for all z such that $|z| > M$, i.e., throughout a deleted neighborhood of infinity. Then the **residue of f at $z = \infty$** is defined by

$$\text{Res}\,[f, \infty] = \frac{1}{2\pi i} \int_C f(z)\, dz,$$

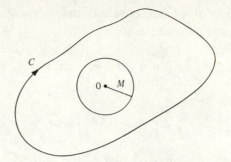

FIGURE 8.4 RESIDUE AT INFINITY

where C is a simple closed path *negatively oriented* and such that $|z| = M$ is in Int (C); see Figure 8.4.

(a) Show that if $\sum_{n=-\infty}^{\infty} a_n z^n$ is the Laurent expansion of a function $f(z)$ at $z = \infty$, then Res $[f, \infty] = -a_{-1}$; in other words, the residue of a function f at infinity is equal to the negative of the coefficient of z^{-1} in the Laurent series expansion of f *with center at infinity.*

(b) Verify that Res $[e^{1/z}, \infty] = -1$, thus illustrating the fact that, unlike the residues at finite points, a function may be analytic at infinity and still have a nonzero residue there.

Section 33
Some Special Transformations

Some simple examples of how physical problems may be formulated in terms of complex function theory were discussed in Appendix 3. In this section we discuss certain special functions as they are employed to represent physical problems in a form that lends itself to analytical treatment.

THE FUNCTION $w = z + 1/z$

The transformation

$$w = z + \frac{1}{z} \tag{1}$$

is used extensively in the study of problems in the theory of flows. The polar form of (1) is

$$w = r(\cos t + i \sin t) + \frac{1}{r}(\cos t - i \sin t).$$

Thus, if we think of this function as a complex potential, then its velocity potential and its stream function are

$$\phi(r, t) = \left(r + \frac{1}{r}\right)\cos t$$

and

$$\psi(r, t) = \left(r - \frac{1}{r}\right)\sin t, \tag{2}$$

respectively. Now, let a long cyclinder of unit radius be placed in a steady, irrotational flow of an incompressible fluid; a planar cross section of the flow perpendicular to the axis of the cylinder is shown in Figure 8.5. Setting (2) equal to a constant yields the streamlines of the flow:

$$\left(r - \frac{1}{r}\right)\sin t = c.$$

In particular, if we let $c = 0$, then either $\sin t = 0$, i.e., $t = 0$ or π, or $r = 1$ and, therefore, we obtain the streamline consisting of the part of the real axis with $|x| \geq 1$ and the unit circle. Of course, as c varies over all real values, we obtain all the streamlines of the flow; see Figure 8.5.

From our discussion on p. 143, we recall that the velocity of the flow is given by the conjugate of the derivative of the complex potential:

$$V = 1 - \frac{1}{(\bar{z})^2}. \tag{3}$$

Clearly, $V = 0$ if and only if $z = \pm 1$; these two points are called the **stagnation points** of the flow. In Exercise 33.1 the reader is asked to prove that the velocity V attains its maximum magnitude at the points $z = \pm i$. For points far away from the cylinder, i.e., points with large $|z|$, it is clear from (1) and (3) that the motion approaches a uniform flow, since then

$$w \simeq z \qquad \text{and} \qquad V \simeq 1.$$

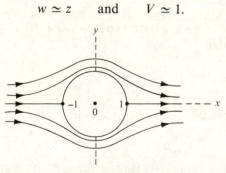

FIGURE 8.5 FLOW AROUND A CYLINDER

We have thus seen that the transformation in (1) may be used to describe the flow around a cylinder.

We continue by investigating some general mapping properties of (1). Earlier, we saw that its real and imaginary component functions are

$$u = \left(r + \frac{1}{r}\right)\cos t \qquad \text{and} \qquad v = \left(r - \frac{1}{r}\right)\sin t.$$

Algebraic manipulation then yields the equation

$$\frac{u^2}{[r + (1/r)]^2} + \frac{v^2}{[r - (1/r)]^2} = 1$$

from which one may derive the following mapping properties of the function under consideration; see the exercises at the end of this section.

(a) Circles $|z| = r$ from the z-plane map onto ellipses in the w-plane each with center at the origin.

(b) All the ellipses referred to in (a) are confocal with foci at $u = \pm 2$.

(c) The circle with $r = 1$ and center at the origin constitutes an exception to (a), for it is mapped onto the segment of the u-axis with $-2 \le u \le 2$. However, as one may easily show, the segment in question is covered twice, once in each direction, so that it may be thought of as a collapsed ellipse; see Figure 8.6(a).

(d) The region of the z-plane which is exterior to the unit circle maps onto the entire w-plane in a one-to-one fashion, upper half onto upper half and lower onto lower; see Figure 8.6(a).

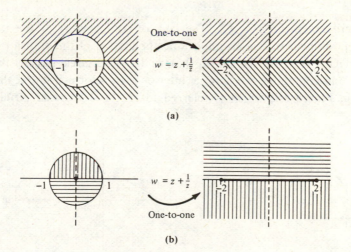

(a)

(b)

FIGURE 8.6 $w = z + 1/z$

(e) The region interior to the unit circle, except $z = 0$, is also mapped onto the w-plane, one-to-one, with the upper half of the disk onto the lower half of the w-plane, and vice versa; see Figure 8.6(b).

(f) Every point of the w-plane, except $w = \pm 2$, is the image of exactly two points from the z-plane. This is easy to see since, from (1), we have

$$z = \tfrac{1}{2}(w \pm (w^2 - 4)^{1/2}).$$

(g) Most of the above facts can also be viewed as follows: Since

$$w(z) = w\left(\frac{1}{z}\right) = z + \frac{1}{z},$$

one may easily argue that any two circles centered at the origin and whose radii are in the relation

$$r_1 = \frac{1}{r_2}$$

map onto the same ellipse in the w-plane. However, the smaller of these two circles is inverted as it is mapped by (1); see property (e), above.

The transformation given in (1) is conformal everywhere except at $z = \pm 1$; this fact is evident since the derivative, given by

$$w' = 1 - \frac{1}{z^2},$$

is nonzero except for $z = \pm 1$. The lack of conformality at these points plays a prominent role in an important application of this transformation in aerodynamics. It can be shown that, under (1), a circle passing through $z = -1$ and containing $z = 1$ in its interior is transformed into a shape as shown in Figure 8.7, which is referred to as the **Joukowski airfoil**. The

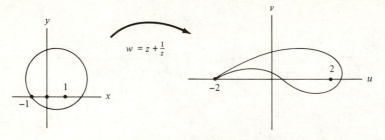

FIGURE 8.7 JOUKOWSKI AIRFOIL

airfoil resembles a cross section of an airplane wing and has a sharp tail edge at the point $w = -2$; note that $w = -2$ is precisely the image of the point $z = -1$, which is one of the two points at which conformality of (1) is destroyed.

THE TRANSFORMATION $w = z + e^z$

We turn now to the transformation

$$w = z + e^z \qquad (4)$$

and we restrict our consideration to the fundamental strip (p. 116) of the z-plane,

$$S : -\pi \le y \le \pi, \ -\infty < x < +\infty;$$

see Figure 8.8(a). The decomposition of (4) is

$$u = x + e^x \cos y \qquad \text{and} \qquad v = y + e^x \sin y. \qquad (5)$$

The images of the horizontal lines $y = c$, in S, represent the streamlines of a fluid flowing through the open channel determined by the rays

$$R_1 : u \le -1, v = 1 \qquad \text{and} \qquad R_2 : u \le -1, v = -1,$$

in the w-plane; see Figure 8.8(b). To see this, let us consider some special cases.

1. If we take the line $y = \pi$, then Equations (5) yield

$$u = x - e^x \qquad \text{and} \qquad v = \pi.$$

As x varies over all the reals, v remains constant at $v = \pi$. More specifically, as x varies from $-\infty$ to 0, u varies from $-\infty$ to -1; on the other hand, as

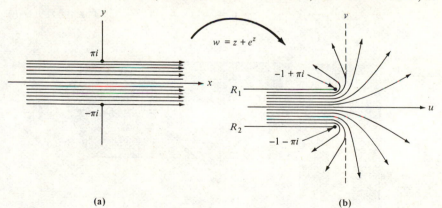

(a)   (b)

FIGURE 8.8 $w = z + e^z$

x varies from 0 to $+\infty$, u retraces the ray R_1 from -1 to $-\infty$. It is not difficult to see that the physical interpretation of this is that a line of flow close to the boundary of the channel will "very nearly turn back" as soon as the flow reaches the opening of the channel into an open field.

2. A similar argument will show that if we take the line $y = -\pi$, then its image will trace the ray R_2 from $-\infty$ to -1 and then back to $-\infty$.

3. If $y = 0$, then

$$u = x + e^x \qquad \text{and} \qquad v = 0$$

and, as x varies from $-\infty$ to $+\infty$, we obtain the entire u-axis, since $v = 0$ and u will range from $-\infty$ to $+\infty$.

The transformation in (4) describes a flow out of a channel or the electrostatic field in the vicinity of the edge of a parallel-plate capacitor.

THE SCHWARZ–CHRISTOFFEL MAP

In potential theory, in connection with boundary value problems, it is often desirable to transform the upper half of the z-plane in a one-to-one fashion onto a region of the w-plane bounded by straight lines, line segments, or rays. The region in question may be a (closed) polygon which may or may not be convex, or, it may be an unbounded region whose boundary, however, consists of lines, segments, or rays; see Figure 8.9. For the remainder of this discussion we shall refer to the boundary of any such region as the "image polygon."

It can be shown that such a general mapping is possible via a transformation $w = f(z)$ such that

$$w' = A(z - x_1)^{-k_1}(z - x_2)^{-k_2} \cdots (z - x_n)^{-k_n}, \qquad A \neq 0, \tag{6}$$

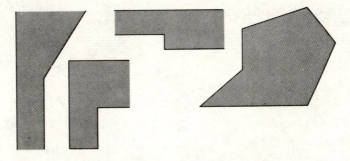

FIGURE 8.9 IMAGE POLYGONS

or, equivalently,

$$w = A \int (z - x_1)^{-k_1}(z - x_2)^{-k_2} \cdots (z - x_n)^{-k_n} \, dz + B, \qquad (7)$$

where A and B are complex constants. The function (7) is called the **Schwarz–Christoffel transformation.**

In order to analyze this mapping, let us first note that, taking the argument of both sides of (6), we obtain

$$\arg w' = \arg A - k_1 \arg(z - x_1) - \cdots - k_n \arg(z - x_n). \qquad (8)$$

The points $x_1, x_2, \ldots, x_n$ are the points on the real axis in the z-plane which, under (7), map onto the "vertices" $w_1, w_2, \ldots, w_n$ of the image polygon in the w-plane; see Figure 8.10.

Now, let us trace a *real point* $z = x$ as it traverses the x-axis. When $x < x_1$, each $\arg(z - x_j)$ in (8) is equal to π; hence $\arg w'$ remains constant until z becomes x_1, at which time the image point w becomes w_1; see Figure 8.10.

When $x_1 < x < x_2$, then $\arg(z - x_1) = 0$ while the other arguments in (8) remain unchanged; hence, at w_1, the argument of w' is changed by $k_1\pi$ and then remains the same until w reaches w_2. Thus, as z traces the segment from x_1 to x_2, w describes a line segment from w_1 to w_2.

Repetition of the process described above will result in mapping the intervals $x_j \leq x \leq x_{j+1}$ onto the sides of the image polygon with vertices $w_1, \ldots, w_n$ and whose exterior angles are $k_1\pi, \ldots, k_n\pi$.

If the angles of deformation are restricted to satisfy

$$-\pi < k_j\pi < \pi, \qquad j = 1, \ldots, n,$$

i.e., if each k_j satisfies $-1 < k_j < 1$, then the function in (7) is continuous for all z with $I(z) \geq 0$. If we further restrict the k_j's to satisfy

$$0 < k_j < 1,$$

then (7) is a single-valued function in the upper half of the z-plane. Moreover, its derivative w' is nonzero except at the points $z = x_j$. It follows, then, that the transformation is conformal in the half-plane $I(z) \geq 0$ except at those points.

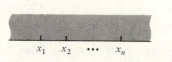
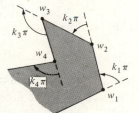

FIGURE 8.10 THE SCHWARZ–CHRISTOFFEL MAP

For an in-depth discussion of conformal mapping in general, as well as of specific transformations, the reader is referred to Copson, *Theory of Functions of a Complex Variable* (New York: Oxford University Press, 1935), Chap. 8, and Titchmarsh, *The Theory of Functions*, 2nd ed. (New York: Oxford University Press, 1939), Chap. 6.

EXERCISE 33

A

33.1. (a) Refer to formula (3) to find the velocity V at $z = \pm i$.

    (b) Use formula (3) to show that $|V| \leq 2$ for every z in the flow around a cylinder of radius 1.

    (c) From (a) and (b) conclude that the $|V|$ is maximum at $z = \pm i$.

33.2. In connection with the transformation $f(z) = z + 1/z$:

    (a) Consider $f(\bar{z})$ and show that conjugate points are mapped onto conjugate points. Conclude that the transformation is symmetric with respect to the real axis.

    (b) Review the concept of inversion in the unit circle, p. 105. Then, consider $f(1/\bar{z})$ and show that, under f, if z is mapped onto w, then $1/\bar{z}$ is mapped onto $\bar{w}$; in short, inverse points with respect to the unit circle map onto conjugate points.

    (c) Consider $f(-\bar{z})$ to show that the transformation is also symmetric with respect to the imaginary axis.

33.3. Review the discussion on pp. 327–29 and then verify properties (a), (b), and (c) on p. 329.

33.4. Show that the function

$$w = z^{1/2},$$

with its domain restricted to the upper half-plane, is a special case of the Schwarz–Christoffel map. Find the image of its domain.

B

33.5. The transformation

$$w = \tfrac{1}{2}(e^z + e^{-z}) = \cosh z$$

is employed in a number of problems in the theory of flows.

    (a) Decompose the function in the form

$$w = u(x, y) + iv(x, y).$$

(b) From u and v, in (a), use algebraic manipulations to eliminate first x and then y to obtain

$$\frac{u^2}{\cos^2 y} - \frac{v^2}{\sin^2 y} = 1 \quad \text{and} \quad \frac{u^2}{\cosh^2 x} + \frac{v^2}{\sinh^2 x} = 1.$$

(c) Consider the horizontal strip of the z-plane between the lines $y = 0$ and $y = \pi$. Then use the formulas in (b) to show that, under the given transformation, horizontal lines in the strip map onto confocal hyperbolas centered at $w = 0$ and vertical line segments in the strip map onto confocal ellipses centered at the origin. Draw the configuration in each plane.

33.6. Show that the function

$$w = iA \sin^{-1} z + B = A \int (z+1)^{-1/2}(z-1)^{-1/2}\, dz + B$$

is a special case of the Schwarz–Christoffel transformation mapping the upper half of the z-plane onto the semi-infinite vertical strip

$$-\frac{\pi}{2} \leq u \leq \frac{\pi}{2}, \qquad v \leq 0,$$

for appropriately chosen constants A and B.

REVIEW EXERCISES — CHAPTER 8

1. Verify that

$$\int_0^{2\pi} \frac{dt}{1 + a \cos t} = \frac{2\pi}{\sqrt{1 - a^2}} \int_0^{2\pi} \frac{dt}{1 + a \sin t}, \quad |a| < 1.$$

2. Verify that

$$\int_{-\infty}^{\infty} \frac{dx}{x^2 + x + 1} = \frac{2\pi}{\sqrt{3}}.$$

3. Verify that

$$\int_0^{\pi} \frac{dt}{1 + \sin^2 t} = \frac{\pi}{\sqrt{2}}$$

by expressing $\sin^2 t$ in terms of the double angle $2t$ and then using the substitution $\theta = 2t$.

4. Verify that

$$\int_{-\infty}^{\infty} \frac{\cos x}{x^2 + 1}\, dx = \pi e^{-1} \qquad \text{and} \qquad \int_{-\infty}^{\infty} \frac{\sin x}{x^2 + 1}\, dx = 0.$$

5. Verify that

$$\int_{-\infty}^{\infty} \frac{x^2}{(x^2 + 1)^3}\, dx = \frac{\pi}{8}.$$

6. Review the mapping properties of the function $w = z + z^{-1}$, which were discussed on p. 329. Examine the inverse (multivalued) function

$$z = \tfrac{1}{2}[w \pm (w^2 - 4)^{1/2}],$$

which maps every point $w \neq \pm 2$ onto two distinct z's and then sketch its Riemann surface, noting that its two sheets are joined along the segment of the real axis between -2 and 2.

CHAPTER 9
Some Theoretical Results

Discussion and proof of a number of theoretical results: the maximum modulus principle; Liouville's theorem; the fundamental theorem of algebra; behavior of an analytic function near a pole and near an essential singularity; the uniqueness of Taylor and Laurent series.

In this chapter we discuss a number of topics that are primarily of theoretical nature, although by no means devoid of practical importance. Some of them are consequences of our development in Part II, while others constitute an extension of that development.

Section 34
The Maximum Modulus Principle

The maximum modulus principle may be expressed in a variety of ways, all of which are essentially equivalent. Simply stated, the principle asserts that if $f(z)$ is analytic and nonconstant on a closed and bounded set S, then $|f(z)|$ attains its maximum value on the boundary of S. We shall begin with a weak version of this result in Theorem 9.1, but first we discuss two other results that we shall need in the proof of that theorem.

The **identity theorem for analytic functions** states that

> if two functions $f(z)$ and $g(z)$ are analytic in a region R and if $f(z) = g(z)$ for all z in a neighborhood N of a point ζ, in R, however small N may be, then $f(z) = g(z)$ at every point of R; in other words, f and g are identical in R.

It is interesting to note that the hypothesis of the identity theorem as stated can be considerably weakened to require only that $f(z) = g(z)$ for an infinity of distinct points having ζ as a limit. This extraordinary property of analytic functions demonstrates once again the strong inner structure possessed by such functions. For further discussion of the identity theorem, see Knopp, *Theory of Functions*, Part 1 (New York: Dover, 1945), Chap. 7.

The second preliminary result that we shall need in the proof of the maximum modulus principle is the following:

> If $f(z)$ is continuous at z_0 and if $|f(z_0)| < M$, then $|f(z)| < M$ for every z in some neighborhood of z_0.

We prove this result as follows. By the continuity of f at z_0, given any $\varepsilon > 0$, the relation

$$|f(z) - f(z_0)| < \varepsilon$$

holds for every z in some δ-neighborhood of z_0. In particular, choosing

$$\varepsilon = M - |f(z_0)| > 0,$$

then for every z in some $N(z_0, \delta)$, we have

$$|f(z)| \leq |f(z) - f(z_0)| + |f(z_0)|$$
$$< \varepsilon + |f(z_0)|$$
$$= M,$$

and the proof of this preliminary result is complete.

Theorem 9.1

Suppose that
1. *$f(z)$ is analytic and nonconstant on a closed region S, i.e., on a set S consisting of a nonempty region R and its boundary.*
2. *$|f(z)|$ attains a maximum on S; i.e., there is at least one point ζ in S such that $|f(z)| \leq |f(\zeta)|$, for all z in S.*
Then ζ is a point of the boundary of S.

Proof:
The proof is by contradiction.

We thus suppose that $|f(\zeta)| = M$ is the maximum of f on S for some interior point of S. Since $f(z)$ is analytic and nonconstant on R (the interior of the set S), it follows that $|f(z)|$ is also nonconstant on R. Then, by the identity theorem for analytic functions, p. 337, there must be points arbitrarily close to ζ for which

$$|f(z)| < M; \tag{1}$$

for, if $|f(z)| = M$ for every point in some neighborhood of ζ, then $f(z)$ would be constant in that neighborhood and, therefore, in all of R, thus contradicting the hypothesis. So, let z_0 be one such point for which (1) holds, chosen in such a way that the circle

$$C : |z - \zeta| = |z_0 - \zeta| = r$$

along with its interior will be in R; see Figure 9.1. Since $|f(z_0)| < M$, it follows from the preliminary result proved above that (1) holds for every z

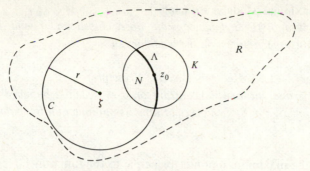

FIGURE 9.1 THEOREM 9.1

in some neighborhood N of z_0 enclosed by a circle, say, K. Now, since $|f|$ attains a maximum on S, there is a number $p > 0$ such that

$$|f(z)| \le M - p$$

for all z in N.

Next, consider the part Λ of C that is enclosed by K and denote the length of Λ by L. Then, using the Cauchy integral formula, p. 198, and Theorem 4.5(5), we have

$$|f(\zeta)| = \frac{1}{2\pi} \left| \int_C \frac{f(z)}{z - \zeta}\, dz \right|$$

$$= \frac{1}{2\pi} \left| \int_{C-\Lambda} \frac{f(z)}{z - \zeta}\, dz + \int_\Lambda \frac{f(z)}{z - \zeta}\, dz \right|$$

$$\le \frac{1}{2\pi} \left| \int_{C-\Lambda} \frac{f(z)}{z - \zeta}\, dz \right| + \frac{1}{2\pi} \left| \int_\Lambda \frac{f(z)}{z - \zeta}\, dz \right|$$

$$\le \frac{1}{2\pi} \frac{M}{r} (2\pi r - L) + \frac{1}{2\pi} \frac{M - p}{r} L$$

$$< M.$$

But since $M = |f(\zeta)|$, we have reached the absurdity $|f(\zeta)| < |f(\zeta)|$ as a result of our assumption that ζ is an interior point of S. It follows, then, that ζ must be a boundary point of S, and the proof is thus complete.

In Appendix 5 it is shown that if a function is continuous on a closed and bounded set S, then the function is bounded on S. As a consequence of this result, one may prove that if a function $f(z)$ is analytic on a closed and bounded set S, then $|f(z)|$ attains a maximum and a minimum on that set. In view of this fact, in Theorem 9.1, hypothesis 1 clearly implies hypothesis 2; we thus have the following sharper form of that theorem.

The Maximum Modulus Principle

If $f(z)$ is analytic and nonconstant on a closed and bounded set S, then $|f(z)|$ attains its maximum on the boundary of S.

Corollary (*The Minimum Modulus Principle*)

If $f(z)$ is analytic and nonconstant on a closed and bounded set S and if $f(z) \neq 0$ for all z in S, then $|f(z)|$ attains its minimum on the boundary of S.

Proof:

Apply the maximum modulus principle to the function $1/f(z)$.

Section 35
Liouville's Theorem. The Fundamental Theorem of Algebra

We recall that a function is called *entire*, provided that it is analytic at every point of the finite plane. Suppose now that $f(z)$ is entire and that c is any point in the plane. Then there is a Taylor series that represents f at every point z:

$$f(z) = \sum_{n=0}^{\infty} a_n(z - c)^n.$$

If $a_n = 0$ for all $n \geq 1$, then, clearly, $f(z)$ is a *constant function*:

$$f(z) = a_0.$$

If $a_k \neq 0$ for some integer $k \geq 1$ but $a_n = 0$ for all $n > k$, then $f(z)$ is a *polynomial* of degree k:

$$f(z) = a_0 + a_1 z + c_2 z^2 + \cdots + a_k z^k.$$

Finally, if $a_n \neq 0$ for an infinite number of values of n, then f is called an **entire transcendental function.**

Now, if we examine the behavior of these three types of entire functions for points z of arbitrarily large magnitude, we see that whereas a constant function remains bounded, the other two are unbounded. Put in different terms, if a function is entire and nonconstant, then it can be made to assume arbitrarily large values by appropriate choice of values for the independent variable z. The first theorem of this section describes this fact in a slightly different but equivalent form.

Theorem 9.2 (*Liouville's Theorem*)
Suppose that
1. $f(z)$ *is an entire function.*
2. *f is bounded on the plane; i.e., there is a positive real M such that, for all z, $|f(z)| \le M$.*
Then f is a constant function.

Proof:
Since f is entire, it possesses a Taylor series expansion

$$f(z) = \sum_{n=0}^{\infty} a_n(z - c)^n$$

at every point c and the series converges for all z. By Corollary 2, p. 200, we know that

$$|a_n| \le \frac{M}{\rho^n}$$

for all $n = 1, 2, 3, \ldots$, where ρ is the radius of any circle centered at c and so chosen that f is analytic on the circle and its interior. But f is entire and, therefore, ρ can be taken arbitrarily large. Hence, letting $\rho \to \infty$, we have

$$|a_n| = 0, \qquad n = 1, 2, \ldots .$$

Thus $f(z) = a_0$, and the theorem is proved.

As a consequence of Liouville's theorem, one may effect a simple proof of the Fundamental Theorem of Algebra, which asserts that every non-constant polynomial with complex coefficients has at least one complex root. The reader is probably aware of the fact that an elementary proof of this theorem is extremely involved.

Theorem 9.3 (*Fundamental Theorem of Algebra*)
Suppose that
$$P(z) = a_0 + a_1 z + \cdots + a_n z^n$$
is a polynomial with $a_n \ne 0$ and $n > 0$.
Then there is a complex number α such that $P(\alpha) = 0$.

Proof:
First, we prove that $P(z)$ is an unbounded function, i.e., that given any real number $M > 0$, there is a point ζ such that $|P(\zeta)| > M$. To that end, we note the following:

(a) Let $M > 0$ be arbitrarily chosen. Since, by hypothesis, $|a_n| > 0$, one can find a point z of sufficiently large modulus so that

$$\frac{|a_n|\,|z|^n}{2} > M.$$

(b) Again, a point z of sufficiently large modulus can be chosen so that

$$\left| \frac{a_{n-1}}{z} + \frac{a_{n-2}}{z^2} + \cdots + \frac{a_0}{z^n} \right| \leq \left| \frac{a_{n-1}}{z} \right| + \left| \frac{a_{n-2}}{z^2} \right| + \cdots + \left| \frac{a_0}{z^n} \right| < \frac{|a_n|}{2}.$$

Then, by use of Exercise 2.14(h), we have

$$\left| a_n + \frac{a_{n-1}}{z} + \cdots + \frac{a_0}{z^n} \right| \geq |a_n| - \left| \frac{a_{n-1}}{z} + \cdots + \frac{a_0}{z^n} \right|$$

$$> |a_n| - \frac{|a_n|}{2}$$

$$= \frac{|a_n|}{2}.$$

(c) Now, choosing ζ of sufficiently large modulus so that all the relations in (a) and (b) will hold for $z = \zeta$, we find that

$$|P(\zeta)| = |a_n \zeta^n + a_{n-1}\zeta^{n-1} + \cdots + a_1\zeta + a_0|$$

$$= |\zeta|^n \left| a_n + \frac{a_{n-1}}{\zeta} + \cdots + \frac{a_1}{\zeta^{n-1}} + \frac{a_0}{\zeta^n} \right|$$

$$> |a_n|\,|\zeta|^n/2$$

$$> M.$$

Therefore, $P(z)$ is an unbounded function.

With this fact at our disposal we now complete the proof by assuming, contrary to the assertion of the theorem, that $P(z) \neq 0$ for every z, and we reach a contradiction. We argue this as follows: $P(z) \neq 0$ for all z implies that the function $1/P(z)$ is entire, since it is a rational function whose denominator never vanishes; see p. 81. On the other hand, $P(z) \neq 0$ for all z also implies that $1/P(z)$ is a bounded function. (Why?) It then follows from Liouville's theorem that $1/P(z)$ is a constant function, which is impossible since $P(z)$ is a nonconstant polynomial. This contradiction establishes the theorem.

Section 36
Behavior of Functions Near Isolated Singularities

In Section 28 we saw that a removable singularity is a singularity only in a superficial sense and that, in fact, a function possessing such a singular point can be defined at that point so as to be analytic there; see Exercise 28.21.

In the present section we will describe the behavior of a function $f(z)$ in the vicinity of its poles and essential singularities. We will thus see that near each of its poles, $f(z)$ is not as well behaved as it is near its removable singularities. Its behavior, however, is still quite predictable, for, as Theorem 9.4 will show, $f(z)$ always approaches infinity as z approaches any one of its poles. As for the behavior of a function near one of its essential singularities, it is completely unpredictable and extremely unstable in the sense described in the Casorati–Weierstrass theorem later in this section; closely associated with that behavior is the property put forth in Picard's theorem.

Theorem 9.4

Suppose that $f(z)$ has a pole of order n at a point z_0.
Then

$$\lim_{z \to z_0} f(z) = \infty.$$

Proof:

Without loss of generality, we prove this theorem for the case $z_0 = 0$.

By hypothesis, there is a Laurent series expansion that converges to f for all z in some deleted neighborhood of zero:

$$f(z) = \frac{a_{-n}}{z^n} + \frac{a_{-n+1}}{z^{n-1}} + \cdots + \frac{a_{-1}}{z} + \sum_{k=0}^{\infty} a_k z^k, \qquad a_{-n} \neq 0. \tag{1}$$

Now, if we rewrite (1) in the form

$$f(z) = \frac{a_{-n}}{z^n}\left[1 + \frac{a_{-n+1}}{a_{-n}}z + \frac{a_{-n+2}}{a_n}z^2 + \cdots \right] = \frac{a_{-n}}{z^n}[1 + S(z)],$$

we note that

$$\text{as } z \to 0, \qquad S(z) \nrightarrow 0.$$

Hence, given $\varepsilon > 0$, say $\varepsilon = \frac{1}{2}$, there is a $\delta > 0$ such that

$$|z| < \delta \qquad \text{implies that} \qquad |S(z)| < \tfrac{1}{2} \tag{2}$$

and, therefore,

$$|1 + S(z)| \geq 1 - |S(z)| > \tfrac{1}{2}.$$

But then,

$$|f(z)| = \frac{|a_{-n}|}{|z|^n}|1 + S(z)| > \frac{|a_{-n}|}{2|z|^n} > \frac{|a_{-n}|}{2\delta^n}.$$

Now, the last relation is true for any positive number less than δ. Hence, letting $\delta \to 0$ which, according to (2), means that $z \to 0$, we see that $\delta^n \to 0$ and, therefore,

$$|f(z)| \to \infty$$

which proves the assertion of the theorem.

We now turn our attention to the case in which $f(z)$ has an essential singularity at, say, z_0 and we demonstrate that as $z \to z_0$, the behavior of f is extremely erratic. We accomplish this demonstration in the next two theorems. The first is a preliminary result that will be used in the proof of the second; it is, however, an interesting and very important result in its own right.

Theorem 9.5   (*Riemann's Theorem*)
Suppose that $f(z)$ has the following two properties:
1. It has an isolated singularity at z_0.
2. It is bounded in some deleted neighborhood N_1 of z_0; i.e., there is a positive number M such that $|f(z)| \leq M$ for all z in the annulus

$$N_1 : 0 < |z - z_0| < \rho.$$

Then z_0 is a removable singularity of f.

Proof:
Without loss of generality, we prove this theorem for the case $z_0 = 0$.

By hypothesis 1, f is analytic in some deleted neighborhood N_2 of zero. Denote by N the smaller of N_1 and N_2. Then f has a Laurent series representation

$$f(z) = \sum_{n=-\infty}^{\infty} c_n z^n$$

at every point of N whose coefficients are given by

$$c_n = \frac{1}{2\pi i} \int_C \frac{f(z)}{z^{n+1}}\, dz,$$

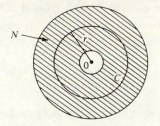

FIGURE 9.2 THEOREM 9.5

where C is the circle $|z| = r$ shown in Figure 9.2. By use of Theorem 4.5(5) we then have

$$|c_n| = \left| \frac{1}{2\pi i} \int_C \frac{f(z)}{z^{n+1}} \, dz \right| \le \frac{1}{2\pi} \frac{M}{r^{n+1}} 2\pi r = \frac{M}{r^n}.$$

Since r can be taken as small as we please, the last quantity can be made to approach zero for all $n = -1, -2, \ldots$. Thus

$$c_{-1} = c_{-2} = \cdots = c_{-n} = \cdots = 0,$$

and, therefore, the Laurent series is actually a Taylor series. It follows then that zero is a removable singularity and the proof is complete.

Theorem 9.6   (*Casorati–Weierstrass*)
Suppose that $f(z)$ has an essential singularity at z_0. Then, for any given number L and any $\varepsilon > 0$, there is a point z in every neighborhood of z_0 such that $|f(z) - L| < \varepsilon$.

Proof:
By hypothesis, f is analytic throughout a deleted neighborhood of z_0. All considerations in this proof which involve analyticity of f are restricted to points of that neighborhood.

Let L be an arbitrary complex number and consider the difference $f(z) - L$. We distinguish two possibilities:

1. $f(z) - L$ has a zero in every neighborhood of z_0.
2. There is a neighborhood $N(z_0, \rho)$ no point of which is a zero of $f(z) - L$.

If (1) is true, then the assertion of the theorem follows. If, on the other hand, (2) holds, we consider the function

$$g(z) = \frac{1}{f(z) - L}$$

which is analytic in the deleted neighborhood $N^*(z_0, \rho)$, since $f(z) - L$ is analytic and nonzero there; at z_0, $g(z)$ has an isolated singularity. Now, if z_0 were a removable singularity or a pole of g, then from

$$f(z) = \frac{1}{g(z)} + L$$

we see that f would then either have a pole or be analytic at z_0, contrary to hypothesis. Therefore, g has an essential singularity at z_0. It follows, as a consequence of Theorem 9.5 that g is unbounded in N. Hence, for any $\varepsilon > 0$, there at least one point z in N for which

$$|g(z)| > \frac{1}{\varepsilon}$$

and, therefore,

$$|f(z) - L| < \varepsilon.$$

This completes the proof.

Section 37
Uniqueness of Taylor and Laurent Expansions

In Theorem 6.13 it was proved that every function analytic at a point c has a Taylor series expansion that converges within some neighborhood of c. Similarly, in Theorem 7.1, the existence of a Laurent expansion was proved for any function that is analytic throughout a circular annulus. In this section we propose to prove that these expansions are unique. As has been our practice in most instances involving series, we shall prove our propositions for the case where the center of the expansion is the origin: $c = 0$; the general case is immediately obtainable by merely substituting $(z - c)$ for z at the appropriate places.

Theorem 9.7

Suppose that the series $\sum_{n=0}^{\infty} a_n z^n$ and $\sum_{n=0}^{\infty} b_n z^n$ have a positive radius of convergence and that both converge to the same function throughout a neighborhood of $z = 0$.

Then

$$a_n = b_n, \quad \text{for all } n = 0, 1, 2, \ldots,$$

and hence the two series are equal.

Proof:

The proof is by induction on n.

1. $n = 0$: Since the two series have the same sum for all z in the neighborhood of zero, in particular this is true for $z = 0$; hence,

$$a_0 = b_0.$$

2. Inductive step: Suppose that

$$a_j = b_j, \qquad \text{for all } j = 0, 1, \ldots, k.$$

Then

$$\sum_{n=k+1}^{\infty} a_n z^n = \sum_{n=k+1}^{\infty} b_n z^n.$$

Now, for $z \neq 0$ but within the neighborhood of coincidence of the two series, we divide both sides of the last equality, above, by z^{k+1} to obtain

$$a_{k+1} + a_{k+2}z + a_{k+3}z^2 + \cdots = b_{k+1} + b_{k+2}z + b_{k+3}z^2 + \cdots.$$

Then, letting $z \to 0$, we have

$$a_{k+1} = b_{k+1}.$$

Hence, by induction,

$$a_n = b_n, \qquad \text{for all } n,$$

and the proof is complete.

Corollary (*Uniqueness of Taylor Series Expansion*)
Suppose that $f(z)$ is analytic at a point c.
Then any two Taylor series expansions of f with center at c are equal.

The following stronger version of Theorem 9.7 is closely related to the Identity Theorem for Analytic Functions; see reference given on p. 337.

Theorem 9.8
Suppose that the two Taylor series

$$\sum_{n=0}^{\infty} a_n z^n \qquad and \qquad \sum_{n=0}^{\infty} b_n z^n$$

have a positive radius of convergence and that their sums coincide at an infinity of distinct points converging to zero. Moreover, suppose that the two series are equal at $z = 0$.
Then the two series are identical within some neighborhood of zero.

We now proceed to prove that the Laurent series expansion of a function is unique in the sense that any two such expansions for one and the same function over the same circular annulus are termwise identical.

Theorem 9.9   (*Uniqueness of Laurent Series Expansion*)
Suppose that
1. $f(z)$ is analytic on the closed annulus

$$A : r \leq |z| \leq R, \qquad where \qquad 0 \leq r < R \leq \infty.$$

2. $f(z) = \sum_{n=-\infty}^{\infty} c_n z^n$, for every z in A.
Then the above series is a Laurent series.

Proof:
The proof consists in showing that, for every integer n,

$$c_n = \frac{1}{2\pi i} \int_C \frac{f(\zeta)}{\zeta^{n+1}} \, d\zeta,$$

where C is any circle $|z| = \rho$, where $r < \rho < R$; see Theorem 7.1. So, let C be as specified above and let ζ be an arbitrary point of C. Then, by hypothesis, for any such ζ, the given series converges to f:

$$f(\zeta) = \sum_{k=-\infty}^{\infty} c_k \zeta^k.$$

Multiplying through by $1/\zeta^{n+1}$, for any n, we obtain

$$\frac{f(\zeta)}{\zeta^{n+1}} = \sum_{k=-\infty}^{n-1} c_k \zeta^{k-n-1} + \frac{c_n}{\zeta} + \sum_{k=n+1}^{\infty} c_k \zeta^{k-n-1}.$$

Next, we integrate the above relation along C, oriented positively, and in view of the fact described in Remark 1, p. 281, we carry the integration term by term. Thus we have

$$\int_C \frac{f(\zeta)}{\zeta^{n+1}} \, d\zeta = \sum_{k=-\infty}^{n-1} c_k \int_C \zeta^{k-n-1} \, d\zeta + c_n \int_C \frac{d\zeta}{\zeta} + \sum_{k=n+1}^{\infty} c_k \int_C \zeta^{k-n-1} \, d\zeta.$$

By the Cauchy integral formulas, p. 199, every integral in the above two summations is zero. Therefore,

$$\int_C \frac{f(\zeta)}{\zeta^{n+1}} \, d\zeta = c_n 2\pi i;$$

hence

$$c_n = \frac{1}{2\pi i} \int_C \frac{f(\zeta)}{\zeta^{n+1}} \, d\zeta.$$

But n was chosen randomly. Therefore, the last relation is true for any n, and the theorem is proved.

Answers to Exercises

CHAPTER 1

Exercise 1, p. 8

1.1. $7 + i.$

1.2. $-4 + 2i.$

1.3. $5 - 12i.$

1.4. $1 - 5i.$

1.5. $1.$

1.6. $a^2 + b^2.$

1.7. $-\dfrac{30}{61} + \dfrac{36}{61}i.$

1.8. $\dfrac{a^2 - b^2}{a^2 + b^2} + \dfrac{2ab}{a^2 + b^2}i.$

1.9. $\dfrac{3}{13} - \dfrac{2}{13}i.$

1.10. $-1, -i, 1, i, \ldots, -1.$

1.11. For $k =$ any integer, if $n = 4k$, $i^n = 1$; if $n = 4k + 1$, $i^n = i$; if $n = 4k + 2$, $i^n = -1$; and if $n = 4k + 3$, $i^n = -i.$

1.14. $z^2 = a^2 - b^2 + 2abi$, $z^3 = (a^3 - 3ab^2) + (3a^2b - b^3)i.$

1.15. (a) $i.$ (b) $-(3 + i)/2.$ (c) $(3 - 5i)/2.$ (d) $-9i.$

1.16. (a) $z = \pm 1.$ (b) $z = 0 + bi$, for any $b.$ (c) $z = a + bi$, with $a^2 + b^2 = 1.$

Exercise 2, p. 21

2.2. $\sqrt{5}.$

2.4. (a) cis $\pi.$

(b) 3 cis 0.

(c) 4 cis $\dfrac{3\pi}{2}.$

(d) $\sqrt{8}$ cis $\dfrac{3\pi}{4}.$

(e) $\sqrt{3}$ cis $\dfrac{\pi}{2}.$

(f) 6 cis $\dfrac{7\pi}{6}.$

(g) $\sqrt{2}$ cis $\left(-\dfrac{\pi}{4}\right).$

(h) $\sqrt{5}$ cis $[\tan^{-1}(-\tfrac{1}{2})].$

(i) 2 cis $\dfrac{5\pi}{4}.$

(j) $\sqrt{13}$ cis $[\tan^{-1}(-\tfrac{3}{2})].$

2.6. (a) 4 cis $\dfrac{\pi}{2}.$

(b) $\sqrt{2}$ cis $\dfrac{3\pi}{4}.$

(c) 8 cis $\dfrac{\pi}{2}.$

(d) $8^{15/2}$ cis $\dfrac{5\pi}{4}.$

2.7. (a) Circle: center at $5, r = 6.$
(b) Circle: center at $-2i, r = 1$, and its exterior.
(c) Vertical line: $x = -3.$
(d) Horizontal line $y = 3.$
(e) The real axis.
(f) Ellipse $3x^2 + 4y^2 + 12x = 0.$
(g) Hyperbola $12x^2 - 4y^2 + 48x + 45 = 0.$

351

    (h) Vertical strip $-1 \le x < 1$.

    (i) The lower half of the z-plane, not including the real axis.

    (j) Horizontal strip $0 < y \le 2\pi$.

2.9. $\pm 1; (1 \pm i\sqrt{3})/2; (-1 \pm i\sqrt{3})/2$.

2.10. $-2; 1 \pm i\sqrt{3}$.

2.12. $z_0 = 1, z_1 = (-1 + i\sqrt{3})/2, z_2 = (-1 - i\sqrt{3})/2; z_1^2 = z_2, z_1^3 = z_0$.

2.13. $z_k = \text{cis } \dfrac{k\pi}{6}, k = 0, 1, 2, \ldots, 11$.

2.20. z must be real in the first case, and pure imaginary or zero in the second case.

Review—Chapter 1, p. 23

1. (a) $3 - 4i$.

    (b) $(5 - 12i)/\sqrt{13}$.

    (c) $4^{1/4} \text{ cis } [\frac{1}{4}(\pi + 2k\pi)], k = 0, 1, 2, 3$.

    (d) -2^{90}.

    (e) $\text{cis } \dfrac{k\pi}{4}, k = 0, 1, 2, \ldots, 7$.

    (f) $(-1 + 3i)/10$.

    (g) $-\frac{7}{5}$.

    (h) 1.

    (i) $2^{1/6} \text{ cis } \left(\dfrac{\pi}{4} + \dfrac{2k\pi}{3} \right), k = 0, 1, 2$.

    (j) $26 - 36i$.

2. T T F F T T F T T F.

3. When z, w, and the origin are collinear.

4. All z such that $R(z) \ge 1$.

5. (a) The $45°$ ray emanating from (but not including) the origin.

    (b) The upper half-plane, not including the real axis.

    (c) The entire z-plane, except the origin.

13. $z_k = 2^{-1/12} \text{ cis } \left[\dfrac{1}{6} \left(2k\pi - \dfrac{5\pi}{12} \right) \right]$.

CHAPTER 2

Exercise 3, p. 35

3.1. Boundary: $|z| = 1$; open; bounded.

3.2. Boundary: $y = 0$ and $y = 1$; neither open nor closed; unbounded.

3.3. Boundary: $x = -2$ and $x = 0$; open, unbounded.

3.4. Boundary: $|z + i| = 2$; open; bounded.

3.5. Boundary: $|z| = 1$ and $|z| = 3$; closed; bounded.

3.6. Boundary: $|z - i| = 3$; closed; unbounded.

3.7. Boundary: $y = 0$ and $y = 1$; neither open nor closed; unbounded.

3.8. Boundary: the rectangle with vertices at $0, i, -2 + i$, and -2; neither open nor closed; bounded.

3.9. Boundary: the upper half of $|z| = 1$ and the segment $-1 \le R(z) \le 1$, $I(z) = 0$; open; bounded.

3.10. Boundary: $|z| = 3$; closed; bounded.

3.11. Boundary: $x = 2$ and $x = 5$; neither open nor closed; unbounded.

3.12. Boundary: $|z - \frac{3}{2}| = \frac{1}{2}$; closed; unbounded.

3.13. Boundary: $|z + (i/6)| = \frac{1}{6}$; open; bounded.

3.14. Closed; it contains "all" its boundary points.

3.15. Neither; each point of the set is a boundary point, hence it is contained in the set. However, zero is also a boundary point, and it is not in the set.

3.17. Open. Boundary: $|z| = 1$ and the segment $0 \le R(z) \le 1$ of the real axis. Complement: $|z| \ge 1$ and the above segment.

3.18. The finite z-plane.

Exercise 4, p. 41

4.1. $2, -6$.

4.2. $-2, 13 - 2i$.

4.3. $-i, i, (4 - 3i)/5$.

4.4. $1, 16$.

4.5. $1, e, -e^2$.

4.6. $(x^2 - y^2 + 3x^3 - 9xy^2) + (2xy + 9x^2y - 3y^3)i$;
$(r^2 \cos 2\theta + 3r^3 \cos 3\theta) + (r^2 \sin 2\theta + 3r^3 \sin 3\theta)i$.

4.7. $\left(y + \dfrac{x}{x^2 + y^2} \right) + xi; \left(\dfrac{\cos \theta}{r} + r \sin \theta \right) + ir \cos \theta$.

4.8. $2 + \pi i$.

4.9. $1, -1, i, 0, 2 + i$; changes nothing.

4.10. $1, 2 + i, 0, -2 + 2i, 1 - i$; shift by one unit to the right.

4.11. $-2i, 0, 1 - 2i, -i, -4i, 1 - i$; shift by two units downward.

4.12. $0, i, 2i, 3i, -1 + i, -2 + 2i, -1, -2, -3$; rotation through $\pi/2$ around the origin.

4.13. $0, 0, 0, 1, -1, 0, 2, 2$; vertical projection onto the real axis.

4.14. $1, i, -1, -i, 1 + i, -1 - i, 2i, -4$; reflection in the line $y = x$.

4.15. $1, 1, 1, e, -e, e^2, -1, e^{-2}i, -e^{-2}i$.

4.16. (a) $z^3 = r^3 (\cos 3\theta + i \sin 3\theta)$.

(c) The point $z = 0$, and all z such that, for any real number α,

$$\alpha \le \arg z < \alpha + \frac{2\pi}{3}.$$

(d) Moduli are equal; arguments differ by $2\pi/3$.

4.17. (a) $-e, -e^2, -e^{-1}, -e^{-2}, -e^3$. For $z = x + \pi i$, $w = -e^x$, and the resulting image is the negative u-axis.

(b) $e(\sqrt{2} + i\sqrt{2})/2, ei, -e, -ei, e$. For $z = 1 + yi$, $w = e$ cis y, and the resulting image is the circle $|w| = e$.

(c) Circle: $|w| = e^{-2}$; ray: $u = 0$, $v < 0$.

(d) Horizontal lines map onto rays emanating from the origin; vertical lines map onto circles centered at the origin (see Section 15).

Exercise 5, p. 48

5.1. $3 + 2i$.   **5.2.** 2.

5.3. $(13 - 51i)/5$.   **5.4.** $3a^2$.

5.5. $1 + i$.   **5.6.** n.

5.7. i.   **5.8.** -1.

5.9. $-e^{2i}$.

5.10. (a) $1 - i$. (b) $\frac{4}{5} - 4i$. (c) 0. Limit, as $z \to 0$, does not exist, since it depends on the path.

5.11. (a) Let $z \to i$ along the imaginary axis ($x = 0$, $y \to 1$) to find the limit $Z = -i$. Then, let $z \to i$ along a horizontal path ($x \to 0$, $y = 1$) to find the limit $Z = 1$.
    (b) $z \to 0$ along the line $y = x$ yields the limit $L = 2\sqrt{2}$, whereas $z \to 0$ along the parabola $y = x^2$ yields $L = 4$.

Exercise 6, p. 56

6.1. (a) $6z^5 + 6z^2$.

    (b) $10(2z + 5)^8(1 - 2z + z^2)^9(-2 + 2z) + 16(2z + 5)^7(1 - 2z + z^2)^{10}$.

    (c) $\dfrac{16(1 - 2z + z^2)^{10}(2z + 5)^7}{(1 - 2z + z^2)^{20}} - \dfrac{10(2z + 5)^8(1 - 2z + z^2)^9(-2 + 2z)}{(1 - 2z + z^2)^{20}}$.

6.2. (a) $2z + 3$. (b) $-z^{-2}$

6.3. (a) $-1 + 6i$. (b) $2 - 6i$. (c) $1 - 2\pi - i$.

6.6. (a) $f(z) = x^2 - y^2 + 2xyi$.

        $f' = u_x + iv_x = 2x + 2yi = 2(x + yi) = 2z$.

Exercise 7, p. 61

7.1. On the line $y = \frac{1}{2}$; $f' = 1$.

7.2. All z; $f' = 3z^2$.

7.3. All z; $f' = 0$.

7.4. On the line $x = -\frac{1}{2}$; $f' = -1$.

7.5. All z; $f' = \cos x \cosh y - i \sin x \sinh y$.

7.6. Nowhere.   **7.7.** Nowhere.

7.8. Nowhere.   **7.9.** For $z = 0$, only; $f' = 0$.

7.10. On the parabola $4x = 9y^2$; $f' = 4x = 9y^2$.

7.11. $f(z) = x + iy$; $f'(z) = 1$ exists. However, the derivatives of $g(z) = x$ and $h(z) = y$ do not exist anywhere.

7.14. (a) $f' = 1/z$. (b) $f' = nz^{n-1}$. (c) $f' = \dfrac{1}{n}r^{1/n-1} \ \text{cis} \left(\dfrac{1}{n} - 1\right)\theta$.

Exercise 8, p. 67

8.1. All z.   **8.2.** Nowhere.

8.3. All z.   **8.4.** All z.

8.5. All $z \neq -1$.   **8.6.** All $z \neq 0, i, -i$.

8.7. All z not on the nonpositive real axis.

8.8. All z.

8.9. All z.

8.10. Nowhere.

8.17. $v = y; f(z) = z + c$.

8.18. $v = \frac{1}{2}(y^2 - x^2); f(z) = -\frac{i}{2}z^2 + c$.

8.19. $v = 2 \tan^{-1} \frac{y}{x}; f(z) = 2(\ln r + i\theta) + c$.

8.20. $v = e^x \sin y; f(z) = e^x \operatorname{cis} y$.

8.27. HINT: Consider each of the four cases resulting from combining $u = \pm(x^2 - y^2)$ with $v = \pm 2xy$. Each case represents a pair of angular sectors of the z-plane; e.g., $u = x^2 - y^2$ and $v = 2xy$ represent half of the first and half of the third quadrant. In each case, check the Cauchy–Riemann equations.

Review—Chapter 2, p. 68

1. F F T F T F F F F F.

3. $f'(z) = -2z^{-3}$.

7. (a) As $z \to 0$ along $x = 0$, $L = 0$; as $z \to 0$ along $y = x^3$, $L = 1$.

    (b) As $z \to 0$ along $x = 0$, $L = 0$; as $z \to 0$ along $y = x$, $L = -i$.

8. f' exists only at $z = 0$; hence f is nowhere analytic.

10. Reflection in the real axis followed by a shift by one unit to the right.

11. If f is to be analytic at a point on the circle $|z| = 1$, then it must be analytic at points outside that circle.

12. $v = \sinh x \sin y + c$.

13. $v = x + c; f(z) = iz + k$.

CHAPTER 3

Exercise 9, p. 77

9.1. Not one-to-one; every z maps onto one and the same point: $3i$.

9.2. One-to-one.

9.3. One-to-one.

9.4. Not one-to-one; e.g., $z = 2$ and $z = -1 + i\sqrt{3}$ both map onto $w = 5$.

9.5. Take as its domain the upper half of the z-plane; or the left half; or the right half. In fact, one may take any one quadrant.

9.6. No. Although the function preserves angles in magnitude, it reverses their direction. From an analytic standpoint, $w = \bar{z}$ is not an analytic function and hence cannot be conformal.

Exercise 10, p. 92

10.1. (a) i. (b) $-e$. (c) -1. (d) $1 + i\sqrt{3}$. (e) e^2.

10.5. (a) $\ln 3 + \left(\frac{3\pi}{2} + 2k\pi\right)i$, $k =$ integer.

    (b) $\ln \sqrt{2} + \left(2k\pi - \frac{\pi}{4}\right)i$, $k =$ integer.

10.7. (In all parts, k = integer.)

(a) $(2k\pi - \pi/2)i$. (b) $2k\pi i$. (c) $\ln \sqrt{2} + [(\pi/4) + 2k\pi]i$.

(d) $\ln 5 + (\text{Tan}^{-1}\frac{4}{3} + 2k\pi)i$. (e) $\ln \sqrt{5} + \left(\text{Tan}^{-1}\frac{-1}{2} + 2k\pi\right)i$.

10.8. Either approach yields the answer $(k\pi - \pi/4)i$, k = integer. The first method yields the answer directly. In the second case, one obtains $(2k\pi - \pi/4)i$ for the one root and $(2k\pi + 3\pi/4)i$ for the other. Note that these two answers combined yield the one obtained earlier.

10.9. -2.

10.10. $(k\pi - \pi/4)i$, k = integer.

10.11. (a) -1. (b) 1. (c) 0. (d) $\cos 2$. (e) $[(e^{-\pi} - e^{\pi})/(e^{-\pi} + e^{\pi})]i$. (f) $i \sinh 1$. (g) $\cosh 1$. (h) $-i \cot 1$. (i) $\sin 1 \cosh 1 + i \cos 1 \sinh 1$.

10.12. The derivative of $\sin z$, $\cos z$, $\tan z$, $\cot z$, $\sec z$, and $\csc z$, is given, respectively, by $\cos z$, $-\sin z$, $\sec^2 z$, $-\csc^2 z$, $\sec z \tan z$, and $-\csc z \cot z$.

10.13. The derivative of $\sinh z$, $\cosh z$, $\tanh z$, $\coth z$, $\text{sech } z$, and $\text{csch } z$, is given, respectively, by $\cosh z$, $\sinh z$, $\text{sech}^2 z$, $-\text{csch}^2 z$, $-\text{sech } z \tanh z$, and $-\text{csch } z \coth z$.

10.29. For $z = -3i$, it is easy to verify that $|\sin z| > 1$ and $|\cos z| > 1$.

10.33. *H i n t* : Recall that the property Log $z^n = n$ Log z holds "give or take a multiple of $2\pi i$."

10.34. (a) $e^{-(\pi/2 + 2k\pi)}$, k = integer.

(b) $e^{(\ln\sqrt{2} + \pi/4 + 2k\pi) + (\pi/4 + 2k\pi - \ln\sqrt{2})i}$, k = integer.

10.35. (a) In the answers to Exercise 10.34, take $k = 0$.

Exercise 11, p. 99

11.1. (a) $\arg w = 5\pi/6$, $|w| = 2$, $v = -1$, $u = -2$.

(b) $\arg (w - 2i) = -\pi/6$, $|w - 2i| = 2$, $v = 3$, $u = 2$.

(c) $\arg w = 13\pi/12$, $|w| = 2\sqrt{2}$, $u - v - 2 = 0$, $u + v + 4 = 0$.

(d) Rotate by 0 radians (no rotation), stretch by a factor of 1 (no stretching), and translate by 1 unit to the right, to obtain $\arg (w - 1) = \pi/3$, $|w - 1| = 2$, $u = 0$, $v = 2$.

(e) Rotate by $-\pi/4$ radians, stretch by a factor of $\sqrt{2}$, and translate through the vector $1 - i$.

(f) Write $w = 2iz - 2 + 2i$. Then, rotate by $\pi/2$, stretch by a factor of 2, and translate through the vector $-2 + 2i$.

11.2. Since the linear map is a similarity transformation, the image of the given angular section is a similar "wedge" obtained through a rotation by π, stretching by a factor of 2, and a translation by $-2i$. One thus obtains $\pi < \arg (w + 2i) < 7\pi/6$.

11.3. $u^2 = 2(2 - v)$.

11.4. $w = -iz + 1 - i$.

11.5. (a) 0. (b) $1 - i$. (c) None. (d) $-2i/3$.

11.6. The identity function $w = z$ and no other.

11.7. $w = iz + 1$.

Exercise 12, p. 103

12.1. $0 < \arg w < \pi$.

12.2. $\alpha = \pi/4$.

12.4. (a) $u = 3$. (b) $u \geq 0, v = 0$. (c) $u \leq 0, v = 0$. (d) $v^2 = 16(4 - u)$. (e) $v^2 = 36(u + 9)$. (f) $u^2 = 1 - 2v$. (g) $|w| > 4$. (h) $|\arg w| < \pi$. (i) The region between the parabolas $v^2 = 4(1 - u)$ and $v^2 = 16(4 - u)$.

12.5. $\alpha = 5\pi/6$.

12.6. The four nonreal roots of unity: $z_k = \text{cis}\,(2k\pi/5)$, $k = 1, 2, 3, 4$.

12.10. An example can be provided by the function $w = z^2$, under which ray $R : x \geq 0$, $y = 0$ maps onto the ray $R' : u \geq 0$, $v = 0$, and ray $Q : x = 0$, $y \geq 0$ maps onto the ray $Q' : u \leq 0$, $v = 0$. R and Q form an angle of $\pi/2$, whereas R' and Q' form an angle of π.

Exercise 13, p. 108

13.1. (a) 1. (b) -1. (c) $-i$. (d) i. (e) $(1 - i)/2$. (f) $(5 + 12i)/169$. (g) $(-3 - 4i)/25$. (h) $i/3$. (i) $1 - i$.

13.3. (a) $u^2 + v^2 + v = 0$. (b) $u^2 + v^2 - u - v = 0$. (c) $u^2 + v^2 + u = 0$. (d) $u^2 + v^2 - u + v = 0$. (e) $2v + 1 = 0$. (f) $u^2 + v^2 + 2u + 4v + 1 = 0$. (g) $v = 0$. (h) $u = 0$.

13.5. The image is the unit circle in the w-plane with the upper and lower halves interchanged.

13.6. $2u + 1 = 0$.

13.7. $z = \pm 1$.

13.9. $\left| w + \dfrac{i}{2k} \right| = \dfrac{1}{2|k|}$.

13.11. The point $w = -i/2$ is the image of the point $z = 2i$ at which L_1 and L_2 intersect. The other point of intersection, $w = 0$, of the two circles is the image of the point at infinity at which L_1 and L_2 also intersect.

13.12. The lower half of the w-plane, which is exterior to the circle $|w + i| = 1$.

Exercise 14, p. 113

14.1. $-2i, (1 - 3i)/2, -(1 + 3i)/2, -3i/2, \infty, -i$.

14.2. (a) $\pm(\sqrt{2} + i\sqrt{2})/2$. (b) $1 \pm i$.

14.3. $w = \dfrac{z - 1}{z + 1}$.

14.4. $|w - 1 + 2i| = 1$.

14.5. $w = \dfrac{z}{2z + 1}$.

14.6. $w = z + 1$.

14.10. $I(w) \leq 0$.

14.11. $|w| < 1$.

Exercise 15, p. 118

15.1. (a) Ray: $\arg w = 1$, $|w| > e^{-2}$.

(b) Ray: $\arg w = 1$, $|w| > e^{-1}$.

(c) Ray: $\arg w = 1$, $|w| > 1$.

(d) Ray: arg $w = 1$, $|w| > e$.
(e) Circular arc: $|w| = e^{-2}$, $-\pi/2 < \arg w < \pi$.
(f) Unit circle: $|w| = 1$.
(g) Half circle: $|w| = e$, $0 \le \arg w < \pi$.
(h) Circle: $|w| = e^2$.
(i) Ray: arg $w = 3$.
(j) Circle: $|w| = e^{-8}$.

15.2. (a) Segment: $u = \ln c$, $-\pi < v \le \pi$.
(b) Line: $v = -\pi/4$, $-\infty < u < \infty$.
(c) Line: $v = 3\pi/4$, $-\infty < u < \infty$.
(d) Line: $v = \pi/6$, $-\infty < u < \infty$.

15.3. The upper part of the circular annulus determined by the circles $|w| = e^{-1}$ and $|w| = e^3$, and the rays arg $w = 0$ and arg $w = 2$.

15.4. The configuration that can be described as follows: from $w = 1$ to $w = e^2$ along a straight line, to $w = e^2$ cis 1 along a circular arc (counterclockwise), to $w = e^{-2}$ cis 1 along a straight line, to $w = e^{-2}$ cis (-2) along a circular arc (clockwise), to $w = $ cis (-2) along a straight line, and, finally, back to $w = 1$ along a circular arc (counterclockwise).

Exercise 16, p. 122

16.4. (b) The first equation fails to hold when $y = 0$; the second equation fails to hold when $x = k\pi/2$, $k = $ integer.

16.5. (a) The interval $-1 \le u \le 1$, $v = 0$, covered infinitely often.
(b) The ray $u \ge 1$, $v = 0$, covered twice.
(c) The ray $u \le -1$, $v = 0$, covered twice.
(d) The v-axis.
(e) The ray in (b) if $k = $ even; the ray in (c) if $k = $ odd.
(f) The v-axis.

16.8. The upper half of the ellipse

$$\frac{u^2}{\cosh^2 1} + \frac{v^2}{\sinh^2 1} = 1$$

along with its horizontal axis. $A': -1$, $B': 0$, $C': 1$, $D': \cosh 1$, $E': i \sinh 1$, $F': -\cosh 1$.

Review—Chapter 3, p. 123

1. Line $4u + 3v = 0$, second quadrant only.
2. Line $v + 6 = 0$.
3. (a) $0 < |w - 1| < 3$, $0 < \arg (w - 1) < \pi/2$.
(b) $0 < |w| < 3$, $-\pi/2 < \arg w < 0$.
(c) $0 < |w| < 9$, $0 < \arg w < \pi$.
(d) $0 < |w| < 27$, $0 < \arg w < 3\pi/2$.
(e) $0 < |w| < 81$, $0 < \arg w < 2\pi$.
(f) The semiinfinite horizontal strip $u < \ln 3$, $0 < v < \pi/2$.
4. The square with vertices at the points $2 + i$, $-1 - 2i$, $-4 + i$, $-1 + 4i$.
5. (a) 1. (b) 2. (c) None. (d) None.

6. 1.

7. $2v + 1 = 0$.

8. $w = (z + i)/(iz + 1)$.

9. (a) $z = (\pi/2 + 2k\pi + 1)i/2, k = $ integer.
(b) $z = \ln(2 \pm \sqrt{3}) + 2k\pi i, k = $ integer.
(c) $z = 2k\pi - i\ln(\sqrt{5} - 2)$ and $z = (2k + 1)\pi - i\ln|\sqrt{5} + 2|, k = $ integer.

12. $e^{x^2 - y^2}$; $2xy$.

13. F F T F T F F T F F.

Appendix 3, p.132

1. (a) ± 1. (b) $\pm\sqrt{2}(1 + i)/2$. (c) $\pm\sqrt{2}(1 - i)/2$. (d) $\pm(\sqrt{3} + i)$.

2. Arg $w = \pi/6, 4\pi/6, 7\pi/6, 10\pi/6$.

3. (a) v-axis. (b) Line $u = \ln(\pi/2)$. (c) line $v = \pi/4$. (d) Line $v = \pi/2$. (e) The infinite strip between $v = \pi/4$ and $v = \pi/2$.

4. (a) For $k = 0, 1, 2, 3, 4,$ and 5, and $z = re^{i\theta}, f_k(z) = r^{1/6} e^{i(\theta + 2k\pi)/6}, \pi/2 < \theta < 5\pi/2$, $r \neq 0$.
(b) For $k = $ any integer, and $z = re^{i\theta}, f_k(z) = \ln\sqrt{r} + i(\theta + 2k\pi), \pi/2 < \theta < 5\pi/2$, $r \neq 0$.

6. For $k = $ any integer, and $z = re^{i\theta}, f_k(z) = \theta + 2k\pi, 0 \le \theta < 2\pi$.

Appendix 3, p.137

1. See the answer to Exercise 12.10.

Appendix 3, p.139

1. (a) $z = (1/a)(w - b)$. Analytic everywhere.
(b) $z' = 1/a$.

2. $w' = 1/(1 - z^2)^{1/2}$.

Appendix 3, p.144

1. Lines $x = c - 1, c = $ constant.

2. Circle $cx^2 + cy^2 - x = 0, c = $ constant.

CHAPTER 4

Exercise 17, p.157

17.1. (a) $x = y^2 - 1$, from $(0, -1)$ to $(0, 1)$. (b) $x^2/9 + y^2/4 = 1, y \ge 0$. (c) $x^2 + (y + 1)^2 = 1$. (d) $x = e^{1-y}$, from $(1, 1)$ to $(e^{-1}, 2)$. (e) $(x - a)^2 + (y - b)^2 = 4$, clockwise from $(a, b + 2)$ to $(a - 2, b)$, where $z_0 = a + bi$.

17.3. $x = -t, y = (1 - t)/2, -1 \le t \le 3$.

17.4. (a) The strip $-10 < I(z) < 25$. (b) All z such that $|z| \ge 2$. (c) All z such that $|R(z)| > 2$. (d) All z such that either $|z| < 1$ or $1 < |z| < 2$.

17.7. (a) $x = 2\tau + 2, y = 4\tau^2 - 4\tau, 0 \le \tau \le 1$.
(b) $x = \sin \tau, y = -2\cos \tau, \pi \le \tau \le 2\pi$.
(c) $z = -i - e^{2i\tau}, 0 \le \tau \le \pi$.

Exercise 18, p.166

18.1. (a) $\frac{1}{28}, \frac{1}{24}$. (b) $\frac{284}{5}, \frac{34}{3}$. (c) $\frac{4}{35}, \frac{4}{21}$. (d) $-\frac{34}{3}, \frac{34}{3}$.
18.5. (a) 0. (b) 0.   **18.6.** (a) $-\frac{4}{3}$. (b) 0.
18.8. 0.   **18.9.** $3\pi/8, 0$.
18.10. $\ln\left(\frac{3}{2}\right)$.   **18.11.** $\ln 4$.
18.12. 0.   **18.13.** π.
18.14. $\frac{5}{3}$.   **18.15.** $(e-1)/3$.

Exercise 19, p.175

19.1. $728/3 + \ln 9 - 728i/81$.   **19.2.** $\frac{9}{2} - 4i$.
19.3. 1.   **19.4.** $e^{1+2i} - e^{-1-2i}$.
19.5. $2\pi i$.   **19.6.** $(14 - 5i)/15$.
19.8. πi.   **19.9.** (a) -4. (b) -2π. (c) 2π.
19.11. (a) 1. (b) $4\pi e^2$. (c) $4\sqrt{5} + 2$. (d) 2. (e) $\pi/4$.
19.12. (a) $8 - 3\pi i$. (b) $8 - 3\pi i$. (c) $8 + 3\pi i$.

Review—Chapter 4, p.177

1. -16.   **2.** 0.
3. 2.   **4.** $2\pi/3$.
5. Because the path of integration, $y = e^x$, passes through neither of the limits of the integral.
6. (a) $x = 1 + 2\cos t, y = 1 + 2\sin t, 0 \leq t \leq 2\pi$.
   (b) $x = t, y = t^2, -2 \leq t \leq 0$.
   (c) $x = -t, y = t^2, 0 \leq t \leq 2$.
   (d) $x = t, y = (4 + t^2)^{1/2}, 0 \leq t \leq 2$.
   (e) $x = t, y = 1 - t, -1 \leq t \leq 2$.
7. $(1 + i)/2$.   **8.** $-(1 + i)/2$.
9. $\frac{13}{3}$.   **10.** $\frac{13}{3}$.
11. 3.

CHAPTER 5

Exercise 20, p.190

20.1. 0.   **20.2.** 0.
20.3. 0.   **20.4.** $-4\pi i$.
20.5. $1 + i$.   **20.6.** 0.
20.7. $e^{-1} - e - 2$.   **20.8.** 0.
20.9. $-\frac{3}{2}$.   **20.10.** $e^\pi - 3$.
20.11. 0.   **20.12.** $-(\sin \pi^2)/2$.
20.13. $(e^{-1} - 1)/2$.   **20.14.** $\pi/2$.
20.15. $2\pi i$.

Exercise 21, p. 196

21.1. $8\pi i$.

21.2. $12\pi i$.

21.3. 0.

21.4. $-\pi i$.

21.5. 0.

21.6. $ie^i - ie^{-i} + e^i + e^{-i} - 5\pi/2$.

21.7. 0.

21.8. $-2\pi i$.

21.9. -2π.

21.10. 0.

21.11. 0.

21.12. 0.

21.13. 0.

Exercise 22, p. 201

22.1. $6^5\pi i$.

22.2. $-2\pi i$.

22.3. $4\pi i e^4 - 2\pi i$.

22.4. $2\pi i \cos 1$.

22.5. 0.

22.6. 0.

22.7. $20\pi^3 i$.

22.8. 0.

22.9. (a) $(2\pi i/a) \sin a$. (b) $(\pi/a)e^{ai}$.

22.10. $\pi i(e^a - 6a)$.

22.11. 0.

22.12. $2\pi i$.

Review—Chapter 5, p. 203

1. $-1 - 8i + \cos 2i$.

2. 0.

3. $-2\pi i/(2 + 11i)$

4. $2\pi i e^{2i}/5! + 24\pi$.

5. $2\pi i \operatorname{Log}(-2i)$.

6. 0.

7. $24\pi i$.

8. $\frac{5}{2} - 5i$.

9. 0, if $|\alpha| > 1$; $\pi e^{\alpha}i$, if $|\alpha| < 1$.

10. $\zeta^2 + 3\zeta - i$.

11. No. $f(z) = z^{-2}$ is a counterexample.

12. $2\pi i$, if both α and $-\alpha$ are in Int (C). πi, if either α or $-\alpha$ is in Int (C) and the other is in Ext (C). 0, if both α and $-\alpha$ are in Ext (C).

13. Take $r < \pi/2$.

CHAPTER 6

Exercise 23, p. 241

23.1. $\{2i, -4, -8i, 16, 32i, \ldots\}$; divergent.

23.2. $\{2i, -2, -2i, 2, 2i, \ldots\}$; divergent.

23.3. $\left\{ \dfrac{1}{1 - i}, \dfrac{1}{-4i}, \dfrac{1}{-6 - 6i}, \dfrac{1}{-16}, \ldots \right\}$; convergent.

23.4. $\left\{ \dfrac{2 - i}{1 + 2i}, \dfrac{4 - i}{2 + 2i}, \dfrac{6 - i}{3 + 2i}, \dfrac{8 - i}{4 + 2i}, \ldots \right\}$; convergent.

23.5. $\{1 - i, 2 - i/2, 3 - i/3, 4 - i/4, \ldots\}$; divergent.

23.6. Convergent for $|z| \leq 1$. Divergent for $|z| > 1$.

23.7. Absolutely convergent.

23.8. Conditionally convergent.

23.9. Divergent.

23.10. Absolutely convergent.

23.11. Absolutely convergent.

23.12. Absolutely convergent.

Exercise 24, p. 247

24.1. $2; |z| = 2.$
24.3. $1; |z| = 1.$
24.5. $\infty; |z| = \infty.$
24.7. $1; |z + 1| = 1.$
24.9. $\frac{1}{4}; |z| = \frac{1}{4}.$
24.11. Diverges at all four points.

24.2. $e^{-1}; |z + 2| = e^{-1}.$
24.4. $3; |z - i| = 3.$
24.6. $1/e; |z + i| = e^{-1}.$
24.8. $0; |z + \pi i| = 0.$
24.10. $1; |z + e| = 1.$

Exercise 26, p. 266

26.3. $\displaystyle\sum_{n=0}^{\infty} (-1)^n z^{n+1}, r = 1.$

26.4. $\displaystyle\sum_{n=0}^{\infty} \frac{(-1)^n}{2^{n+1}} z^{n+2}, r = 2.$

26.5. $\displaystyle\sum_{n=0}^{\infty} \frac{e^2}{n!}(z - 1)^n, r = \infty.$

26.6. $\displaystyle\sum_{n=0}^{\infty} \frac{(-1)^n}{i^{n+1}}(z - i)^n, r = 1.$

26.7. $\displaystyle\sum_{n=0}^{\infty} \frac{(-1)^n}{4^{n+1}}(z - 4)^n, r = 4.$

26.8. $\displaystyle\sum_{n=0}^{\infty} \frac{(-1)^n}{n!} z^n, r = \infty.$

26.9. $\displaystyle\sum_{n=0}^{\infty} \frac{z^{2n+1}}{(2n + 1)!}, r = \infty.$

26.10. $\displaystyle \text{Log } i + \sum_{n=0}^{\infty} \frac{(-1)^n}{i^{n+1}(n + 1)}(z - i)^{n+1}, r = 1.$

26.11. $\displaystyle\sum_{n=0}^{\infty} (-1)^n 2^n z^n - \sum_{n=0}^{\infty} (-1)^n 2^n z^{n+1}, r = \frac{1}{2}.$

26.12. $\displaystyle\sum_{n=0}^{\infty} (-1)^{n+1} \frac{2^n}{3^{n+1}}(z - 1)^{n+1}, r = \frac{3}{2}.$

26.13. $\displaystyle\sum_{n=0}^{\infty} \frac{z^{2n}}{(2n)!}, r = \infty.$

26.14. $z + \frac{1}{3}z^3 + \frac{2}{15}z^5 + \cdots.$
26.15. $1 + z - \frac{1}{3}z^3 - \frac{1}{6}z^4 + \cdots.$
26.16. $z + z^2 + \frac{5}{6}z^3 + \frac{5}{6}z^4 + \cdots.$

26.17. $\displaystyle e \sum_{n=0}^{\infty} \frac{(z - 1)^{n+2}}{n!}.$

26.18. $\rho = 1.$
26.19. $\rho = 2.$

26.20. $\displaystyle\sum_{n=0}^{\infty} (-1)^{n+1} \frac{(z - \pi)^{2n}}{(2n + 1)!}.$

26.21. $\displaystyle\frac{3 - i}{10} \sum_{n=0}^{\infty} \left[\frac{1}{i^{n-1}} - \frac{(-1)^n}{3^{n+1}} \right] z^n.$

26.22. (b) $(1 - z)^{-2}.$ (c) $z^3 e^z.$ (d) $\dfrac{\sin z}{z}.$ (e) $\dfrac{z \cos z - \sin z}{z^2}.$

Review—Chapter 6, p. 268

1. T F F F F F.
2. (a) 2. (b) 0. (c) 1. (d) e.

3. (a) $e^i \sum\limits_{n=0}^{\infty} \dfrac{(z-i)^n}{n!}$.

(b) $\sum\limits_{n=0}^{\infty} \dfrac{(-1)^{n+1}}{(2n+1)!}\left(z-\dfrac{\pi}{2}\right)^{2n+1}$.

(c) $\sum\limits_{n=0}^{\infty} \dfrac{(-1)^n}{(2i)^{n+1}}z^n$.

(d) $-\sum\limits_{n=0}^{\infty} z^{2n}$.

(e) $\sum\limits_{n=0}^{\infty} \tfrac{1}{2}(n+2)(n+1)z^n$.

(f) $e\sum\limits_{n=0}^{\infty} \dfrac{(z-2)^n}{n!}$.

4. $z\cos z + \sin z$.
5. $e(1 + z + \tfrac{3}{2}z^2 + \tfrac{13}{6}z^3 + \cdots)$.

6. $\sum\limits_{n=1}^{\infty} \dfrac{n}{2^{n+1}}z^{3n-3}$.

7. Because its general term, $\dfrac{n-1}{n} + \dfrac{1}{n^2}$, does not tend to zero as $n \to \infty$; see Theorem 6.5.

8. $e^{-1} \sum\limits_{n=0}^{\infty} \dfrac{(z+1)^{2n}}{n!}$.

9. $\sum\limits_{n=0}^{\infty} \dfrac{(-1)^n}{4^{n+1}}z^{2n}$.

10. (a) $\sum\limits_{n=0}^{\infty} \dfrac{z^{2n+1}}{(2n+1)n!}$.

(b) $\sum\limits_{n=0}^{\infty} (-1)^n \dfrac{z^{4n+1}}{(4n+1)(2n)!}$.

(c) $\sum\limits_{n=0}^{\infty} (-1)^n \dfrac{z^{2n+1}}{(2n+1)(2n+1)!}$.

(d) $\sum\limits_{n=1}^{\infty} \dfrac{z^n}{n \cdot n!}$.

12. $1 + \dfrac{z^2}{2} + \dfrac{5z^4}{24} + \dfrac{61z^6}{720} + \cdots$.

Appendix 6, p. 278

1. (a) $\{n^2\}$. (b) $\{1 + (-1)^n\}$. (c) $\{(-n)^n i\}$. (d) $\{n^{[(-1)^n]}\}$. (e) $\left\{\dfrac{n-1}{n}i^n\right\}$.

2. (a) Yes. (b) No. The sequence $\{1 + (-1)^n\}$ has two limit points, 0 and 2, neither of which is a cluster point.

CHAPTER 7

Exercise 27, p. 286

27.1. $\sum\limits_{n=0}^{\infty} \dfrac{(-1)^n 3^n}{z^{n+1}}$.

27.2. $\dfrac{1}{3} \displaystyle\sum_{n=0}^{\infty} (-1)^n \left[\dfrac{1}{(z-2)^{n+1}} - \dfrac{(z-2)^n}{4^{n+1}} \right].$

27.3. $\displaystyle\sum_{n=0}^{\infty} \dfrac{(-1)^n}{(2n+1)!} \cdot \dfrac{1}{z^{2n+1}}.$

27.4. $\dfrac{1}{3} \displaystyle\sum_{n=0}^{\infty} (-1)^n \dfrac{1-4^n}{(z-2)^{n+1}}.$

27.5. $\displaystyle\sum_{n=0}^{\infty} \dfrac{z^{n-1}}{(n+2)!}.$

27.6. $\displaystyle\sum_{n=0}^{\infty} \dfrac{(-1)^n}{(2n)!}(z-1)^{2n-1}.$

27.7. $\displaystyle\sum_{n=0}^{\infty} \dfrac{1}{n!z^{2n}}.$

27.8. $\displaystyle\sum_{n=0}^{\infty} \dfrac{z^{2n-1}}{(2n+1)!}.$

27.9. $\displaystyle\sum_{n=0}^{\infty} (-1)^n(z-1)^{n-2}.$

27.10. $\dfrac{1}{z} + \displaystyle\sum_{n=0}^{\infty} \left(\dfrac{1}{i^{n-1}} - 1 \right) z^n.$

27.11. $\displaystyle\sum_{n=0}^{\infty} \dfrac{c^n}{z^{n+1}}; \quad \sum_{n=0}^{\infty} \dfrac{(n+1)c^n}{z^{n+2}}.$

27.12. $\dfrac{1}{2z^2} + \dfrac{3}{4z} + \displaystyle\sum_{n=0}^{\infty} \left(1 - \dfrac{1}{2^{n+3}} \right) z^n, \, 0 < |z| < 1.$

$\dfrac{1}{2z^2} + \dfrac{3}{4z} - \displaystyle\sum_{n=0}^{\infty} \dfrac{1}{z^{n+1}} - \sum_{n=0}^{\infty} \dfrac{1}{2^{n+3}} z^n, \, 1 < |z| < 2.$

$\dfrac{1}{2z^2} + \dfrac{3}{4z} - \displaystyle\sum_{n=0}^{\infty} \dfrac{1}{z^{n+1}} + \sum_{n=0}^{\infty} \dfrac{2^{n-2}}{z^{n+1}}, \, 2 < |z| < \infty.$

27.13. $\displaystyle\sum_{n=0}^{\infty} (-1)^n(z-1)^{n-2}.$

27.16. (a) $\displaystyle\sum_{n=1}^{\infty} \dfrac{z^{n-1}}{n!}.$ (b) $\displaystyle\sum_{n=0}^{\infty} \dfrac{(-1)^n}{(2n+1)!}z^{2n}.$

(c) $\displaystyle\sum_{n=1}^{\infty} \dfrac{(-1)^n}{(2n)!}z^{4n-2}.$ (d) $\displaystyle\sum_{n=2}^{\infty} \dfrac{z^{n-2}}{n!}.$

27.17. (a) 1. (b) 1. (c) 0. (d) $\frac{1}{2}$.

Exercise 28, p. 293

28.1. Pole of order 1 at $z = 0$.
28.2. Essential at $z = 0$.

28.3. Pole of order 1 at $z = 1$.
28.4. Pole of order 1 at $z = 1$; pole of order 2 at $z = 2$.
28.5. Removable at $z = 0$; $f(0) = -\frac{1}{2}$.
28.6. Pole of order 2 at $z = -i$.
28.7. Pole of order 2 at $z = 0$; pole of order 1 at $z = \pi$.
28.8. Removable at $z = 2$; $f(2) = 1$.
28.9. Removable at $z = -1$; $f(-1) = -1$.
28.10. Pole of order 3 at $z = 0$.
28.11. Order 1.
28.12. Order 3.
28.13. Order 1.
28.14. Order 1.
28.18. Pole of order 3.
28.19. f has a removable singularity at $z = 0$.

Exercise 29, p. 300

29.1. 0.
29.2. 2.
29.3. $-\frac{2}{3}$.
29.4. 1.
29.5. $1/(2 - \pi)$; 0; $(\pi - 1)/(\pi - 2)$.
29.6. 0.
29.7. Res $[f, 0] = -1$, Res $[f, -1] = e^{-1}/2$, Res $[f, 1] = e/2$.
29.8. 0.
29.9. Res $[f, 0] = -1$, Res $[f, -1] = -e^{-1}/2$, Res $[f, 1] = e/2$.
29.10. $-\frac{1}{24}$.
29.11. 0.
29.12. 0.
29.13. $e\pi(1 + 3i)$.
29.14. $-2i/\pi(1 + \pi i)$.
29.15. $2\pi i$.
29.16. 0.
29.17. $2\pi i$.
29.18. $2\pi(e^3 + \cos 2 - 2)i$.
29.19. 0.
29.20. 0.
29.22. $2\pi i$.

Review—Chapter 7, p. 301

1. (a) $\displaystyle\sum_{n=0}^{\infty} z^{n-1}$.
(b) $\displaystyle -\sum_{n=0}^{\infty} \frac{1}{z^{n+2}}$.

(c) $\displaystyle\sum_{n=0}^{\infty} (-1)^{n+1}(z-1)^{n-1}$.
(d) $\displaystyle\sum_{n=0}^{\infty} \frac{(-1)^{n+1}}{(z-1)^{n+2}}$.

2. (a) $\displaystyle\sum_{n=0}^{\infty} \left(1 - \frac{1}{2^{n+1}}\right) z^n$.
(b) $\displaystyle -\sum_{n=0}^{\infty} \frac{1}{z^{n+1}} - \sum_{n=0}^{\infty} \frac{z^n}{2^{n+1}}$.

(c) $\displaystyle\sum_{n=0}^{\infty} (2^n - 1)\frac{1}{z^{n+1}}$.
(d) $\displaystyle -\sum_{n=0}^{\infty} (z-1)^{n-1}$.

(e) $\displaystyle\sum_{n=0}^{\infty} (-1)^n(z-2)^{n-1}$.

3. $2\pi i$.
6. (b) 1.
7. 0.
8. $2\pi i$.

9. πe^{-1}.   **10.** Pole of order 1.

11. -1.

12. $z = 2k\pi, k = $ integer; all isolated.

13. $\dfrac{1}{2}\displaystyle\sum_{n=0}^{\infty} \dfrac{(-1)^n(n+1)(n+2)}{(z-1)^{n+3}}$.   **14.** -1.

15. 0.

17. (a) $2\pi i$.   (b) $4\pi i$.

     (c) $2\pi i$.   (c) 0.

CHAPTER 8

Exercise 30, p. 320

30.1. $6\pi/\sqrt{21}$.   **30.2.** $\pi/\sqrt{2}$.

30.3. $12\pi/\sqrt{3}$.   **30.4.** π.

30.5. 0.   **30.6.** $2\pi/\sqrt{3}$.

30.7. $\sqrt{2}\pi/4$.   **30.8.** 0.

30.9. π **30.10.** $\pi(e-1)/6\,e^2$.

Exercise 31, p. 323

31.1. $\frac{2}{3}R^{3/2}(e^{3\pi i} - 1); \frac{2}{3}R^{3/2}(1 - e^{3\pi i})$.

31.2. $\frac{4}{5}R^{5/4}(e^{10\pi i/4} - 1); \frac{4}{5}R^{5/4}(e^{20\pi i/4} - e^{10\pi i/4})$;

     $\frac{4}{5}R^{5/4}(e^{30\pi i/4} - e^{20\pi i/4}); \frac{4}{5}R^{5/4}(e^{40\pi i/4} - e^{30\pi i/4})$.

Exercise 32, p. 326

32.2. (a) Analytic. (b) Analytic. (c) Analytic. (d) Not analytic.

32.3. (a) Point of analyticity; pole of order 1; point of analyticity.

     (b) Point of analyticity; pole of order 1; essential singularity.

     (c) Pole of order 1; essential singularity.

Exercise 33, p. 334

33.1. (a) $V = 2$ at both points.

33.5. (a) $w = \cos y \cosh x + i \sin y \sinh x$.

INDEX

A

Absolutely convergent series, 238
Accumulation point, 215
Alternating series test, 240
Analytic function, 62, 63
  at infinity, 325
  inverse of, 137
Annular region, 157
Annulus, 157, 192
  circular, 157
  closed, 192
  of convergence, 280
  multiple, 194
  theorem, 192
Argument of z, 10
Argument principle, 302

B

Bilinear function, 81
Bilinear transformation, 110
Bolzano–Weierstrass theorem, 215
Boundary of a set, 34
Boundary point, 34
Bounded
  above, 214
  below, 214
  set, 35, 214, 219
Branch
  cut, 128
  of a function, 128, 131
  point, 128, 131

C

Casorati–Weierstrass theorem, 345
Cauchy–Hadamard theorem, 243, 277
Cauchy–Riemann equations, 59
  polar form, 61
Cauchy's
  convergence principle, 235
  inequalities, 200
  integral formulas, 198, 199
  integral theorem, 183, 223
Center of a power series, 242

Chain rule, 54
Chordal distance, 31
Circle of convergence, 244
Circular annulus, 157
Closed
  annulus, 192
  multiple annulus, 194
  path, 154
  region, 35
  set, 34, 215
Cluster point, 215
Comparison test, 240
Complement of a set, 33
Complex
  conjugate, 5
  extended plane, 30, 80, 323
  form of equations in the plane, 14
  function, 37
  integral, 169
  number, 3, 25
  plane, 10
  potential, 140
  variable, 37
Composite function, 37
Conditional convergence, 238
Conformal mapping, 77, 134, 136
Conformal transformation, 136
Conjugate, complex, 5
Conjugate harmonic, 66
Conjugation, 5
Connected
  multiply, 156
  set, 156
  simply, 156
Constant
  function, 49, 79
  sequence, 232
Continuous function
  at a point, 47
  on a set, 47
  uniformly, 220
Convergence
  absolute, 238